Blueprint reading

for industry

WRITE-IN TEXT

By

WALTER C. BROWN

Professor, Division of Technology
Arizona State University, Tempe

South Holland, Illinois

THE GOODHEART-WILLCOX COMPANY, INC.

Publishers

STEPS OF PROCEDURE IN READING A BLUEPRINT

Step 1 - Read the Title

Step 2 - Check Drawing Number

Step 3 - Read Title Block and Notes

Step 4 - Read Callouts

Step 5 - Read Changes

Step 6 - Analyze Part or Assembly

Library of Congress Catalog Card Number 82–20949.
International Standard Book Number 0–87006–429–0.

123456789–83–98765432

Library of Congress Cataloging in Publication Data

Brown, Walter Charles.
Blueprint reading for industry.

Includes index.
1. Blue-prints. I. Title.
T379.B76 1983 604.2'5 82–20949
ISBN 0–87006–429–0

INTRODUCTION

The term BLUEPRINT READING, as used in this write-in text, means interpreting and visualizing drawings whether the drawings are actually blueprints or not.

BLUEPRINT READING FOR INDUSTRY is a training course for those who desire a knowledge of basic blueprint reading, or increased knowledge in this area. It is a combination text and workbook. The text tells and shows HOW, and the workbook provides space for meaningful blueprint reading and sketching activities. Variations from standard practices occurring in the numerous actual industry blueprints used in this course, have been retained to provide realistic experiences.

BLUEPRINT READING FOR INDUSTRY, is intended for technical school students, apprentices and adult workers. It is intended also as a self-help course for those who are unable to attend classes.

Walter C. Brown

CONTENTS

Contents

Fig. 1-1. Industry photo. Engineering group at Boeing Company discussing a technical drawing.

PART 1
INTRODUCTION

Unit 1
Blueprints: The Language of Industry

You have heard the saying, "A picture is worth a thousand words." This is certainly true when referring to a drawing of an industrial product.

It would be next to impossible for an engineer or designer to describe in words the shape, size and relationship of the various parts of a machine in sufficient detail for a skilled workman to produce the object. Drawings are the universal language used by engineers, designers, technicians, also skilled craftsmen to communicate quickly and accurately the necessary information to fabricate, assemble and service industrial products. See Fig. 1-1.

The original drawing is seldom used in the plant or field but copies, commonly called "blueprints," are made and distributed to those who need them.

What is a Blueprint?

A BLUEPRINT is a COPY of a DRAWING which tells the mechanic or technician who can read it, what the object will look like when it is completed. Regardless of the color, the terms "drawing," "print" and "blueprint" mean the same thing when referring to copies of engineering drawings. Blueprints provide workmen with the details of size and shape description, tolerances (allowable variation) to be held, materials used, finish and other special treatment.

How Prints are Made

The original drawing, called a "tracing," is usually made on a transparent paper or polyester film. The tracing is placed over a sheet of sensitized paper and exposed to ultraviolet light, Fig. 1-2. The exposed paper then goes through a developing process and becomes a print.

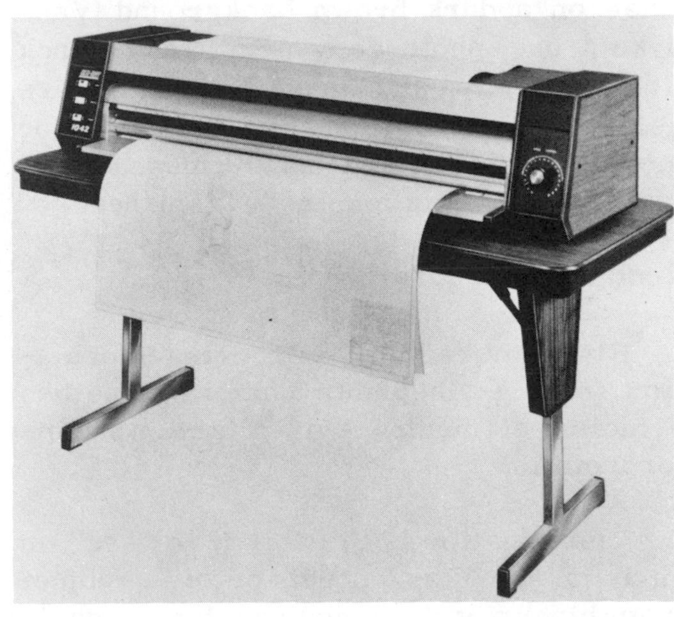

Fig. 1-2. A print making machine. (Blu-Ray, Inc.)

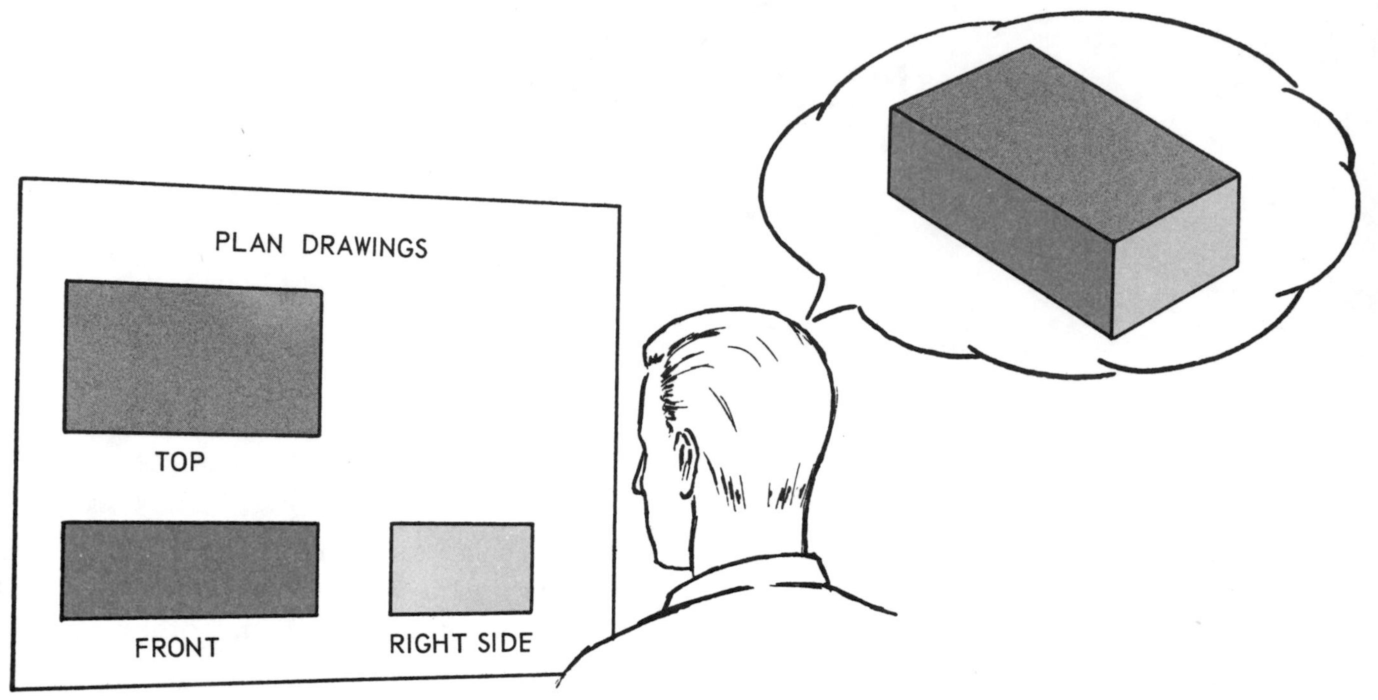

Fig. 1-3. Blueprint drawings show the size and shape of objects by means of a number of views. Each view shows how the object appears from a different location. On a blueprint, the views are arranged systematically so you can mentally connect them together to form an imaginary picture of the part or product.

The process of making blueprints used for many years produced a print with white lines on a blue background, hence the term blueprint. Today, numerous processes are used for making prints - - blue, black or maroon lines with white background; white lines on a dark brown background (Vandyke); and photo-copy prints with black lines on a white background. Tracings are sometimes photographed on microfilm for storage purposes and then, by means of enlargement, prints are made from these.

Reading a Blueprint

Blueprint reading is securing information from a blueprint. This involves two principal elements - - visualization and interpretation.

Visualization is the ability to "see" or envision the size and shape of the object from blueprint drawings which show various views. A study of drafting principles, and learning to do freehand sketching, as presented in this text will help you gain the ability to visualize objects from the views shown. See Figs. 1-3 and 1-4.

The ability to interpret lines, symbols, dimensions, notes and other information on the print is also an important factor in blueprint reading. These factors will be presented to you in a logical order in this text along with actual industrial prints to help you learn to read blueprints used in industry today.

Care of Blueprints

Blueprints are valuable records and, with proper care, can be used many times. Rules you should observe are:

1. Never write on a print unless you have been authorized to make plan changes.

2. Keep prints clean and free of oil and dirt. Soiled prints are difficult to read and contribute to errors.

3. Fold and unfold prints carefully to avoid tearing.

4. Do not lay sharp tools, machine parts or similar objects on prints.

Blueprints should be folded with the print number showing, Fig. 1-5. Blueprints are usually stored in a filing cabinet in a Blueprint Control Area and are issued by signing a "charge-out" card. In most organizations rigid control is maintained over all prints.

Fig. 1-5. Folding a blueprint with the title block and print number showing.

Importance of Blueprint Reading

The blueprint is the primary means of communication in industrial work. Many industrial products such as automobiles and aircraft consist of thousands of component parts, which may be manufactured

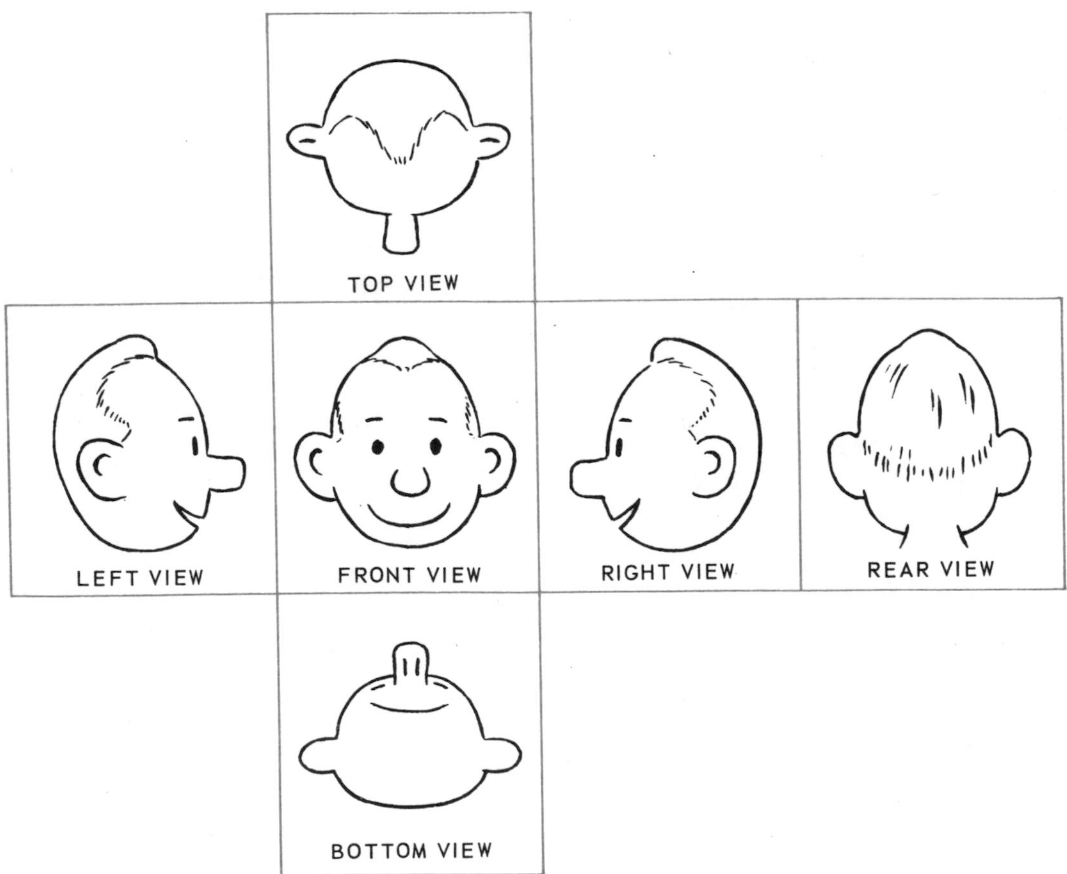

Fig. 1-4. To understand how plan views, as in Fig. 1-3, are obtained, visualize a clown's head to be enclosed in a glass box. You are on the outside looking in, and shift your position and your line of sight around the box to obtain the various views. Since the box has six sides, six views could be obtained and used if required to provide complete details.

in widely separated factories. The "moment of truth" in the manufacture of these products comes when the parts are placed in the final assembly, or a spare part is shipped for installation in the field. These parts must fit. This requires workmen who can read and understand blueprints used in modern industry.

In addition to developing the ability to read blueprints, the ability to sketch industrial objects will be helpful to you in industrial work. This text is designed to assist you in reading blueprints and sketching your ideas and solutions to industrial problems. These two factors are important to your progress in industrial work.

Blueprint Reading Activity 1-BPR-1
INDUSTRIAL BLUEPRINTS

1. Study the GLOSSARY OF TERMS page 316 and COMMON ABBREVIATIONS page 322 to become familiar with these and their location for future reference.

2. Look over a few of the blueprints in this text to gain an idea of the nature of industrial blueprints.

3. Be prepared to tell about any experience you have had in using blueprints at home, school or at work.

4. How many different kinds of prints have you seen or used? Describe these.

Unit 2
How to Read The Steel Rule

While proceeding with this course on BLUEPRINT READING, it is important that you have a 6-inch steel rule and be able to read the divided graduations in 64ths (fractional) and 100ths (decimal) of an inch.

In Fig. 2-1, an enlarged portion of a rule is shown with one edge divided into the common fractional parts of 64 to the inch and the other edge with 100 parts to the inch.

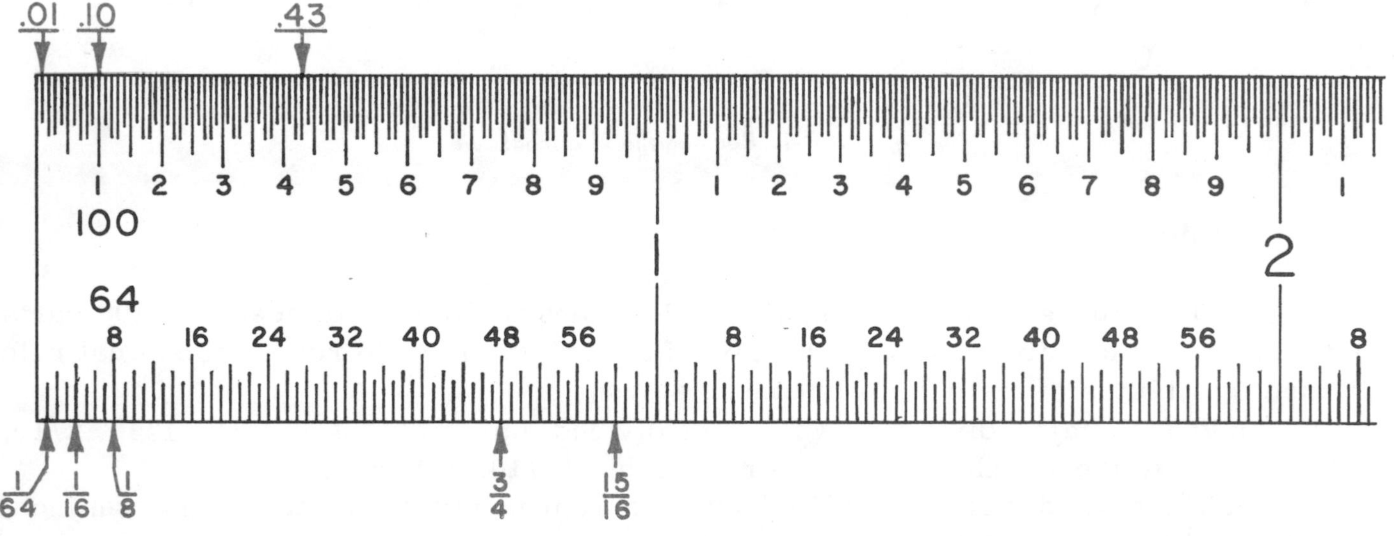

Fig. 2-1. Drawing which shows steel rule graduations (enlarged and with both fractional and decimal graduations starting at same end of rule for ease of reading).

Fractional Rule

The 64 on the end of the rule means the inch is divided into 64 parts and that each small division is 1/64 of an inch in size. To read the fractional rule, start with the edge divided into 64ths and follow these steps:

1. Study the major divisions of the inch numbered 8, 16, 24, etc. These represent

64/64ths or 1 inch. There are eight of these major divisions; therefore, each one is equal to 1/8 of an inch.

2. Note that there are 8 small divisions in each major division. Each small division equals 1/64 of an inch (1/8 x 1/8 = 1/64). If each small division is 1/64 inch in size, then the fourth line (next in length to the lines numbered 8, 16, etc.) represents 4/64 or 1/16 inch.

3. Further study and application of fractional parts will enable you to locate any common fraction which is a multiple of 64ths such as

$$\frac{48}{64} = \frac{3}{4} \text{ and } \frac{60}{64} = \frac{15}{16}.$$

Measurement Activity 2-BPR-1
READING THE FRACTIONAL RULE

Complete the readings called for in the activity below. Reduce your answers to the lowest terms and place in the spaces provided.

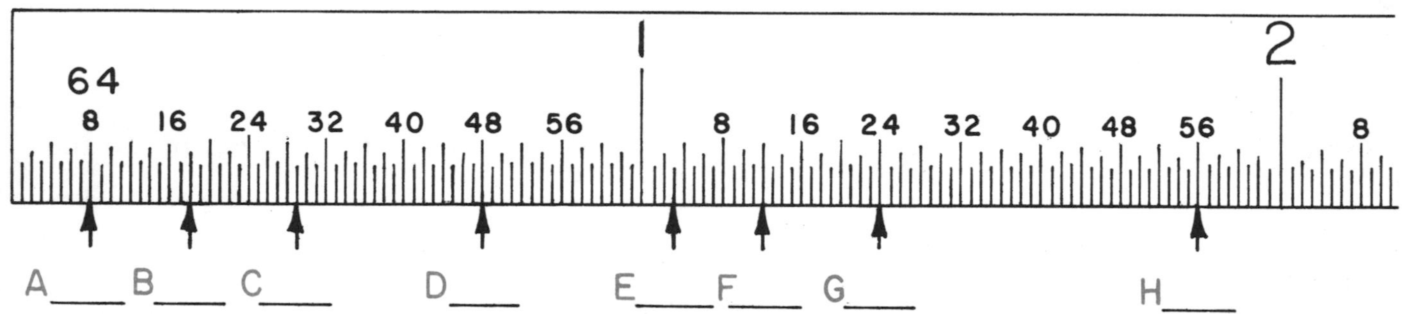

Fig. 2-BPR-1. Reading the fractional rule.

Decimal Rule

The 100 on the end of the rule, Fig. 2-1, means the inch is divided into 100 parts and that each small division is 1/100 (.01) of an inch in size. To read the decimal rule divided into 100ths of an inch, follow these steps:

1. Study the major divisions which are divided into tenths of an inch: 1/10, 2/10, 3/10 - - - on to the 1 inch mark which represents 10/10 or 1 inch.

2. Each major division of 1/10 inch also represents 10/100ths (.10) of an inch; therefore, these major divisions may be read .20, .30, .40 etc.

3. To get a measurement of .43, start at the major division marked 4 and count three additional lines, or small divisions, beyond and this mark represents .43 inch.

Measurement Activity 2-BPR-2
READING THE DECIMAL RULE

Complete the readings called for in the activity on the top of page 13. Place your answers in the spaces provided.

How to Read the Steel Rule

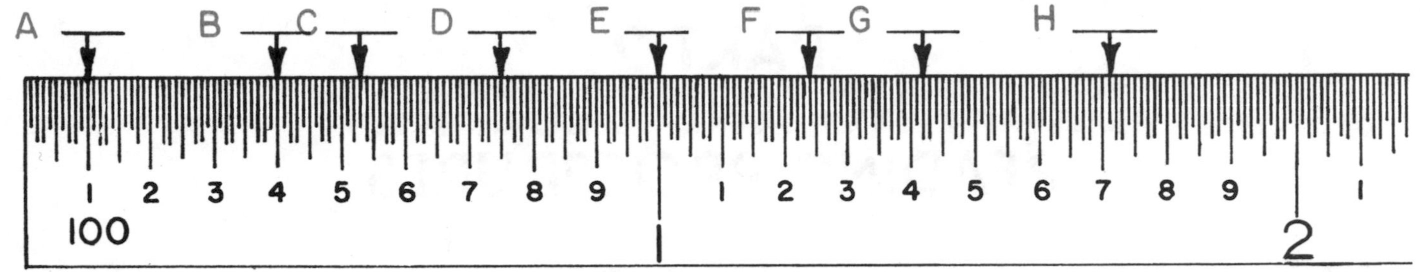

Fig. 2-BPR-2. Reading the decimal rule.

Metric Rule

Metric rules are used in conjunction with blueprints on which the dimensions are given in centimetres and millimetres. A metric rule is shown in Fig. 2-2 and it is read in the same manner as the fractional and decimal rules.

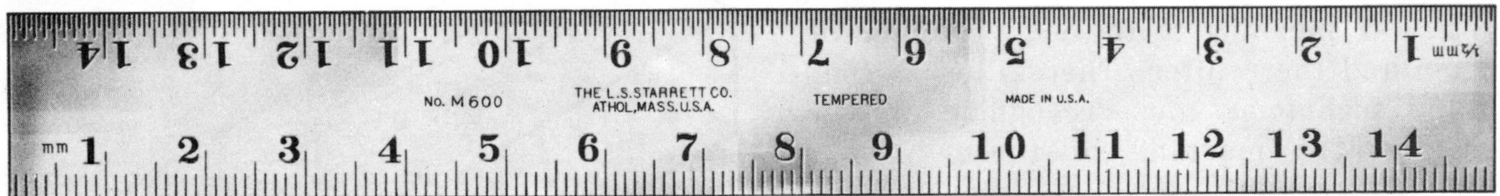

Fig. 2-2. Metric rule. (L. S. Starrett Co.)

When metric dimensions are given on blueprints used in the metalworking industries, they are usually given in millimetres (see blueprint on page 236). Conversion tables for millimetres and inches are shown on page 326.

Measurement Activity 2-BPR-3
READING THE METRIC RULE

Complete the readings called for in the activity below. Place your answers in the spaces provided.

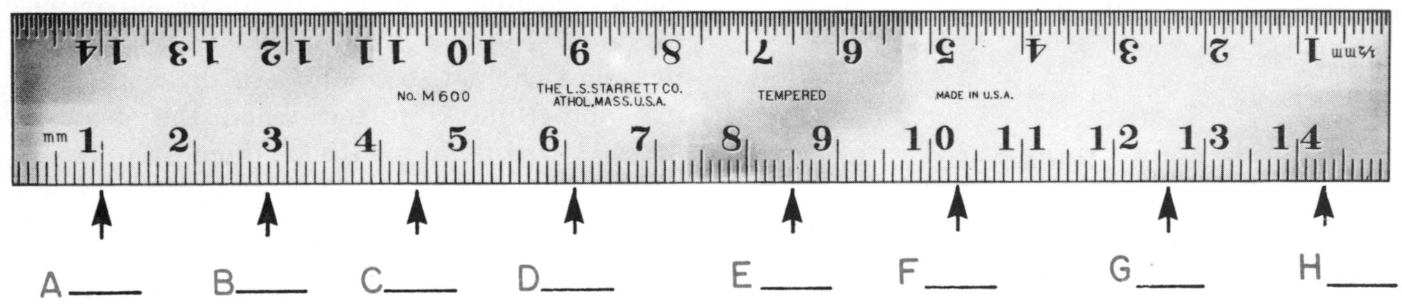

Fig. 2-BPR-3. Reading the metric rule.

PART 2
DRAFTING AND BLUEPRINT READING PROCEDURES

Unit 3
The Alphabet of Lines

In engineering type drawings 10 kinds of lines are commonly used. Each has a particular meaning to the engineer, the designer and the drafter. The skilled mechanic and technician must recognize and understand the meaning of these lines in order to correctly interpret the blueprint in manufacturing or servicing a part or assembly.

These lines, known as the ALPHABET OF LINES, are universally used throughout industry. Each line has a definite form and weight (width - - thick, medium or thin) and when combined in a drawing they convey information essential to understanding the blueprint.

Therefore, to understand the blueprint you must know and understand the alphabet of lines. The following lines make up the alphabet:

THICK

1. Visible Line.
The visible line is a thick, continuous line that represents all edges and surfaces of an object that are visible in the view, Fig. 3-1.

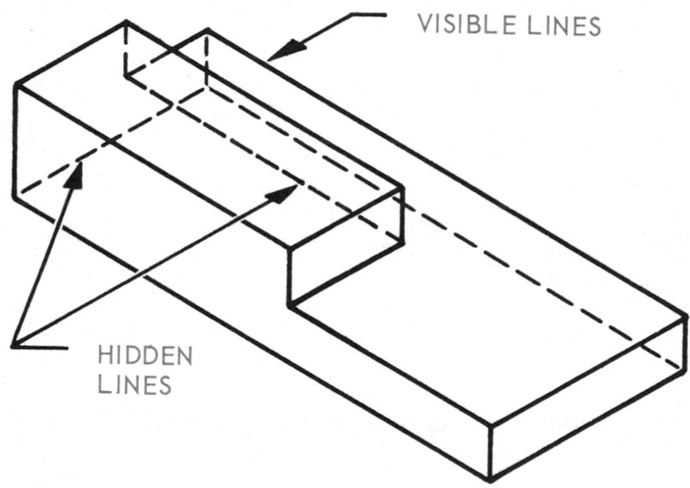

Fig. 3-1. Visible and hidden lines.

MEDIUM

2. Hidden Line.
Hidden lines are medium weight, short dashes used to show edges, surfaces and corners which are not visible in a particular view, Fig. 3-1. They are used when their presence helps to clarify a drawing and are sometimes omitted when the drawing seems to be clearer without them.

THIN

3. Section Line.
Section lines are thin lines usually drawn at an angle of 45 degrees. They indicate

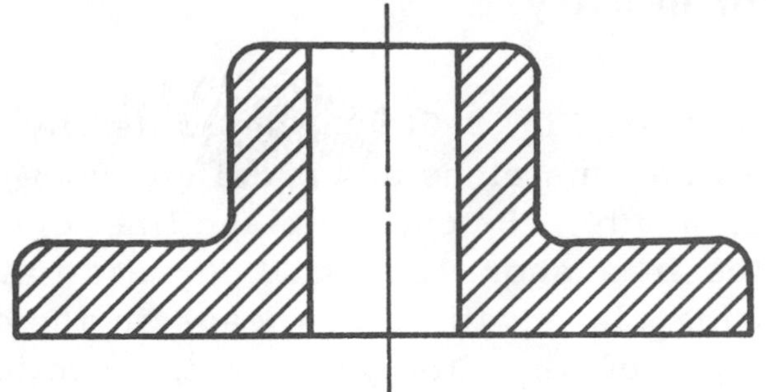

Fig. 3-2. Sectional view.

the cut surface of an object in a sectional view. The section lining for cast iron is shown, Fig. 3-2. This type of section lining is commonly used for other materials in section unless the draftsman or designer wants to indicate the specific material in section. See symbols for other materials on page 329.

THIN

4. Center Line.

Center lines are thin lines used to designate centers of holes, arcs, and symmetrical objects, Fig. 3-3(a). On some drawings, only one side of a part is drawn and the letters SYM are added to indicate the other side is identical in dimension and shape, Fig. 3-3(b). Center lines are also used to indicate paths of motion, Fig. 3-7(a).

THIN

5. Dimension Line.

6. Extension Line.

7. Leader.

Dimension and extension lines are thin lines used to indicate the extent and direction of dimensions of parts. Dimension lines are usually terminated by arrowheads against extension lines, Fig. 3-4.

Leaders are also thin lines and are used to indicate the drawing area to which a dimension note applies, Fig. 3-4.

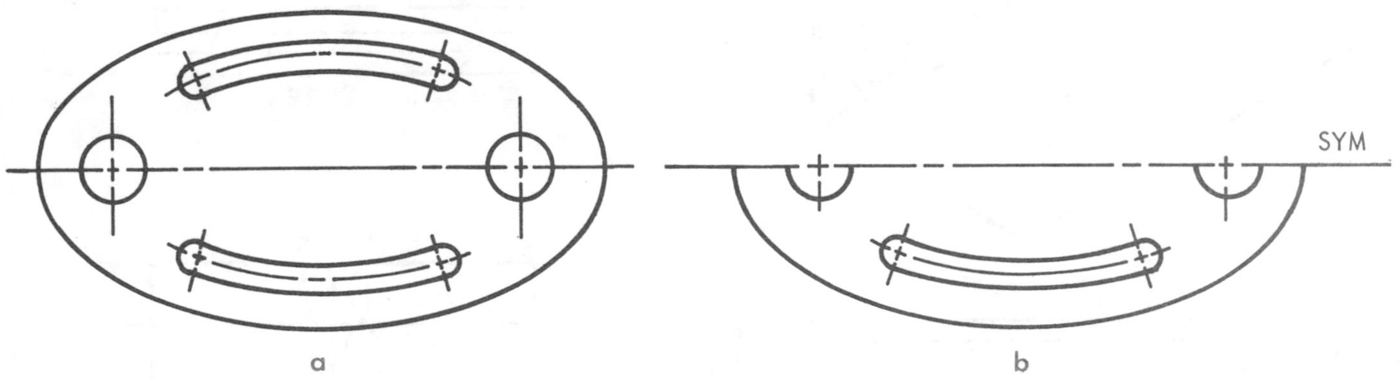

Fig. 3-3. Center lines and symmetrical center line.

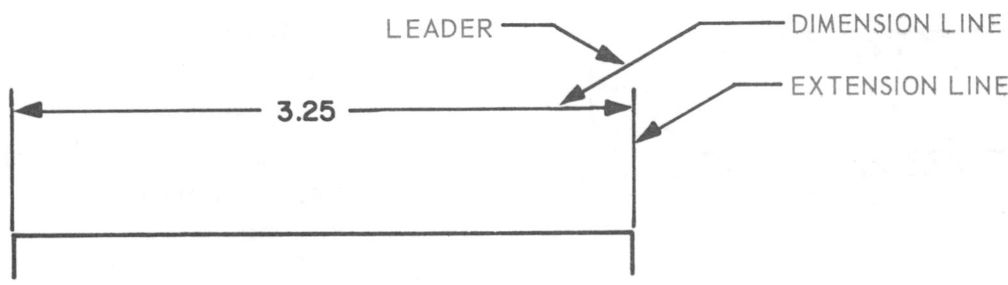

Fig. 3-4. Lines used in dimensioning.

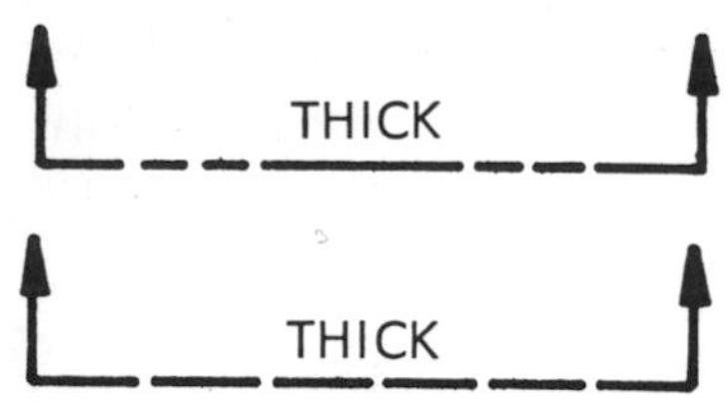

8. Cutting Plane Line
or
Viewing Plane Line

Sometimes it is hardly possible to indicate clearly on the regular views of a drawing the external or internal shape of a part or assembly. The cutting plane or viewing plane line is a thick broken line (two forms approved) located at the point where we "mentally" cut through the part in question and show what is exposed by the cutting plane. See Fig. 3-5(a). A separate view is drawn at this plane and is called a section, Fig. 3-5(b). The cutting plane line terminates in a short line at 90 degrees to the cutting plane with arrowheads in the direction of sight for viewing the section. Letters are used to indicate the section.

9. Break Lines.

There are three types of break lines used in drawings, all of which are for the purpose of breaking out sections for clarity or shortening parts of objects which are constant in detail and would be too long to place on the drawing. When the part to be broken requires a short line, the thick, wavy short break line is used, Fig. 3-6(a). If the part to be broken is longer, the draftsman may use the thin long break line, Fig. 3-6(b). In round stock such as shafts or pipe, the thick "S" break is used, Fig. 3-6(c).

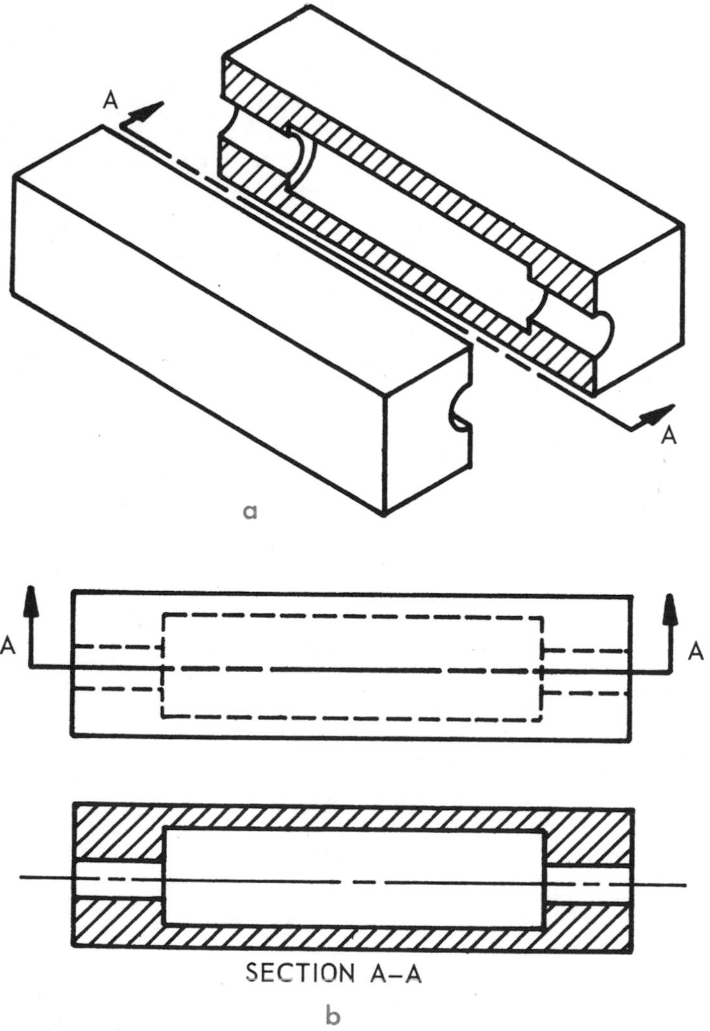

Fig. 3-5. Cutting plane line and section.

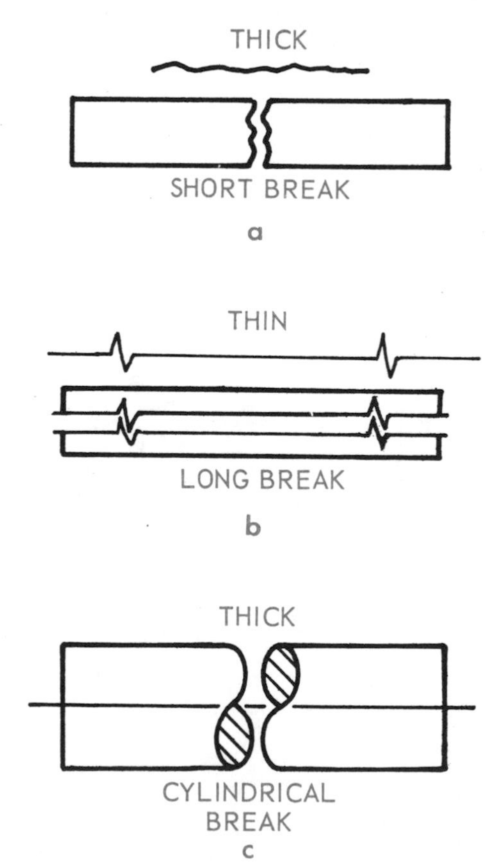

Fig. 3-6. Break lines and their uses.

THIN

10. Phantom Line.
Phantom lines are thin lines composed of long dashes alternating with pairs of short dashes. These are used primarily for three purposes in drawings. To indicate: (1) Alternate positions of moving parts such as a machine arm, Fig. 3-7(a); (2) Adjacent positions of related parts such as an existing column, Fig. 3-7(b); and, (3) Repeated detail, Fig. 3-7(c). They may also be used to indicate datum (line, point or surface from which dimensions are measured) lines and wing and fuselage stations in aircraft blueprint reading.

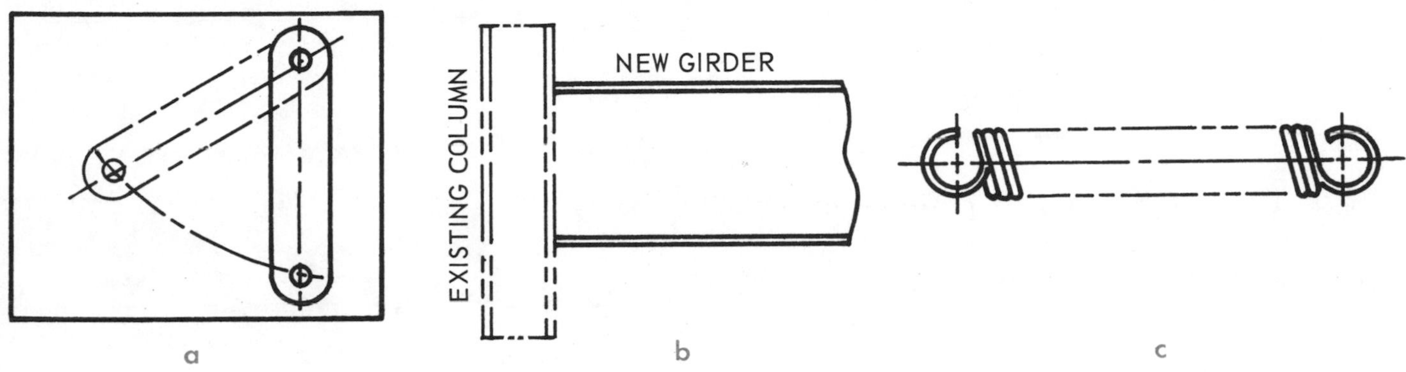

Fig. 3-7. Phantom line uses.

Line Activity 3-BPR-1
ALPHABET OF LINES

Draw freehand, in spaces provided below, the various lines used in blueprints. Be sure to pay close attention to the FORM and WEIGHT of each line.

Example:

1. Visible Line.
2. Hidden Line.
3. Section Line.
4. Center Line.
5. Dimension Line.
6. Extension Line.
7. Leader.
8. Cutting Plane Line.
9. Break Lines.
   a.
   b.
   c.
10. Phantom Line.

Unit 4
Freehand Technical Sketching

Freehand technical sketching is a process used by mechanics, technicians and draftsmen to convey ideas quickly concerning mechanical parts, assemblies and electrical diagrams. Most ideas of a mechanical or design nature find their initial expression in the form of sketches.

You will find that the ability to do freehand sketching will be very helpful to you in learning to read and interpret blueprints.

Sharpen your pencil to a conical point for sketching as shown in Fig. 4-1.

Several types of papers are suitable for sketching, including plain note paper and cross section paper. In applying the sketching techniques discussed in this text, plain paper without ruling is to be used. When you have developed the basic techniques of sketching on plain paper, you will have no difficulty using cross section paper.

Sketching Technique

The manner in which the pencil is held in freehand sketching is important. It should be held with a grip firm enough to control the strokes but not so tight as to stiffen your movements. Your arm and hand should have a free and easy movement. The point of the pencil should extend approximately 1-1/2 inches beyond your fingertips, Fig. 4-2.

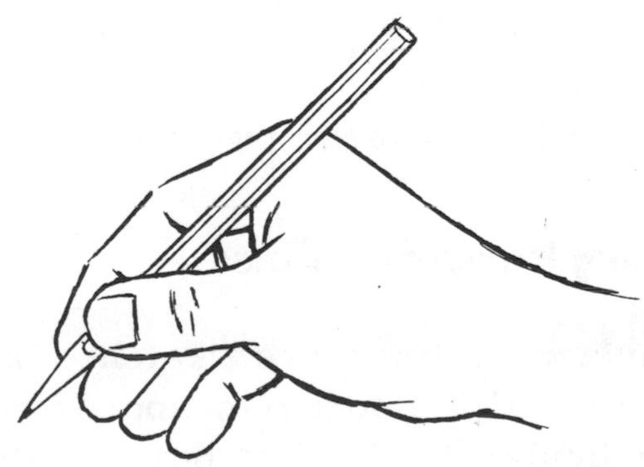

Fig. 4-2. Holding pencil for sketching.

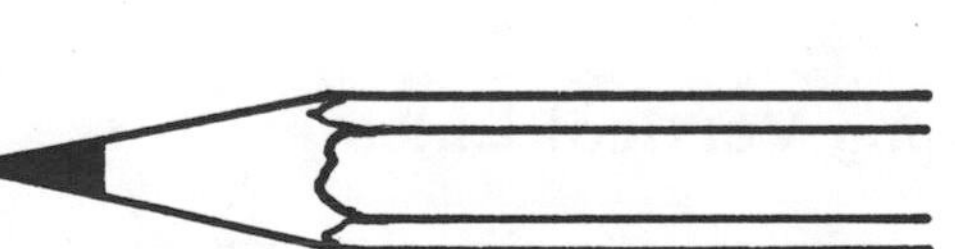

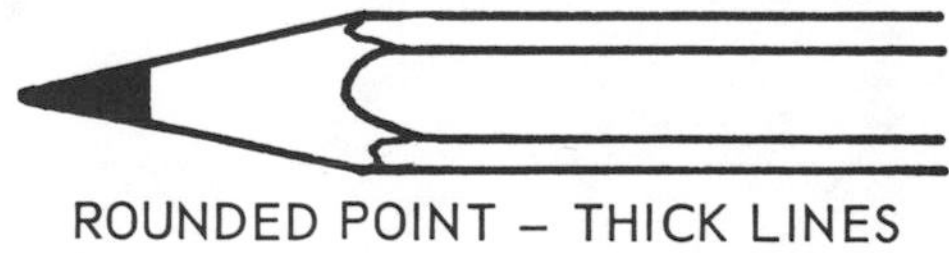

Fig. 4-1. Pencil points for sketching.

Rotate your pencil slightly between strokes to maintain the point. Initial lines should be firm and light but not fuzzy. Avoid making grooves in your paper caused by using too much pressure. In sketching straight lines, the eye should be on the point at which the line is to terminate and a series of short strokes (lines) made, rather than one continuous stroke, Fig. 4-3.

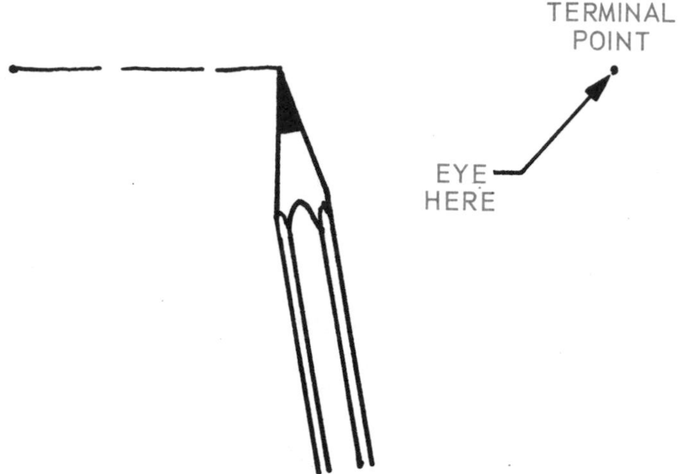

Fig. 4-3. Sketching straight lines.

Sketching Horizontal Lines

Horizontal lines are sketched with a movement of the forearm approximately perpendicular to the line being sketched, Fig. 4-4.

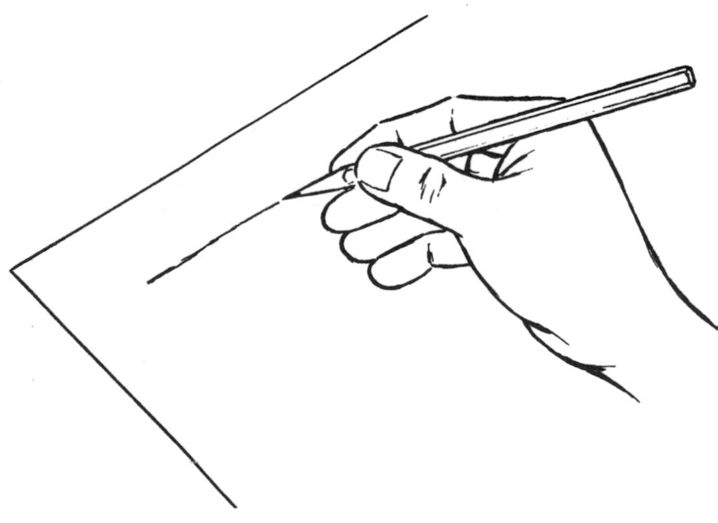

Fig. 4-4. Position for sketching horizontal lines.

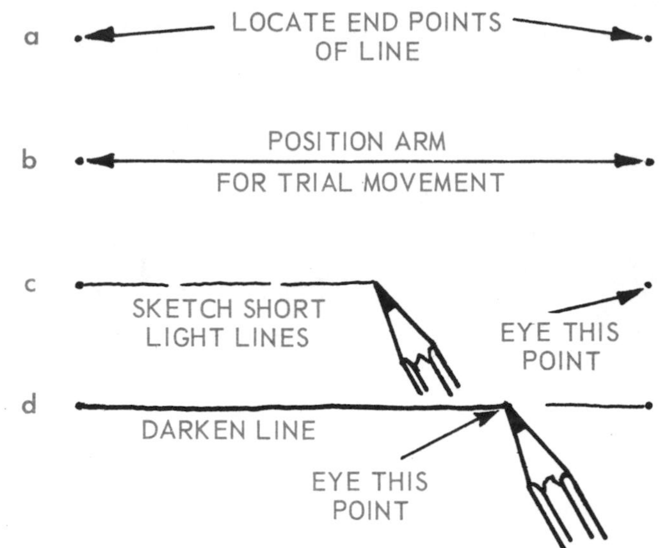

Fig. 4-5. Steps in sketching horizontal lines.

Four steps are essential in the sketching of horizontal lines:

1. Locate and mark the end point of line to be sketched, Fig. 4-5(a).

2. Position the arm by making trial movements from left to right (left-handers, from right to left) without marking the paper, Fig. 4-5(b).

3. Sketch short, light lines between the points, Fig. 4-5(c). Keep your eye on the point where the line is to end.

4. Darken the line to form one continuous line of uniform weight. The eye should lead the pencil along the lightly sketched line, Fig. 4-5(d).

Sketching Assignment

Turn to page 25 and sketch the series of horizontal lines called for in Sketching Assignment 4-BPR-1. Follow the suggested steps of procedure closely and work for improvement with each line.

Sketching Vertical Lines

Vertical lines are sketched from top to bottom by using the same short strokes in series as for horizontal lines. When making the strokes, position your arm comfortably at approximately 15 deg. with the vertical line, Fig. 4-6. A finger and

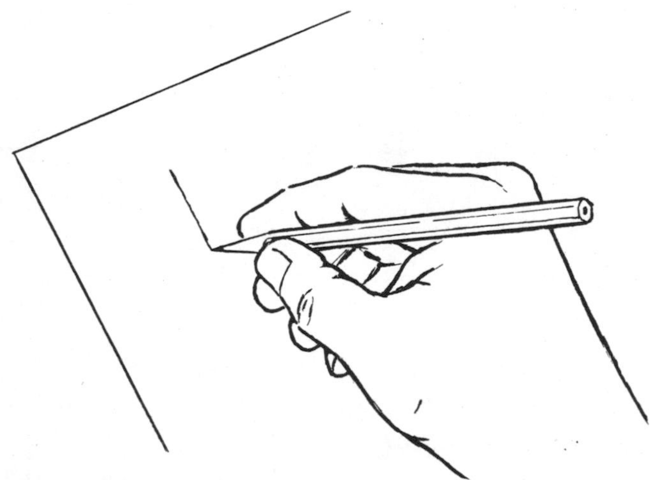

Fig. 4-6. Position for sketching vertical lines. Paper may be rotated slightly counterclockwise for greater ease.

wrist movement together with a pulling arm movement are best for sketching vertical lines.

These four steps should be used in sketching vertical lines:

1. Locate and mark the end points of the line to be sketched, Fig. 4-7(a).
2. Position the arm by making trial movements from top to bottom without marking the paper, Fig. 4-7(b).
3. Sketch short, light lines between the points, Fig. 4-7(c). KEEP YOUR EYE ON THE POINT WHERE LINE IS TO END.
4. Darken the line to form one continuous line of uniform weight. In this step, the eye should lead the pencil along the lightly sketched line, Fig. 4-7(d).

You may find it easier to sketch horizontal and vertical lines if the paper is rotated slightly counterclockwise, Fig. 4-6.

Sketching Assignment

Turn to page 25 and sketch the series of vertical lines called for in Sketching Assignment 4-BPR-2. Follow the suggested steps closely and work for improvement with each line.

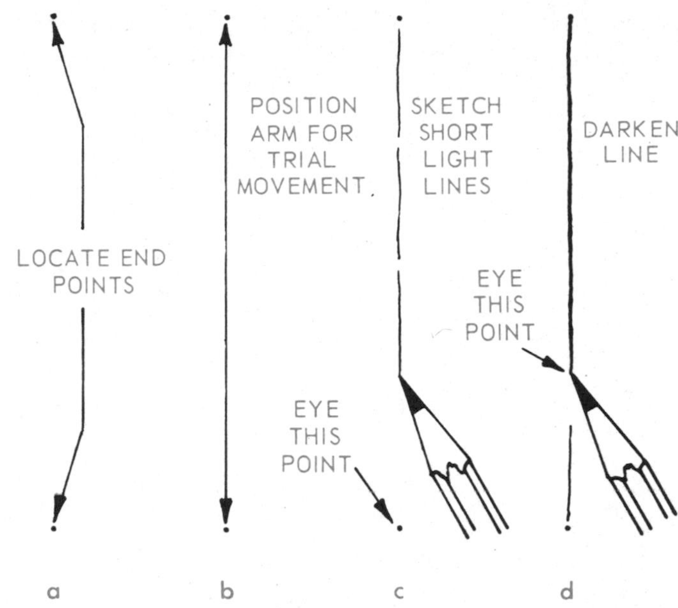

Fig. 4-7. Steps in sketching vertical lines.

Sketching Inclined Lines and Angles

All straight lines which are neither horizontal nor vertical are called INCLINED LINES. The inclined lines are usually sketched between two points or at a designated angle. The same strokes and techniques used for sketching horizontal and vertical lines are used for inclined lines or angles, depending on their position. The paper may be rotated to sketch

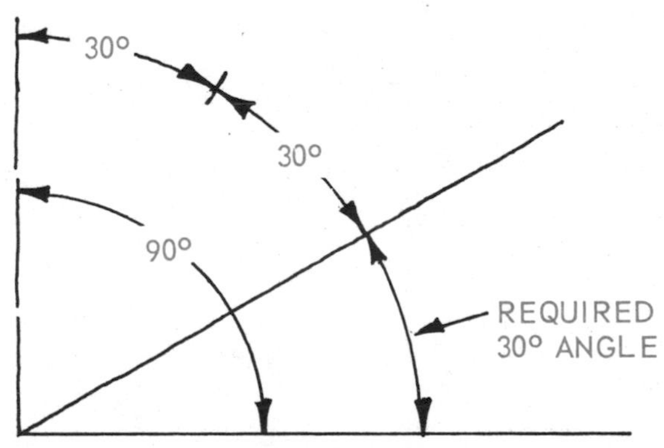

Fig. 4-8. Estimating angle sizes in sketching.

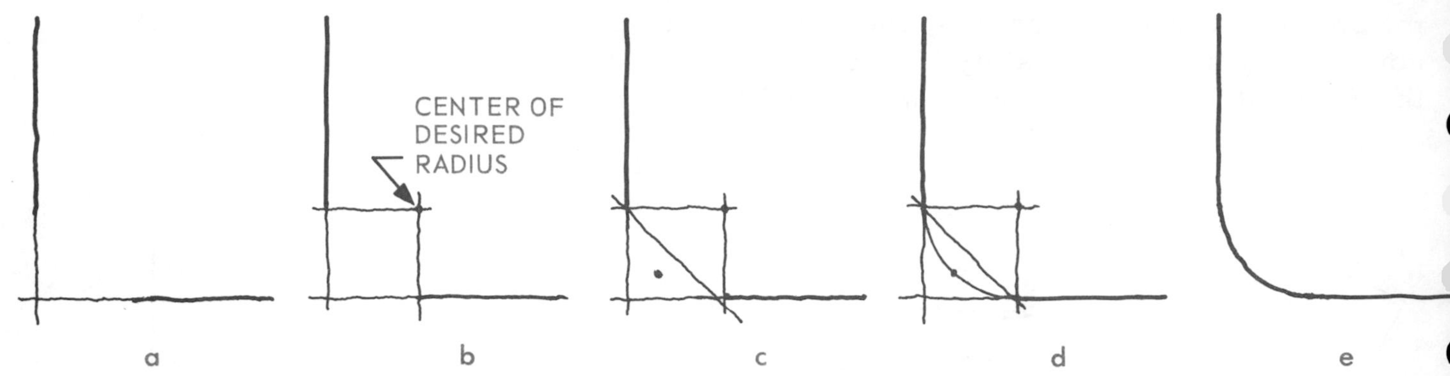

Fig. 4-9. Steps in sketching an arc.

these lines as horizontal or vertical lines if preferred.

Angles may be estimated quite closely by first sketching a right angle (90 deg.) and then subdividing its arc to get the desired angle. Fig. 4-8 illustrates how this is done to get an angle of 30 deg.

Sketching Assignment

Turn to page 26 and sketch the angles in Sketching Assignment 4-BPR-3.

Sketching Arcs, Circles and Ellipses

There are several methods of sketching arcs and circles but the one which is usually most satisfactory for sketching is the triangle-square method.

To SKETCH AN ARC connecting two straight lines, follow these steps:

1. Project the two lines until they intersect, Fig. 4-9(a).
2. Lay out desired arc radius from the point of the intersecting lines, Fig. 4-9(b).
3. Form a triangle by connecting these two points and locate the center point of the triangle, Fig. 4-9(c).
4. Sketch short, light strokes from point where arc is to start on vertical line through center point to point on horizontal line where arc ends, Fig. 4-9(d).
5. Darken the line to form one continuous arc which should join smoothly with each straight line and erase construction lines, Fig. 4-9(e).

To SKETCH A CIRCLE of a certain diameter, follow these steps:

1. Locate the center of the circle and sketch the center lines; lay off half the diameter on each side of the two diameters,

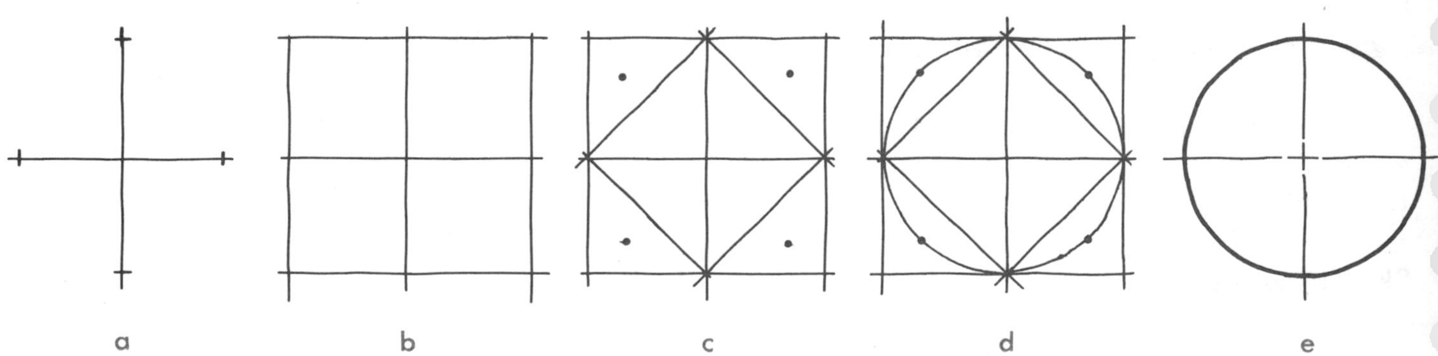

Fig. 4-10. Steps in sketching a circle.

Fig. 4-10(a).

2. Sketch a square lightly at the diameter ends, Fig. 4-10(b).

3. Across each corner, sketch diagonal line forming a triangle and locate center point of each triangle, Fig. 4-10(c).

4. Sketch short, light strokes through each quarter of the circle making sure the arc passes through the triangle center point and joins smoothly with the square at the diameter ends, Fig. 4-10(d).

5. Darken the line to form a smooth, well-formed circle and erase construction lines, Fig. 4-10(e).

Sketching Assignment

Turn to page 27 and sketch the arcs and circles called for in Sketching Assignment 4-BPR-4. Follow the steps suggested above for sketching arcs and circles.

To SKETCH AN ELLIPSE of a certain major and minor axis, follow these steps:

1. Locate center of ellipse and sketch center lines, Fig. 4-11(a).

2. Lay off major axis of ellipse on horizontal center line and minor axis on vertical center line, Fig. 4-11(b).

3. Sketch rectangle through points on axis, Fig. 4-11(c).

4. Sketch tangent arcs at points where center lines cross rectangle, Fig. 4-11(d).

5. Complete the ellipse and darken, then erase the construction lines, Fig. 4-11(e).

Proportion in Sketching

Proportion in sketching is the relationship of the size of one part to another and to the object as a whole. The width, height and depth of your sketch must be kept in the same proportion so the sketch conveys an accurate description of the object being sketched.

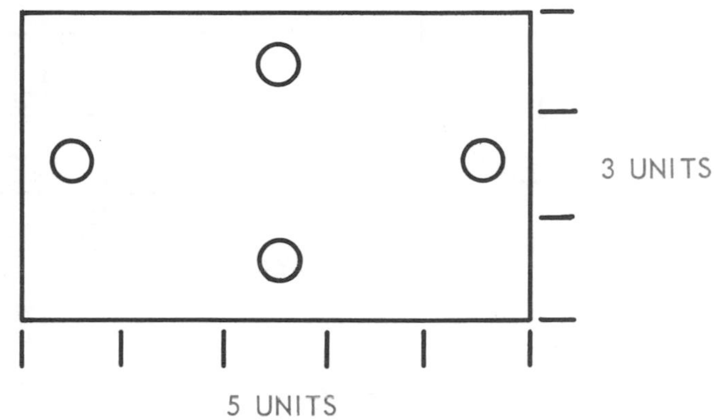

Fig. 4-12. Unit method of proportioning.

One technique useful in estimating proportions is the UNIT method. This involves establishing a relationship between measurements on the object by breaking each of the measurements into units. Compare the width to the height and select a unit that will fit each measurement, Fig. 4-12. Lengths laid off on your sketch should be in the same proportion although the units on the sketch may vary in size from those of the actual object.

Proportion is a matter of estimating

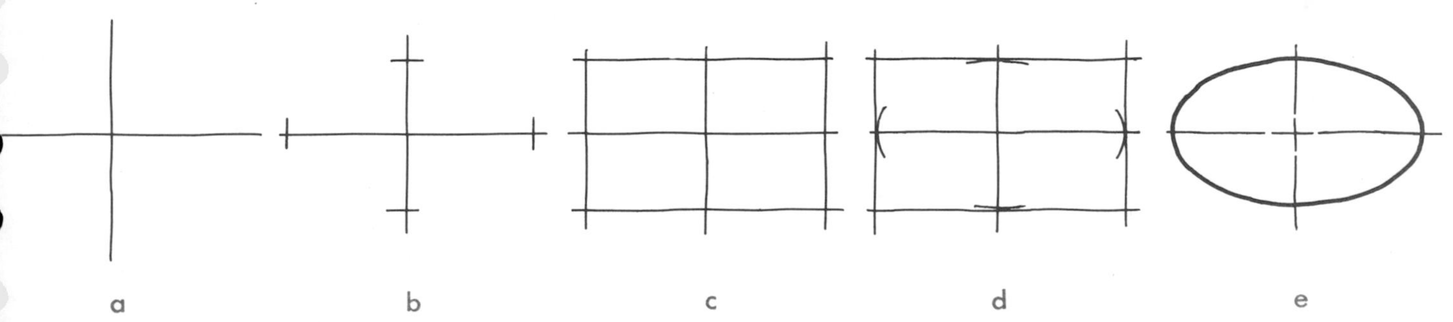

Fig. 4-11. Steps in sketching an ellipse.

lengths on a part or assembly and setting these down on your sketch in the same ratio of units. Practicing this method of establishing proportion in sketches will assist you in developing skill in accurately representing objects you sketch. Actual measurement of an object would of course be necessary if your sketch is to convey actual size dimensions.

Aids to Freehand Sketching

Several aids to freehand sketching have been mentioned in this unit, such as cross section paper for making your sketches and enclosing squares and rectangles for drawing circles and ellipses. There are other aids such as using a piece of folded paper, cardboard or 6-inch rule for a straightedge, and a scrap of paper for measuring proportion or in laying off the radius of a circle.

The aim in freehand sketching, however, should be to develop your skill to a point where aids will no longer be necessary. Sketching is a means for the mechanic or technician to quickly and effectively communicate the shape and size descriptions of an object to another person; or, to record information in the field for his own use at a later time.

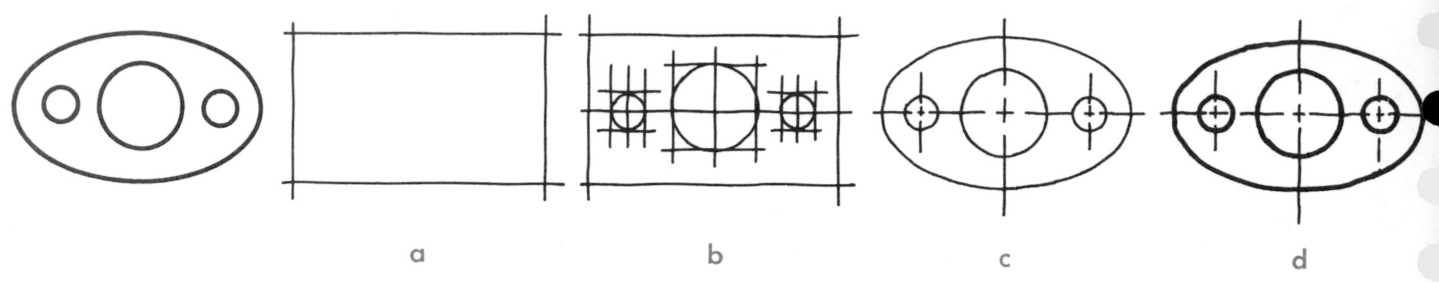

Fig. 4-13. Steps in sketching an object.

Steps in Sketching an Object

The following steps will help you in laying out and completing your freehand sketches:

1. Sketch a rectangle, square, etc., of the correct proportion, Fig. 4-13(a).
2. Sketch major subdivisions and details of the object, Fig. 4-13(b).
3. Remove unnecessary lines with a soft eraser, Fig. 4-13(c).
4. Darken lines to right weight, Fig. 4-13(d).

Sketching Assignment 4-BPR-1
HORIZONTAL LINES

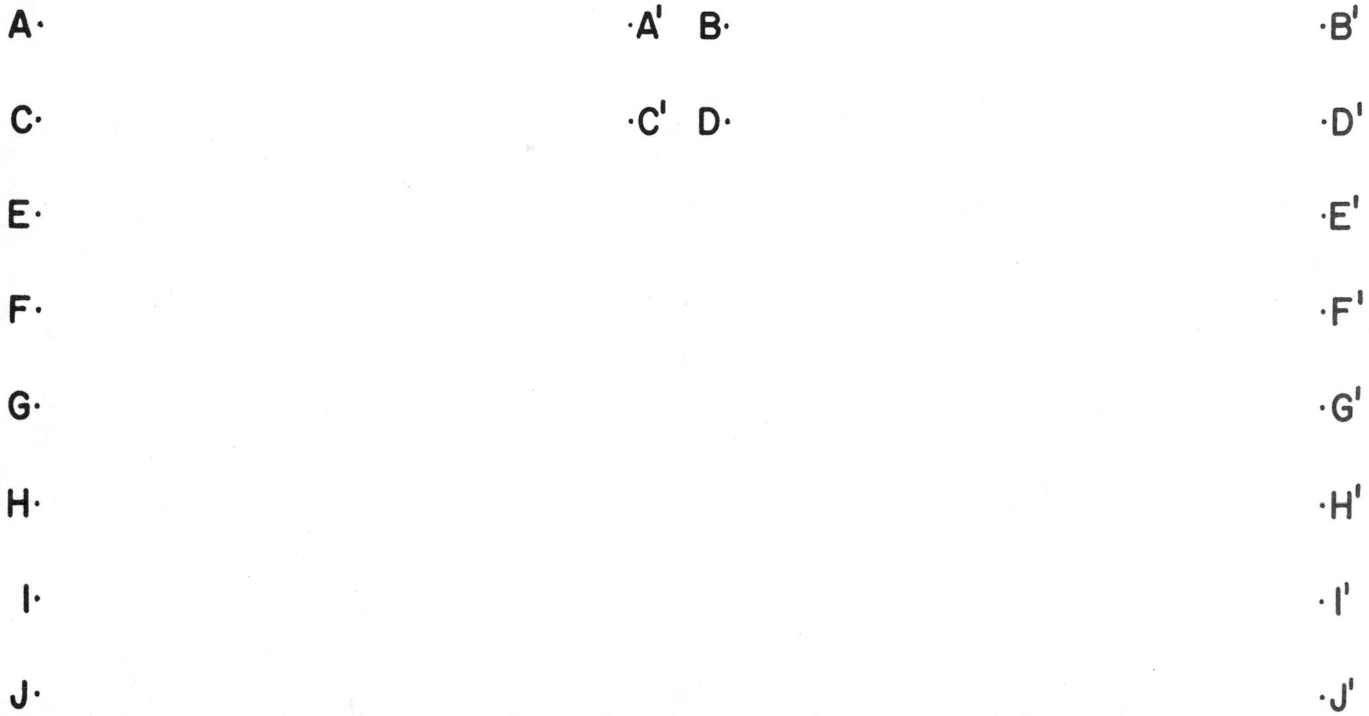

A· ·A' B· ·B'

C· ·C' D· ·D'

E· ·E'

F· ·F'

G· ·G'

H· ·H'

I· ·I'

J· ·J'

Fig. 4-BPR-1. Sketch horizontal lines between points A–A' through J–J'.

Sketching Assignment 4-BPR-2
VERTICAL LINES

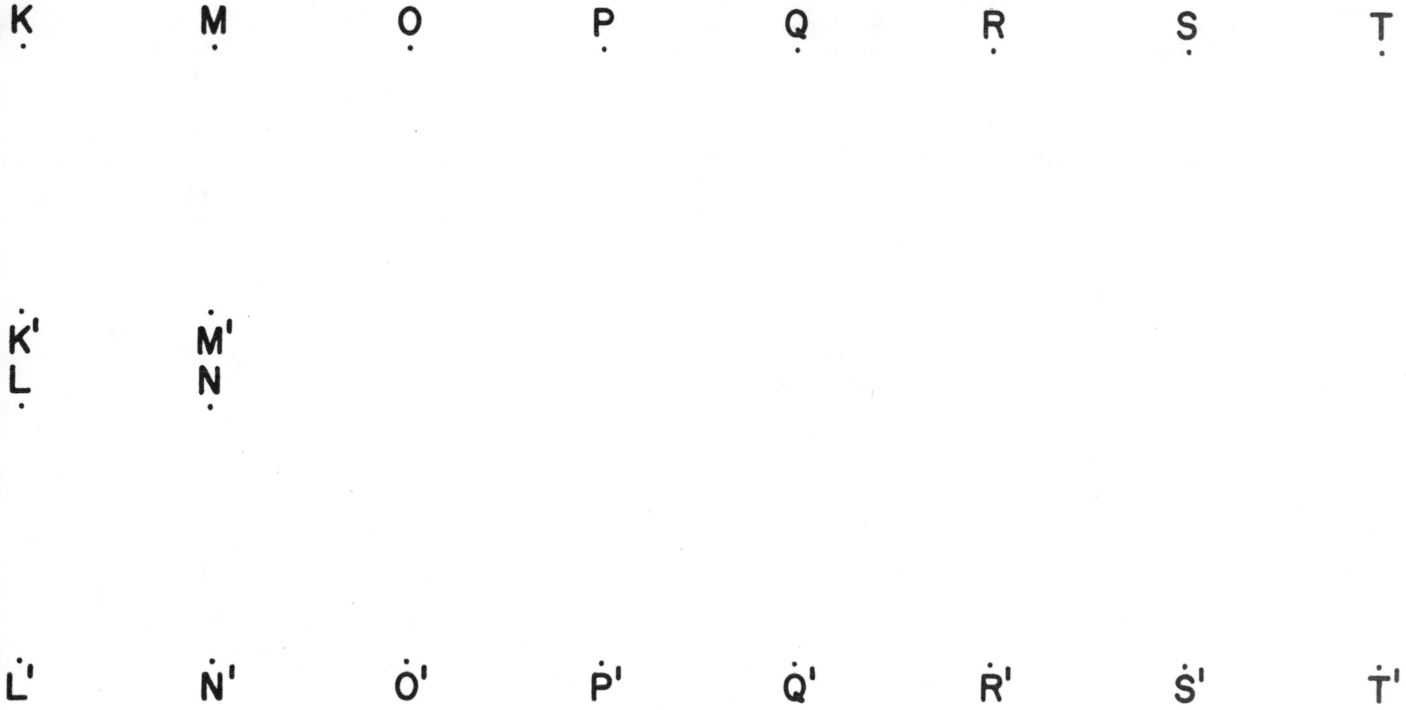

K M O P Q R S T

K' M'
L N

L' N' O' P' Q' R' S' T'

Fig. 4-BPR-2. Sketch vertical lines between points K-K' through T–T'.

Sketching Assignment 4-BPR-3
ANGLES

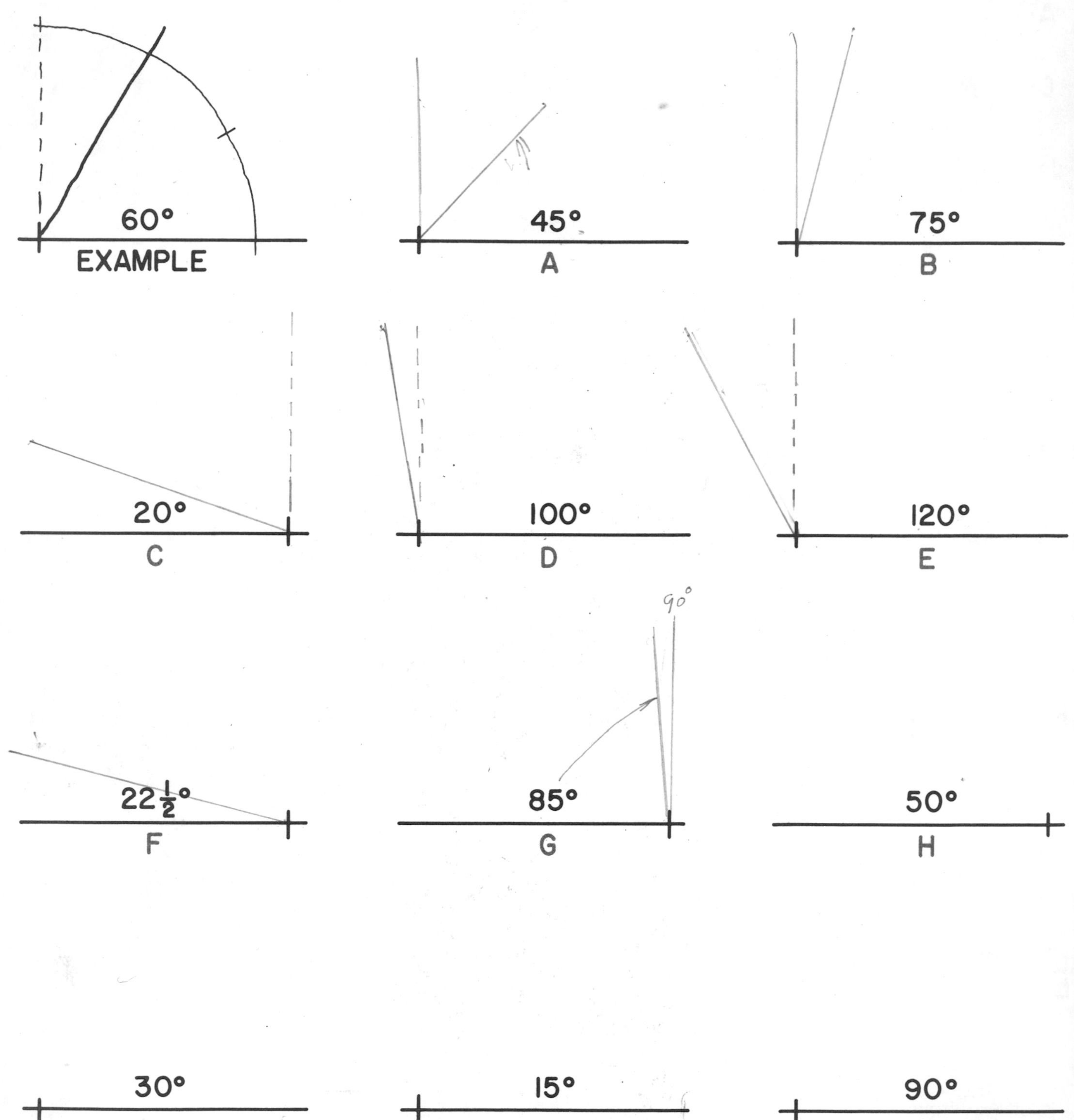

Fig. 4-BPR-3. Sketch the required angles starting at the indicated point.

Sketching Assignment 4-BPR-4
ARCS AND CIRCLES

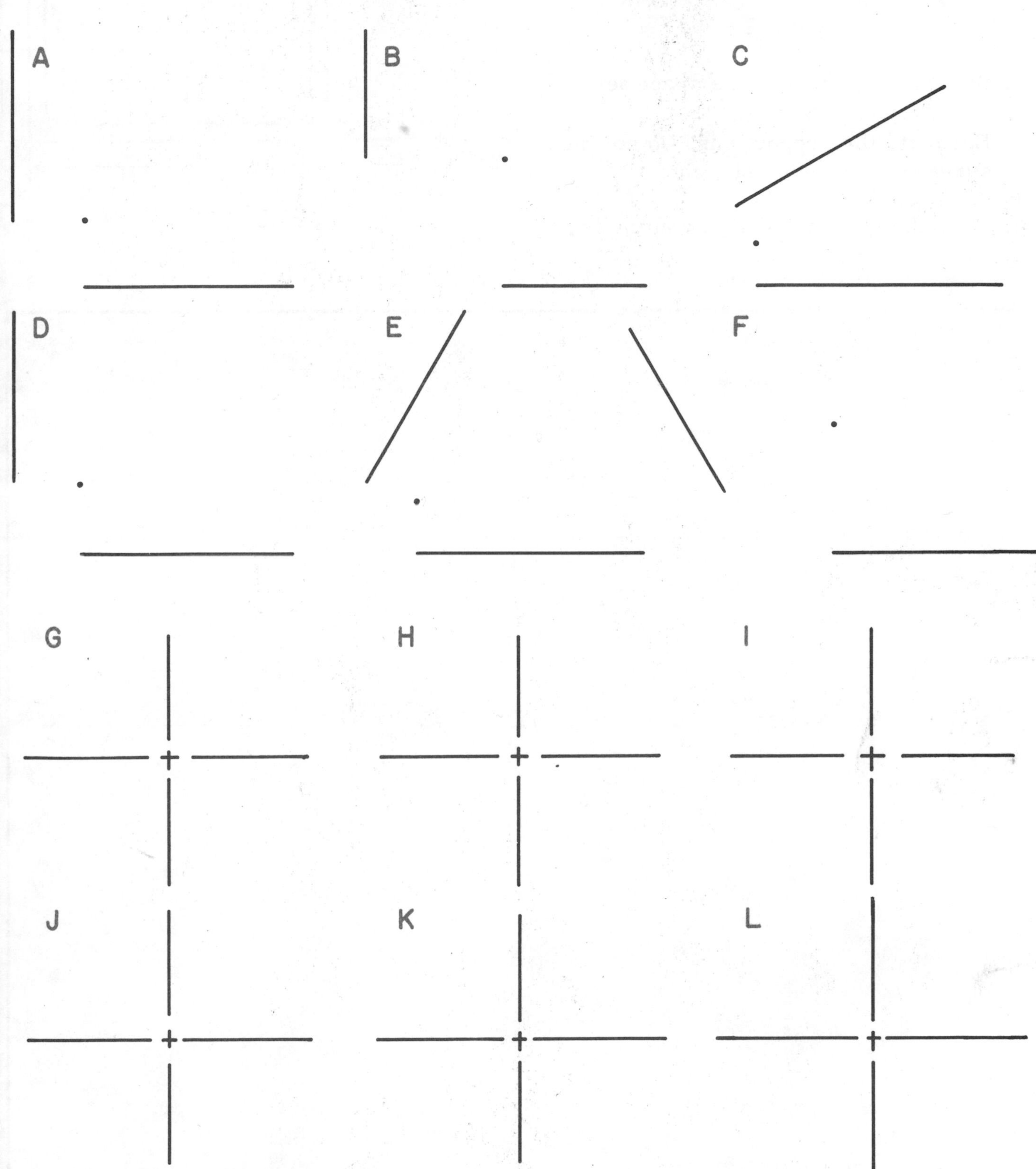

Fig. 4-BPR-4. Sketch arcs joining the sets of lines A through F. Show construction lines for A through C. Erase construction lines for D through F. Sketch circles G through L. Show construction lines for circles G through I. Erase construction lines for J through L.

Sketching Assignment 4-BPR-5
GASKET

1. Sketch the gasket in the space below.

2. Estimate the proportions. Do not measure nor dimension sketch.

3. Date the sketch and sign your name.

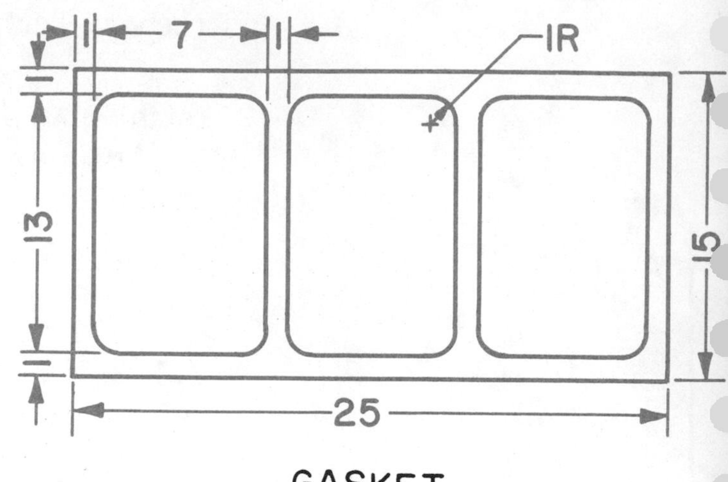

GASKET

NO.4-5	DATE	YOUR NAME

Sketching Assignment 4-BPR-6
BASE PLATE

1. Sketch the base plate in the space below.

2. Estimate the proportions. Do not measure nor dimension the sketch.

3. Date the sketch and sign your name.

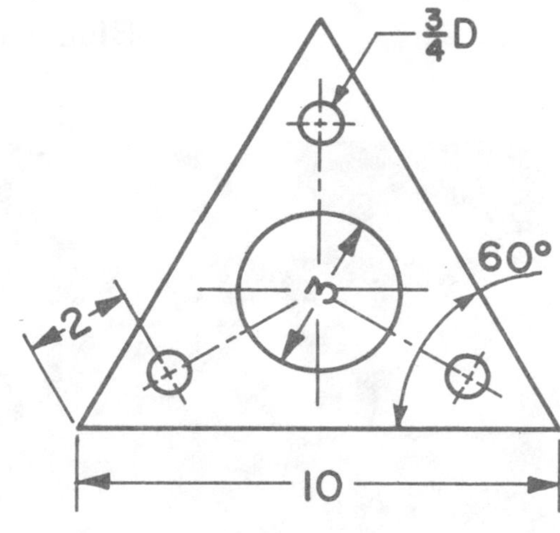

BASE PLATE

NO.4–6	DATE	YOUR NAME

Industry photo. Machining in a turret lathe. (Sheldon Machine Co., Inc.)

Unit 5
Understanding Orthographic Projection Drawings

The purpose of a drawing is to show the size and shape of the object and to provide certain information on how it is to be made. Various methods are available to the draftsman, but the best way to show every feature of the object in its true size and shape is to use a multiview orthographic (ortho-graf-ick) projection drawing.

The different views on an orthographic projection drawing are arranged in a systematic way so the plan user may mentally connect them together, thus forming a mental picture, Fig. 5-1.

Projection of Views

The draftsman's job is one of "dividing" or separating the views of an object so its size and shape may be clearly shown. The skilled mechanic or technician reading the blueprint must be able to visualize (form a mental picture) the object as a whole - -

Fig. 5-1. Forming a mental picture of an object from an orthographic projection drawing.

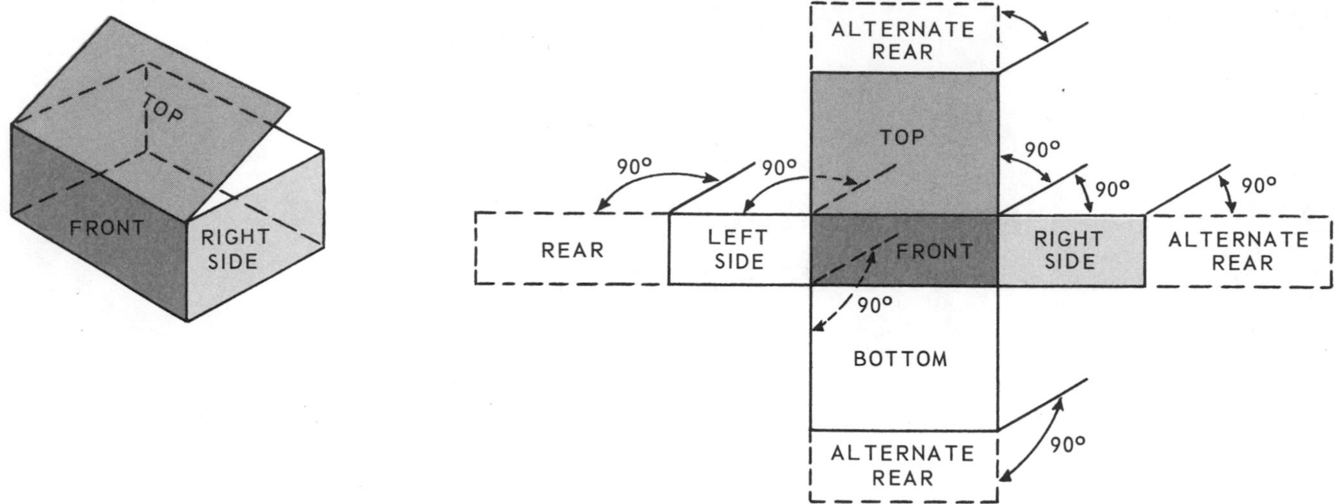

Fig. 5-2. Projection of orthographic views shown by unfolded box.

that is, mentally put the views "back together." Understanding how the draftsman projects and arranges the views will help you later in the visualization process.

The views of an orthographic drawing are projected at right angles (90 deg.) to each other and have a definite relationship. This can best be visualized by cutting and unfolding a cardboard box, Fig. 5-2.

The front view has remained in position. The four adjoining views have revolved on their "hinges" 90 deg. from the adjoining view; the top - up, bottom - down, right side - right, and the left side - left. The rear view, which is possible to have been shown in three alternate positions, is to the far side of the left-side view and was revolved 90 deg. with that view to be in the same plane, Fig. 5-2.

Let us now imagine our box is a "glass box" and we have an object inside the box for which we want to project the views, Fig. 5-3(a). Imaginary projection lines are used to bring the separate views to each projection plane, Fig. 5-3(b). If we further visualize the unfolding of the glass box, the six views are shown in their true size and shape in orthographic projection, Fig. 5-3(c).

After you have fixed in your mind the method of projecting orthographic views, you will be able to visualize the views in their proper position by viewing the object directly, as the draftsman does, without the imaginary "glass box," Fig. 5-4.

Selection of Views

The draftsman draws only those views that ARE NECESSARY to clearly describe the object. This seldom requires all six views. Usually, the necessary details can be shown in three views or less. The following rules, used in combination, guide him in the selection of views:

1. Only those views which clearly describe the shape of the object should be drawn.
2. Views which show the least hidden lines should be selected. Compare the two side views in Fig. 5-3(c).
3. When possible, the object is drawn in its functioning (operating) position.
4. When possible, the view that best describes the shape of the object is selected as the front view.

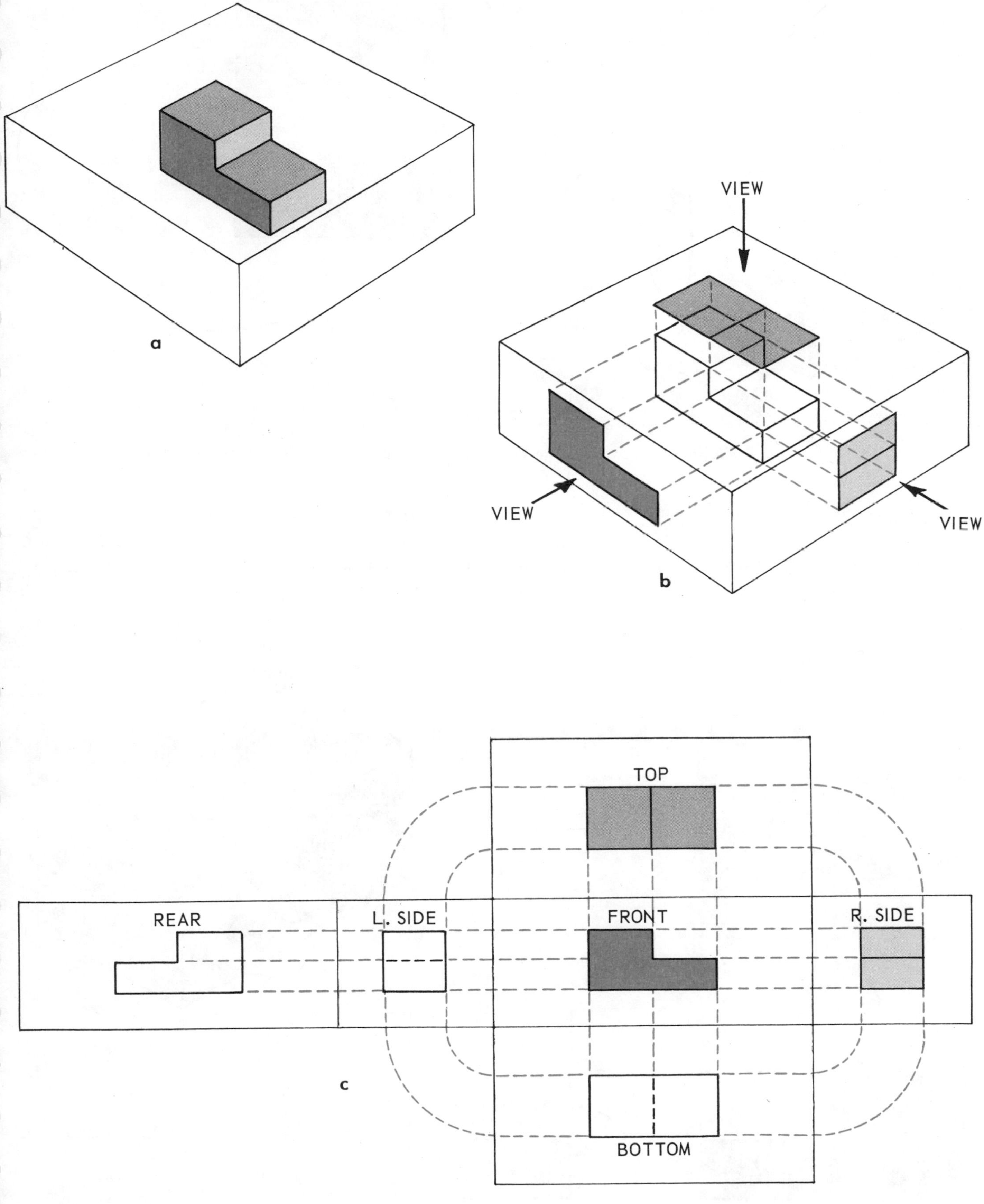

a

VIEW

VIEW

VIEW

b

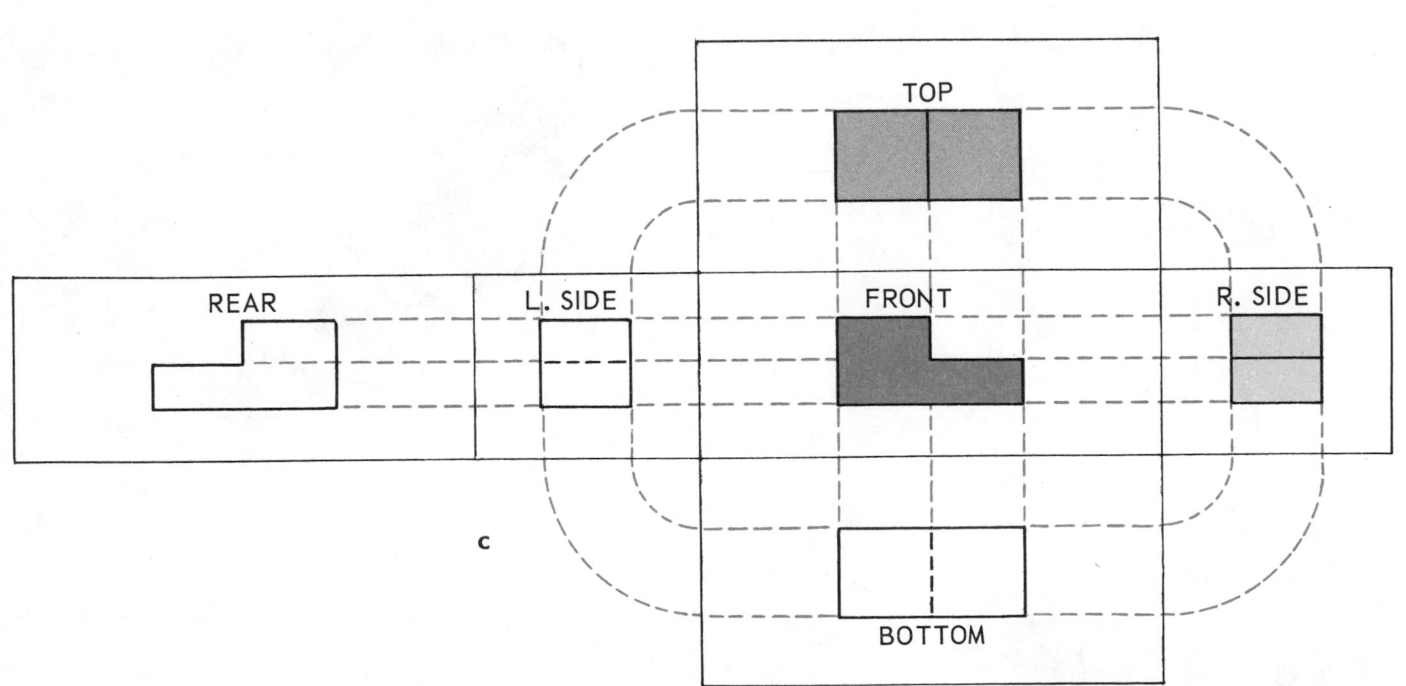

Fig. 5-3. "Glass Box" used to show views in orthographic projection.

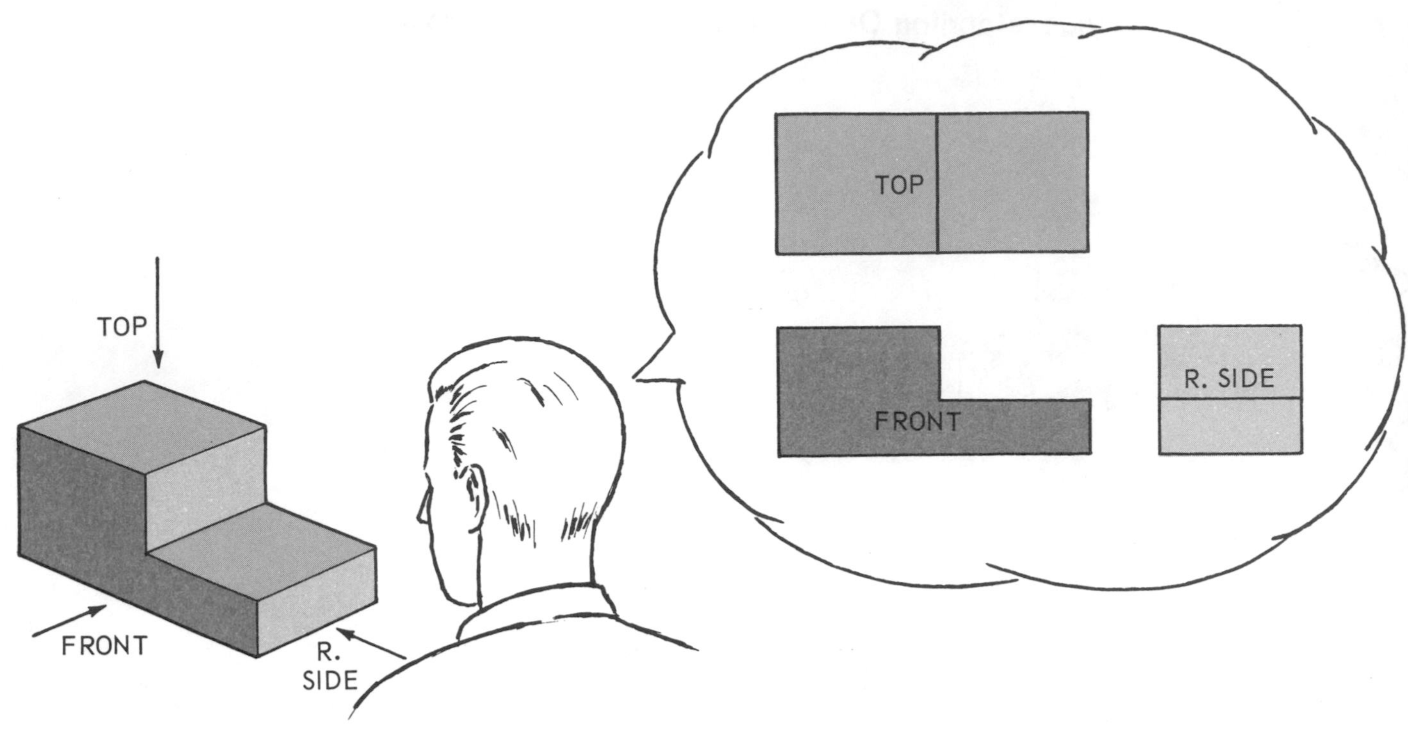

Fig. 5-4. Visualizing the views in orthographic projection.

Visualization of Objects

The preceding description of how the draftsman selects and projects the views of a drawing will help you in the process of visualizing the views as a "whole" and, therefore, better understanding the object shown on the blueprint.

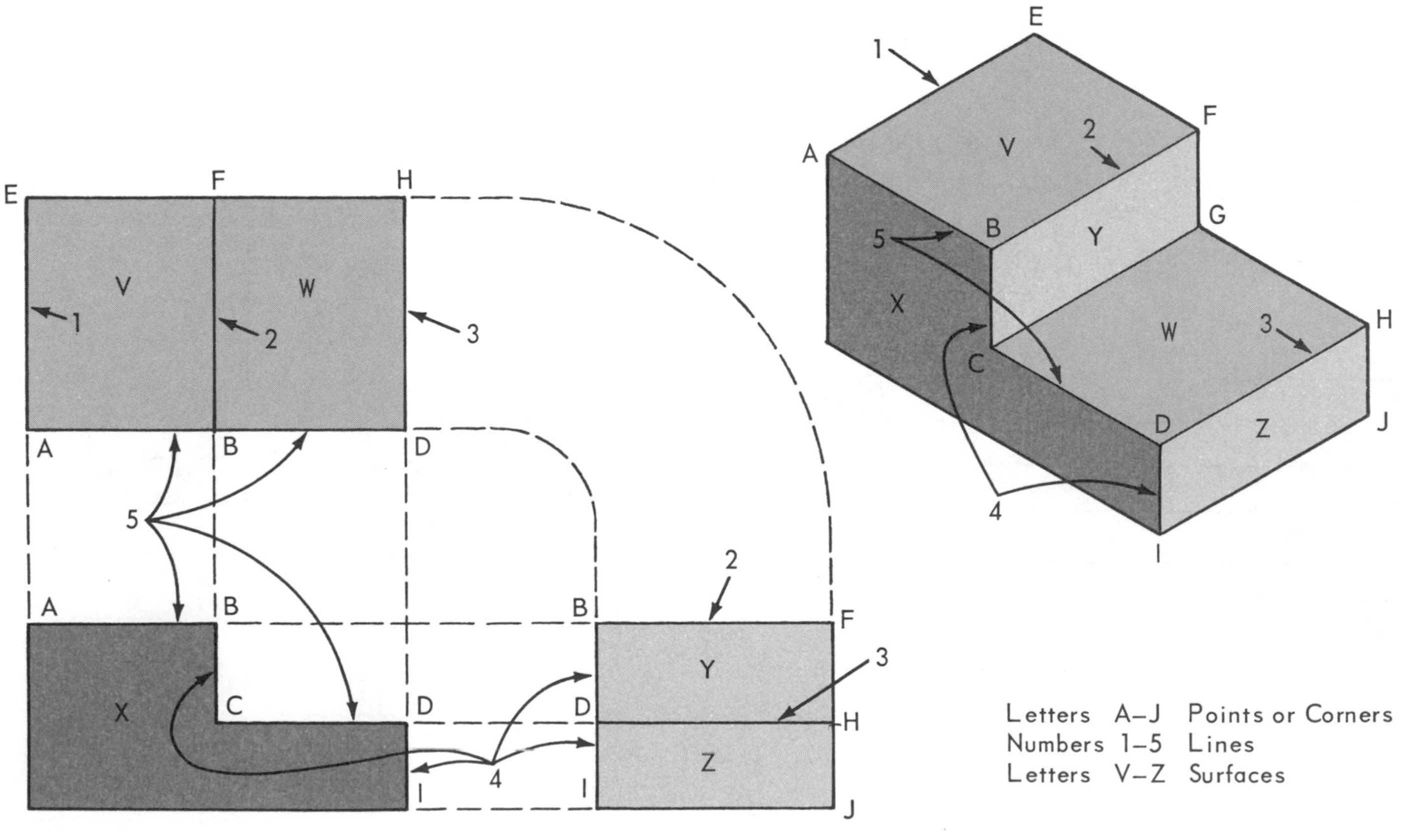

Letters A–J Points or Corners
Numbers 1–5 Lines
Letters V–Z Surfaces

Fig. 5-5. Point, line and surface identification in orthographic projection.

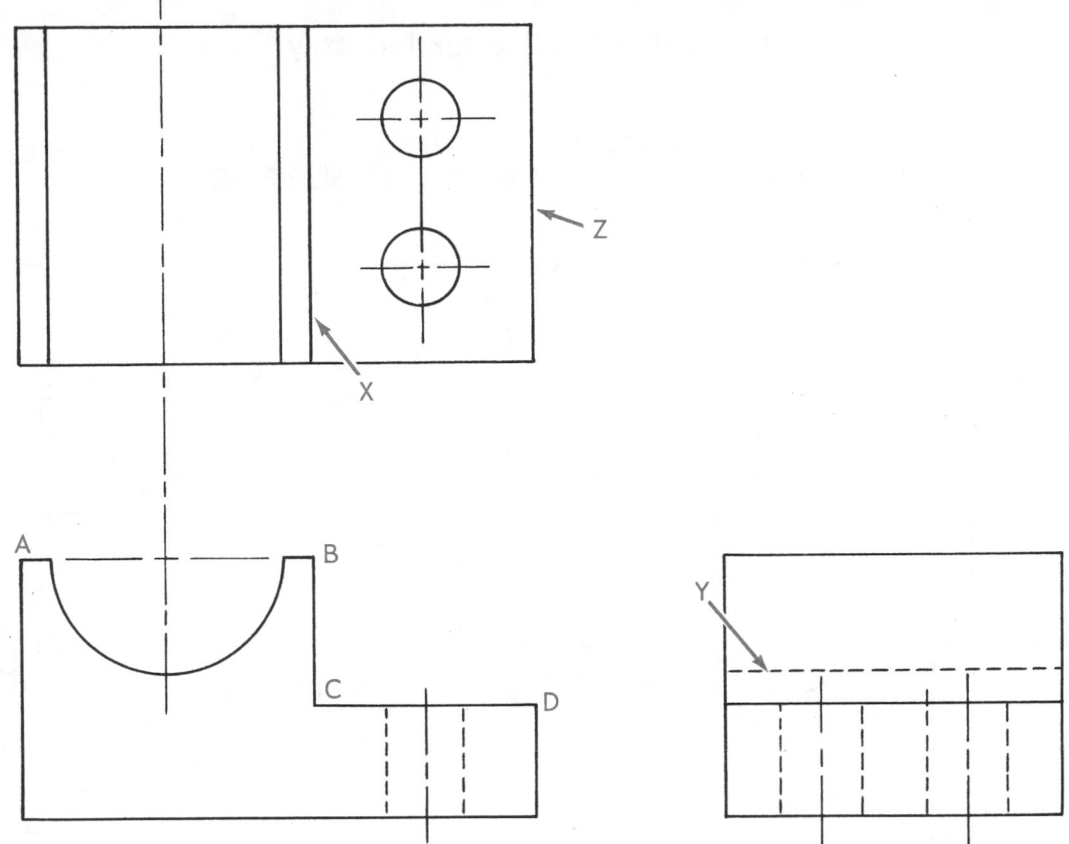

Fig. 5-6. Visualizing an object from an orthographic projection.

You may also find the system of identification of point, line and surface helpful in the visualization process. This system is described in Fig. 5-5. Letters A thru J have been assigned to points or corners of the object; numbers 1 thru 5 to lines; and, letters V thru Z to surfaces.

By comparing the top, front and right side views with the pictorial view, points, lines and surfaces can easily be identified in the various views. For example, surface X in the front view is represented by line 5 in the top view. Point A in the front view represents line 1 in the top view; line 2 in the side view represents surface V in the top view. Points E, F and H in the top view are behind points A, B and D in the front view, but F and H appear in the side view as does J, Fig. 5-5.

Can you identify which line in the front view represents surface W in Fig. 5-5? Surface W is represented by which line in the side view? If you answered line 5 for the front view and line 3 for the side view, you answered correctly.

Projection lines have been drawn in Fig. 5-5 to illustrate how the views are projected from each other. These lines usually do not appear on the finished drawing.

Visualization Rules

In studying an orthographic projection drawing, certain rules of procedure will be helpful in visualizing an object:

1. Scan briefly all views shown, Fig. 5-6.
2. Study carefully the front view for shape description.
3. Move from the front view to the other views and look for lines that describe the intersections of surfaces (line X in the top view, Fig. 5-6), the limits of a surface (line Y in side view) or the edge view of a surface (line Z, top view).
4. Study one feature at a time in the several views and begin to picture in your mind the shape of the real object.

Blueprint Reading Activity 5-BPR-1
IDENTIFICATION OF LINES AND SURFACES

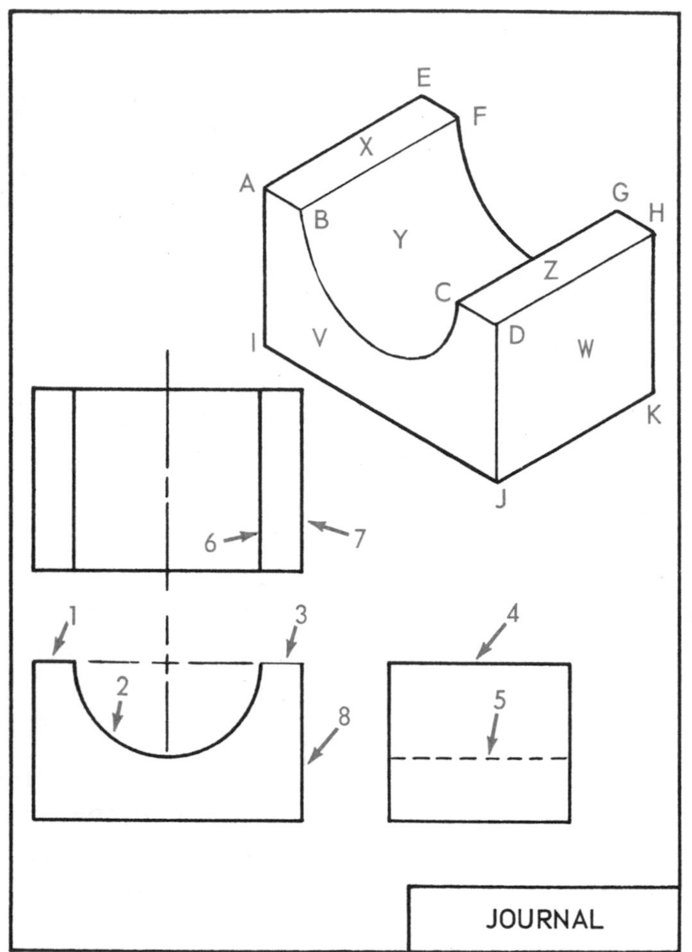

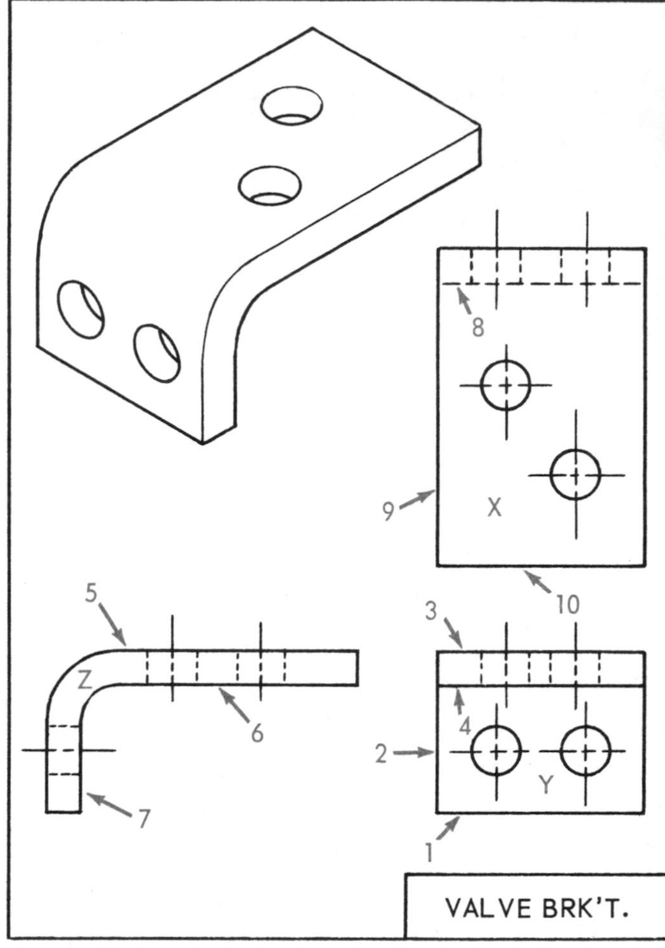

Fig. 5-BPR-1. Study the pictorial drawing in each section and identify the information called for in the questions below.

1. Which line in the top view represents surface W in the pictorial view? ______
2. Line 4 represents which surface in the pictorial view? ______
3. The lower limit of surface Y in the top view is represented by which line in the side view? ______
4. Surface Z is represented by which line in the front view? ______
5. Lines 7 and 8 represent which surface? ______
6. Line 6 extends between which two points? ______

1. Line 2 front view represents which line in top view? ______
2. Surface Z represents which line in the top view? ______
3. Line 3 represents which surface? ______
4. Line 5 represents which surface? ______
5. Line 6 represents which line in front view? ______
6. Line 8 represents which surface? ______
7. Line 7 represents which line in the top view? ______

Blueprint Reading Activity 5-BPR-2
MATCHING VIEWS

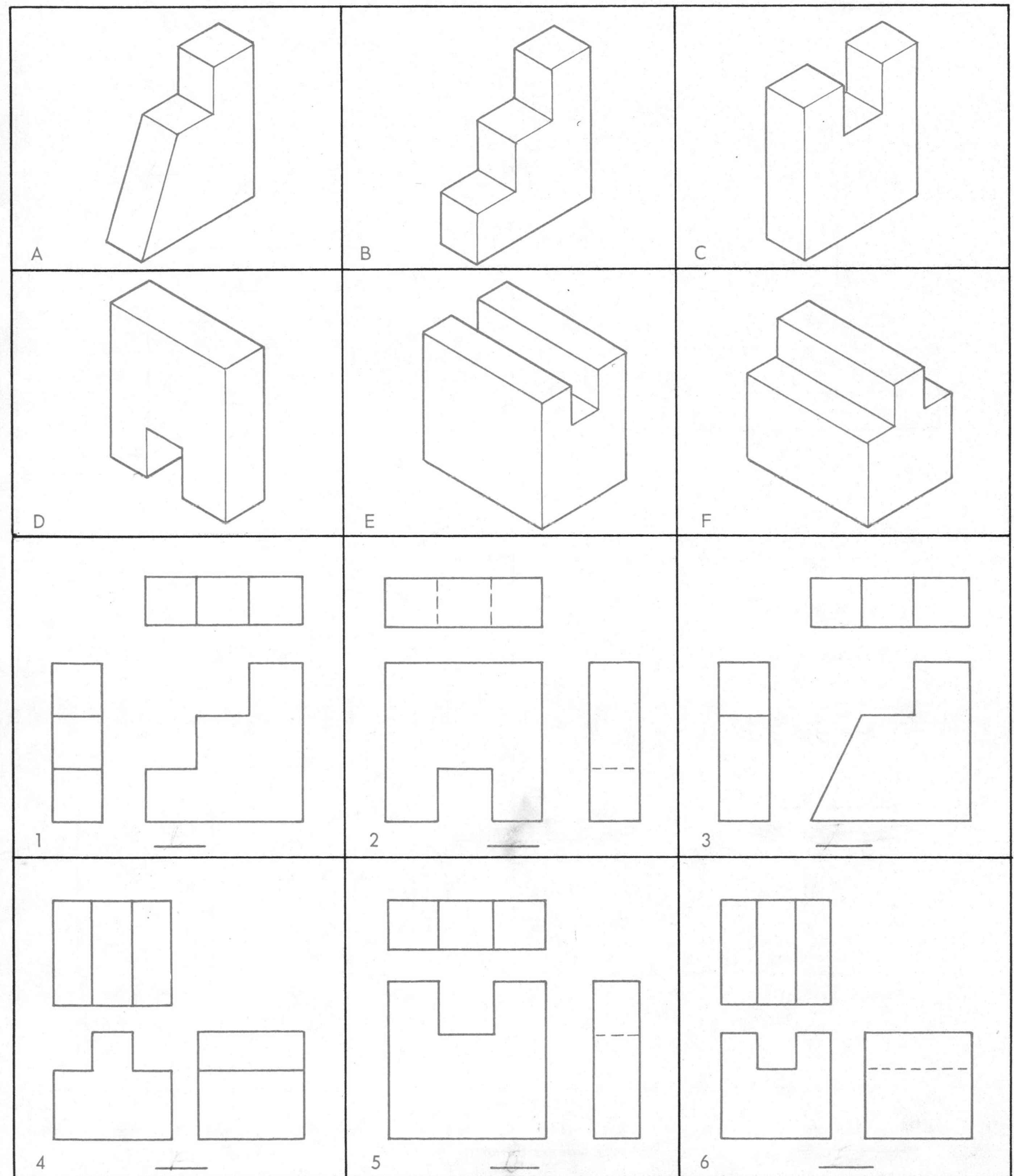

Fig. 5-BPR-2. Study the pictorial views and match each orthographic drawing with its pictorial drawing by inserting the correct letter in the space provided.

Blueprint Reading Activity 5-BPR-3
SKETCHING MISSING LINES

Fig. 5-BPR-3. Sketch the missing lines in the following orthographic projection drawings. Use the sketching technique studied in Unit 4.

Blueprint Reading Activity 5-BPR-4
SKETCHING ORTHOGRAPHIC VIEWS

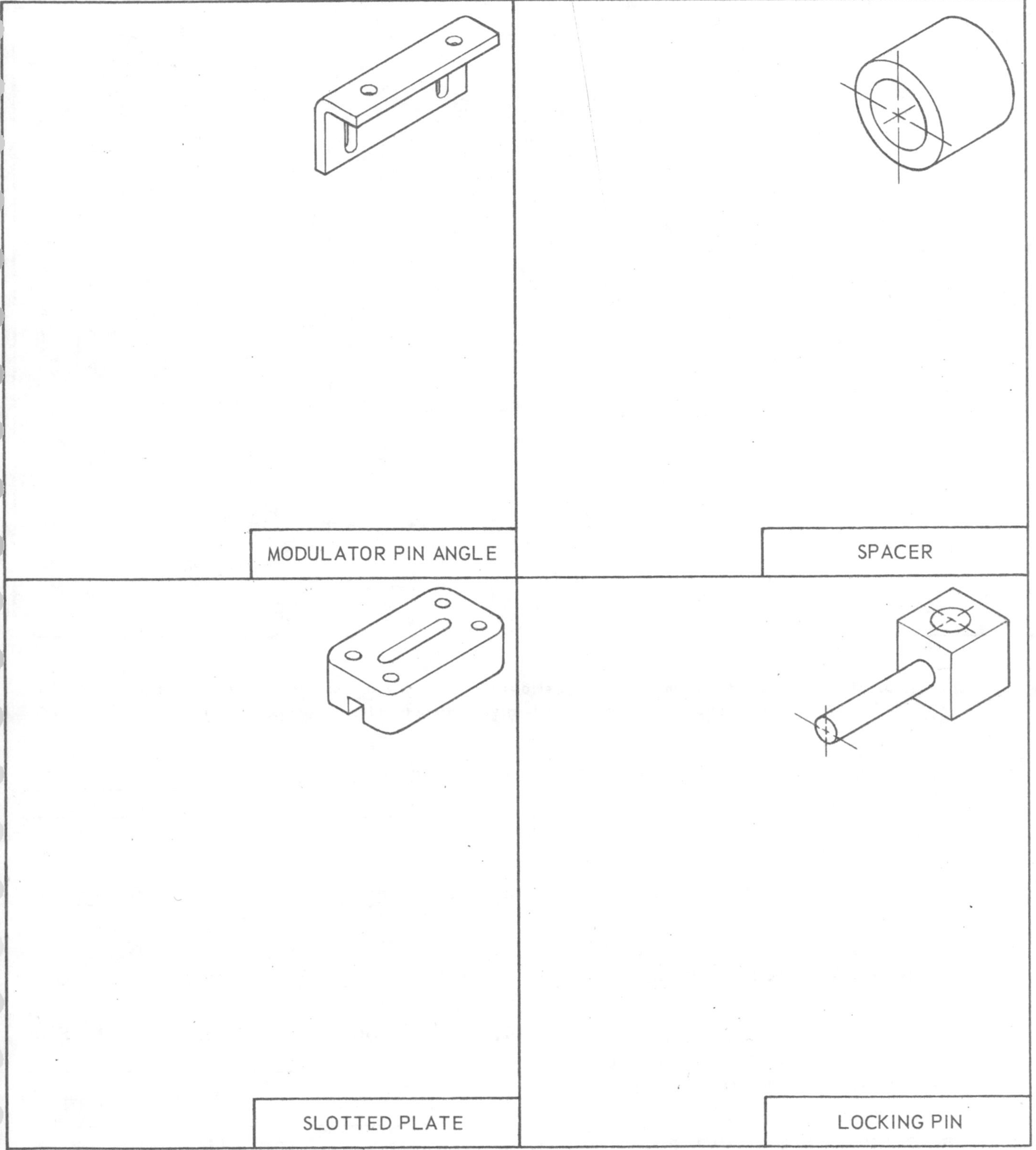

Fig. 5-BPR-4. Sketch the necessary orthographic projection views of the following objects. Select the views carefully and sketch only the views required to fully describe the object.

Blueprint Reading Activity 5-BPR-5
CHILL BAR

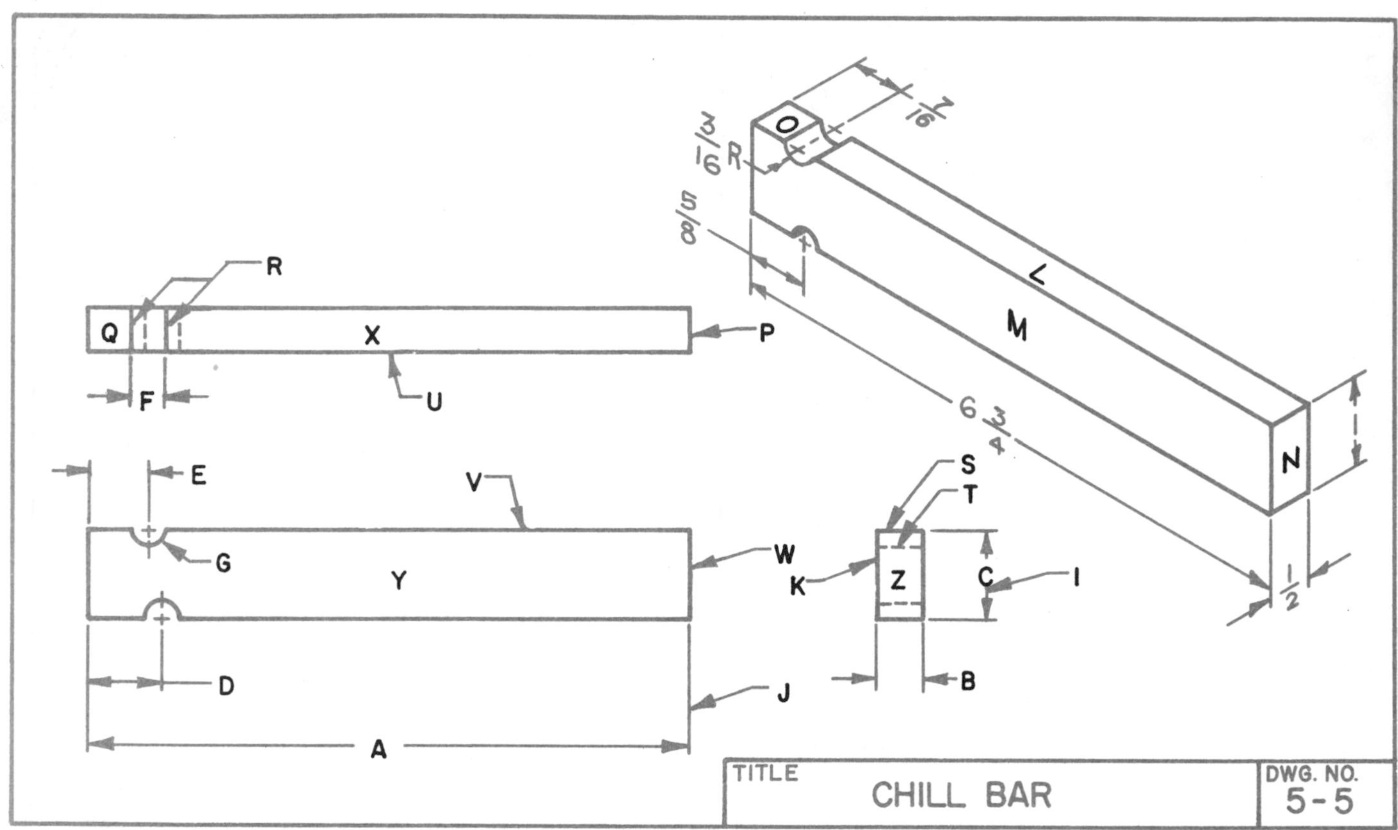

Fig. 5-BPR-5. Study the blueprint and answer the questions by placing your answers in the spaces provided. An example of a correct response is given in each set of answers.

1. Dimensions for:

   1. A ___6 3/4___ B _______ C _______
   D _______ E _______ F _______
   G _______

2. Name orthographic projection views:

   2. X ___Top___ Y _______ Z _______

3. Identify surfaces in the orthographic projection represented by the following letters in the pictorial view:

   3. L ___X___ M _______ N _______
   O _______

4. Identify, in the other views, the line or surface represented by the letters given. Place your response on the same line under the appropriate view.

4.	Top	Front	R. Side
a.	X	V	S
b.		Y	
c.			Z
d.		G	

5. Identification of kinds of lines:

   5. P ___Visible___ J _______
   I _______ T _______

Blueprint Reading Activity 5-BPR-6
PISTON

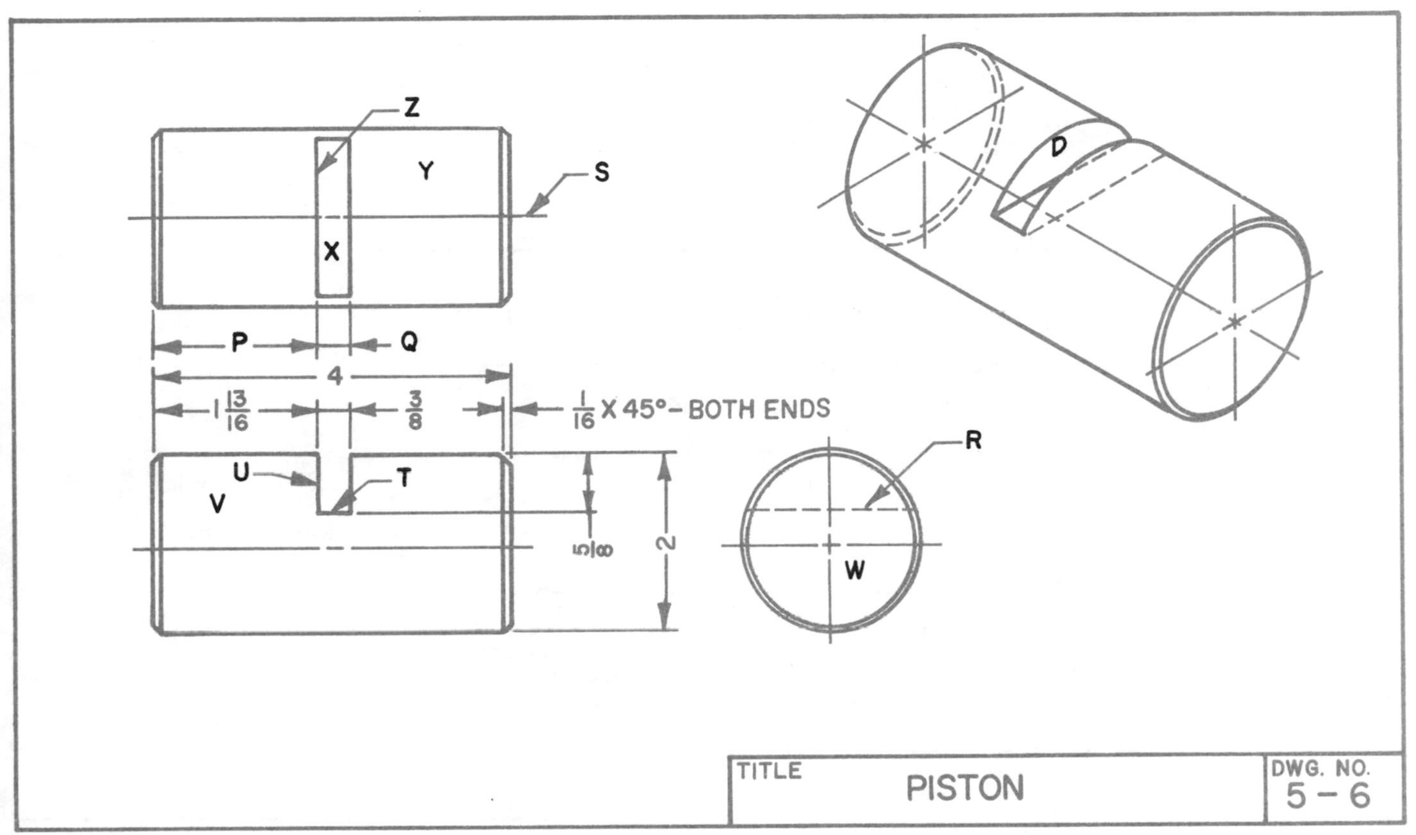

Fig. 5-BPR-6. Study the blueprint and answer the following questions.

1. Name the views represented by: — 1. Y__________ V__________ W__________

2. What kinds of lines are: — 2. Z__________ S__________ R__________

3. Line T represents: — 3. a. In the top view __________.
   b. In the right side view __________.

4. Give the following dimensions: — 4. a. Length-front view __________.
   b. Depth of groove X __________.
   c. Dimension Q __________.
   d. Dimension P __________.

5. Line U represents: — 5. a. In the top view __________.
   b. In the pictorial view __________.

Blueprint Reading Activity 5-BPR-7
BOTTOM CAP

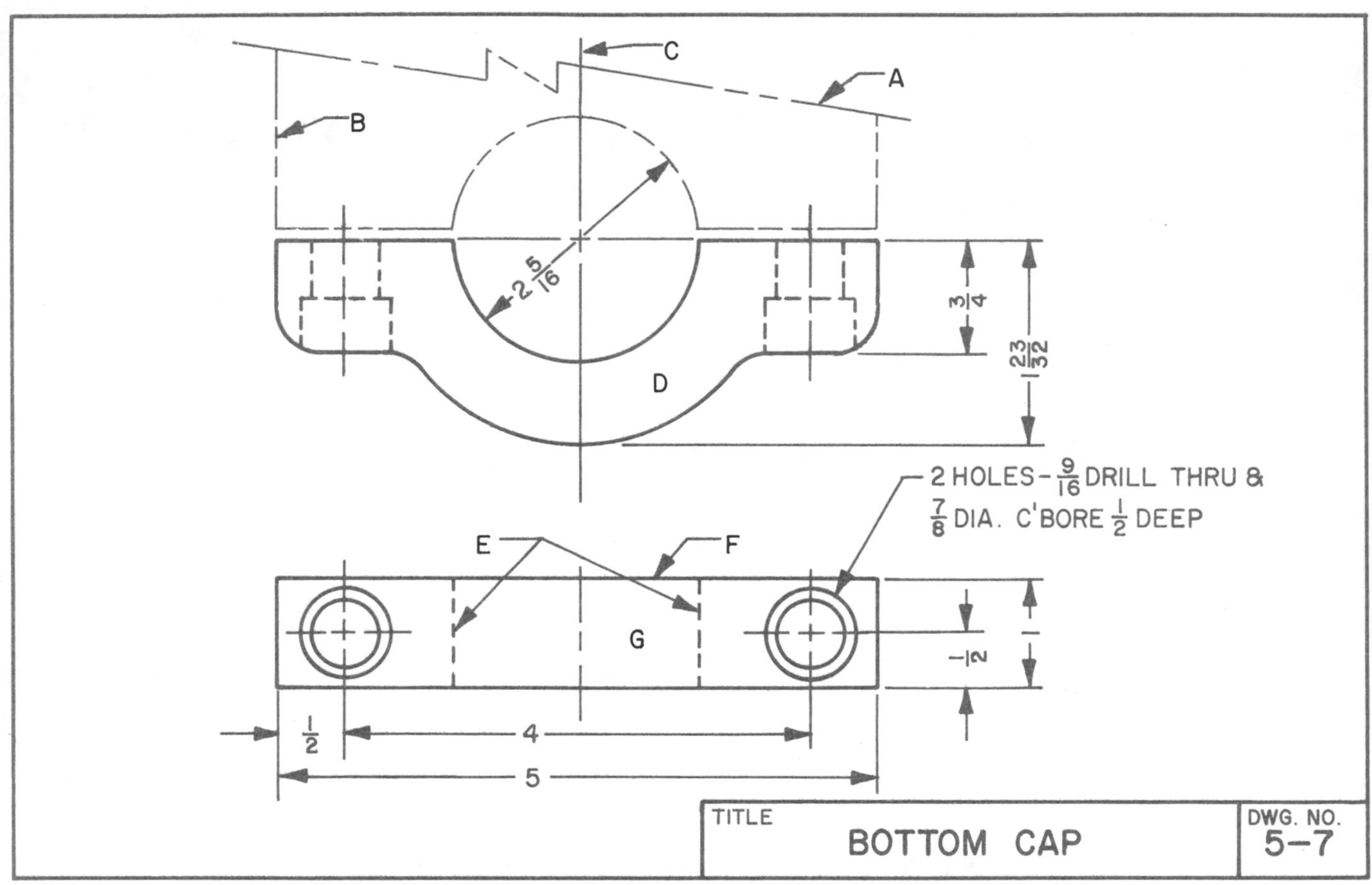

Fig. 5-BPR-7. Study the blueprint and complete the items required.

1. Name the views D and G.	1. D __________ G __________
2. Surface D is represented by what letter in the other view?	2. __________
3. What is the length, width and heighth of the part?	3. l ________ w ________ h ________
4. What is the hole diameter formed by the Bottom Cap and adjoining member?	4. __________
5. Two holes are drilled through the part. What is the diameter of these holes?	5. __________
6. The same two holes are counterbored to what diameter and depth?	6. ________ Dia. ________ Depth
7. Kinds of lines represented by:	7. A ________ B ________ C ________ E ________ F ________
8. What is the distance between the two lines marked E?	8. __________

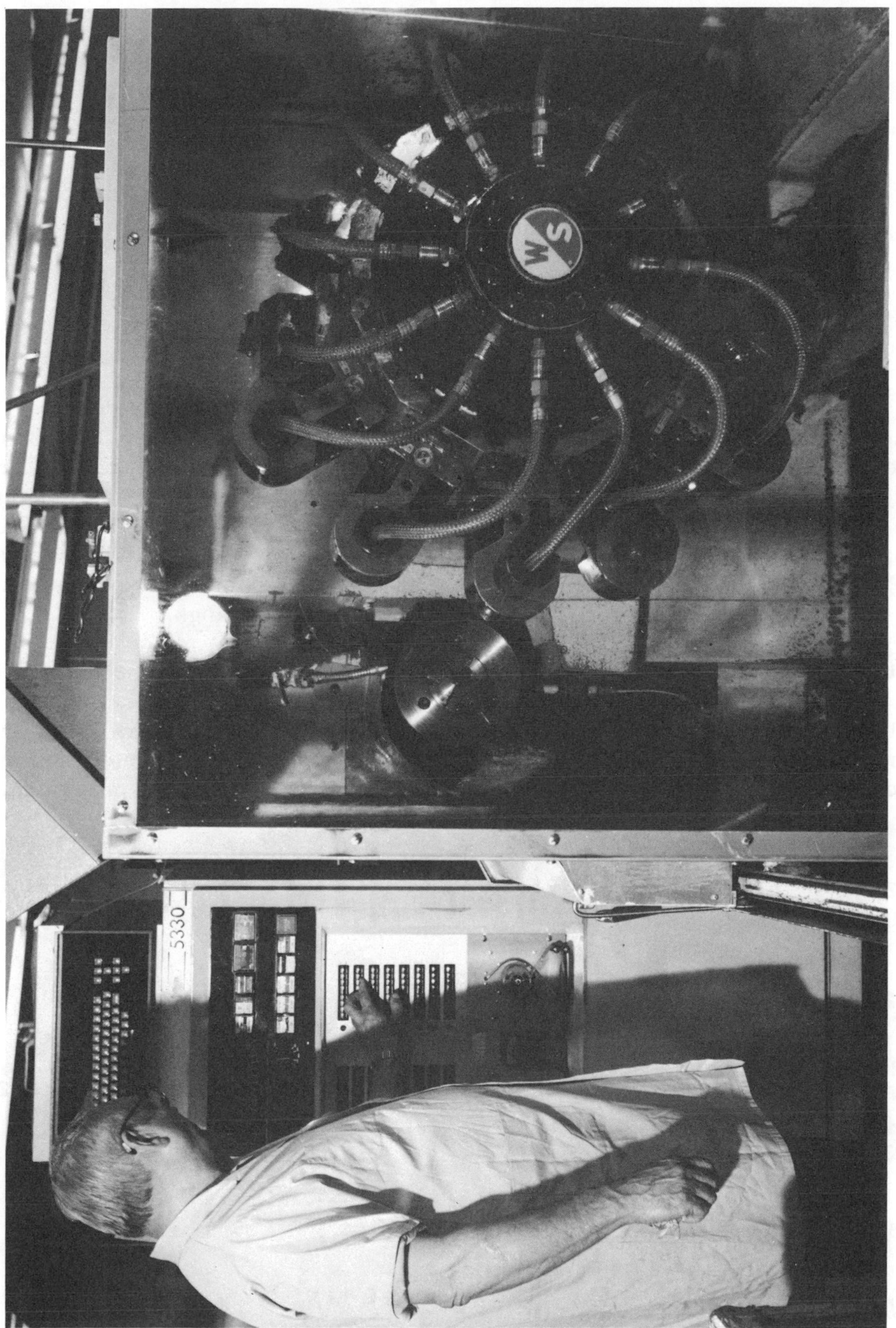

Industry photo. ***Blueprint reading skills are needed by those who program and operate N/C machines such as this N/C lathe.*** *(Perkin-Elmer Corp.)*

Unit 6
Lettering and Dimensioning Freehand Sketches

This unit is planned to give you some experience in forming the style of letters and numerals used on industrial drawings and sketches. The application of dimensions on drawings and freehand sketches is also discussed and illustrated.

Lettering

Style of Lettering

The lettering used on most industrial drawings and sketches is of the Gothic style because it is easy to read and construct. Gothic lettering is done vertically (90°), Fig. 6-1, or inclined ($67\ 1/2^{\circ}$), Fig. 6-2. Vertical upper case (capital) letters are widely used in industry with some industries preferring the inclined upper case letters. Lower case (small letters) are used for notes on some drawings, particularly on map drawings.

Forming Letters and Numerals

The Gothic style refers to letters and numerals which are formed with single strokes of the pencil or pen rather than

A B C D E F G H I J K L M
N O P Q R S T U V W X Y Z
1 2 3 4 5 6 7 8 9 0 8
a b c d e f g h i j k l m n
o p q r s t u v w x y z

Fig. 6-1. Vertical Gothic lettering strokes and proportions.

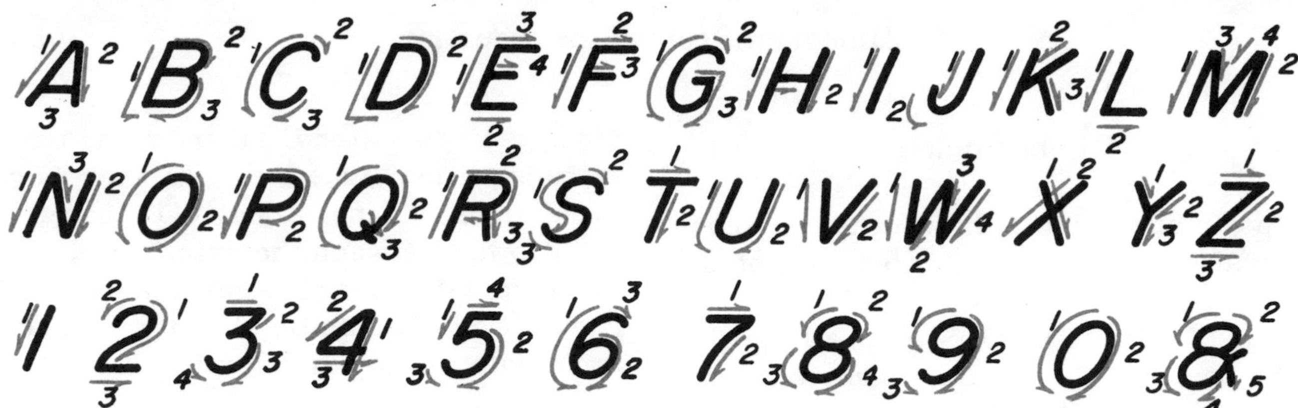

Fig. 6-2. Inclined Gothic lettering strokes and proportions.

letters that vary in thickness. One to four strokes are required to properly construct the letters as shown in Figs. 6-1 and 6-2. By following these stroke patterns you will soon be able to construct neat, well-shaped letters and numerals. Letters are usually made 1/8 inch in height for notes and dimensions with the title of the part 1/4 inch in height. Numerals are equal in height to capital letters and fractions are twice the height of whole numbers, Fig. 6-3. A medium to soft pencil is recommended for lettering.

H
2H
BORE $2\frac{1}{2}$

Fig. 6-3. Height of letters, numerals and fractions.

Guide Lines

Horizontal lines and vertical or inclined guide lines will help you achieve quality in height and slant of letters, Fig. 6-4. These should be constructed lightly so as not to detract from the notes or dimensions.

HORIZONTAL GUIDE LINES
90°
VERTICAL GUIDE LINES
$67\frac{1}{2}$°
INCLINED GUIDE LINES

Fig. 6-4. Guide lines.

Spacing

Good lettering is always compact but not crowded, Fig. 6-5(b).

(a) TOO CROWDED — T O O O P E N

(b) CORRECTLY SPACED LETTERING

Fig. 6-5. Spacing of letters.

The space between letters and words should be pleasing to the eye. Some letters, such as A, C, L and T, because of their shape, can be spaced more closely than letters like B, D, H and M. The spacing between words is equal to about one letter space and about two letter spaces are used between sentences, Fig. 6-6.

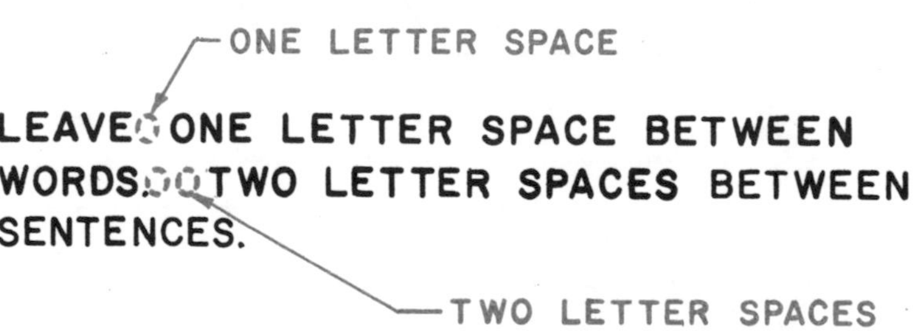

Fig. 6-6. Spacing of words and sentences.

Dimensioning

In dimensioning a drawing the engineer or draftsman provides two types of dimensions:

1. SIZE DIMENSIONS that tell the size of the part and the size of its various geometric features such as arcs, circles, rectangles and angles.
2. LOCATION DIMENSIONS which tell the location of features such as holes, slots and grooves.

Dimension and Extension Lines

Dimension lines are always parallel to the distance dimensioned, Fig. 6-7. Extension lines extend out from the part or feature dimensioned to 1/8 inch beyond the dimension line. About 1/16 inch should be allowed between the extension line and the object to clearly separate the object. Arrowheads should be sharply formed and terminate at the extension line.

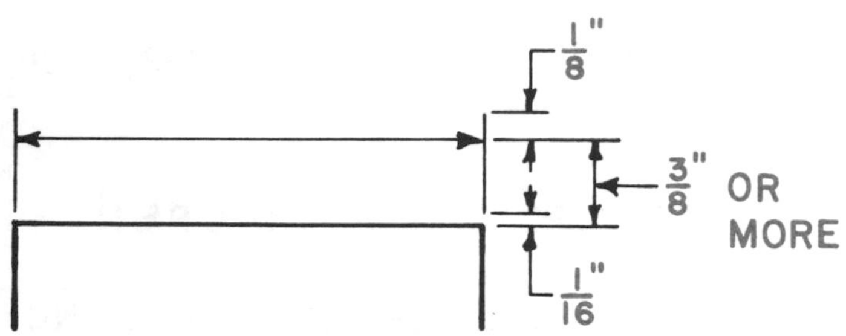

Fig. 6-7. Dimension and extension lines and arrowheads.

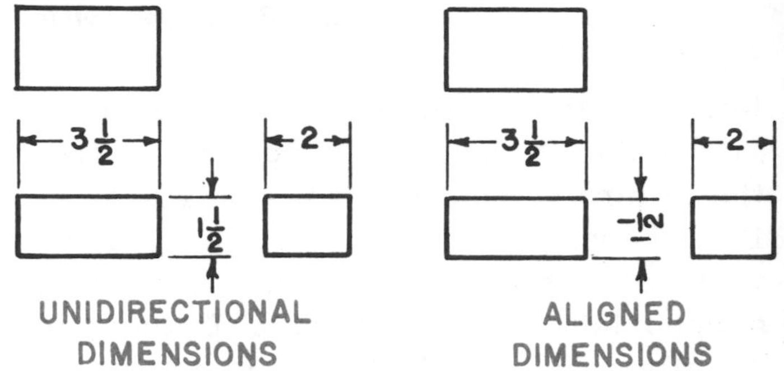

Fig. 6-8. Unidirectional and aligned dimensional systems.

Placement of Dimensions

Dimensions may be placed so they read from the bottom of the drawing (unidirectional dimensions) or read from the bottom and right side (aligned dimensions). See Fig. 6-8. The unidirectional system is

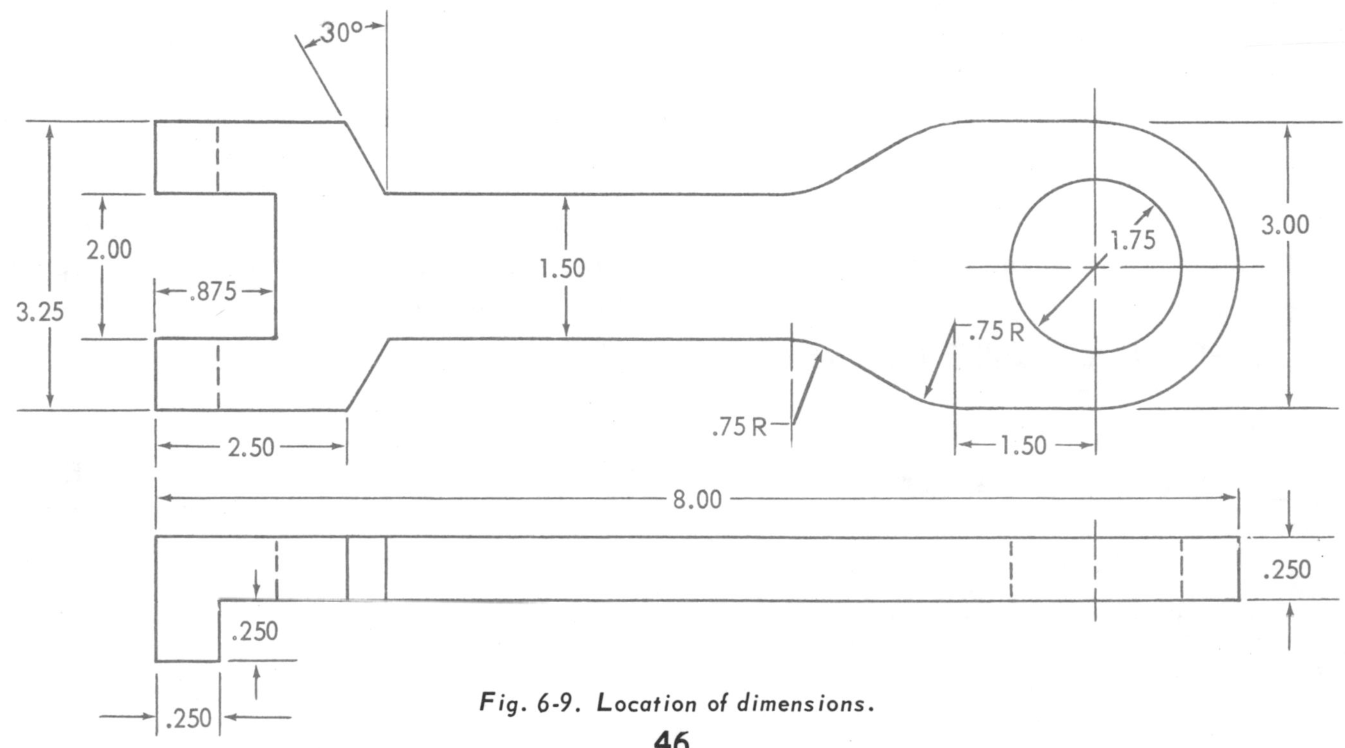

Fig. 6-9. Location of dimensions.

gaining in favor because the dimensions are more easily read by the workmen. Only one system should be used on a single drawing or sketch.

In so far as possible, all dimensions should be placed outside the views, Fig. 6-9. Exceptions are permitted for clarity and where a number of lines would be crossed by extension lines in carrying the dimension outside the view. Diameters and radii may be placed on the view if space and clarity permit. It is considered good practice to dimension between the views and next to the view where the feature dimensioned is best described. Dimensions for narrow spaces, such as slots or thin parts, may be placed on either side of the extension lines or directed to the dimensioned space with a leader.

Lettering Activity 6-BPR-1
GOTHIC LETTERING AND NUMERALS

In the spaces below, practice forming the letters and numerals as directed. Follow the lettering charts and guides in forming your letters and numerals.

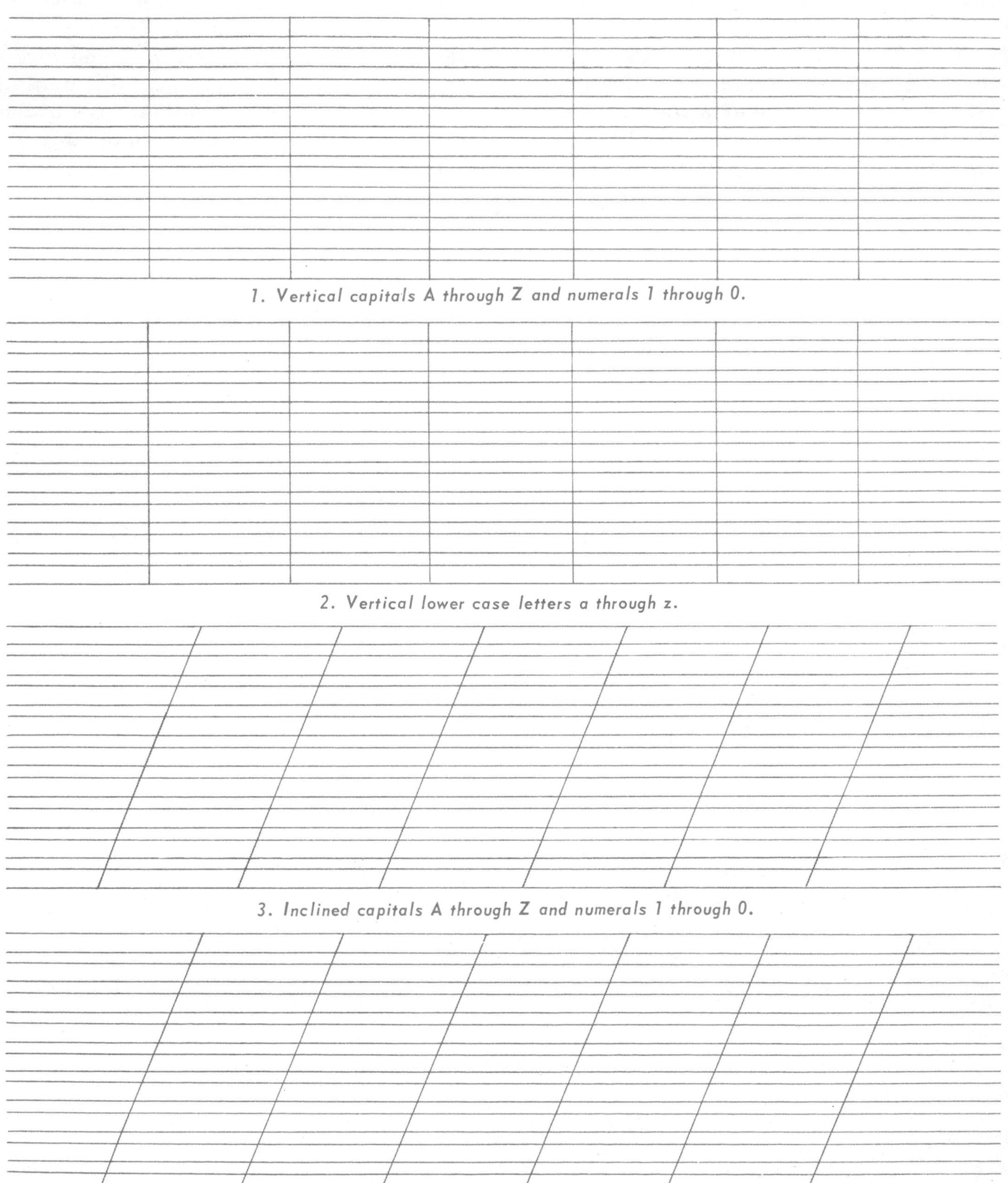

1. Vertical capitals A through Z and numerals 1 through 0.

2. Vertical lower case letters a through z.

3. Inclined capitals A through Z and numerals 1 through 0.

4. Inclined lower case letters a through z.

Unit 7
Auxiliary Views

Draftsmen sometimes find it necessary to use special or AUXILIARY views of an object to show the shape and size of the surfaces which cannot be shown in the regular views.

See Fig. 7-1 Details (a) and (b). You will notice that the three views of the block given in Detail (b) do not show the true size and shape of the inclined surface (surface cut at an angle). An auxiliary view, included in Detail (c), is required to provide the additional information needed. In the auxiliary view, the inclined surface is viewed from a position perpendicular to the surface. Auxiliary views may be projected from any view in which the inclined surface appears as a line.

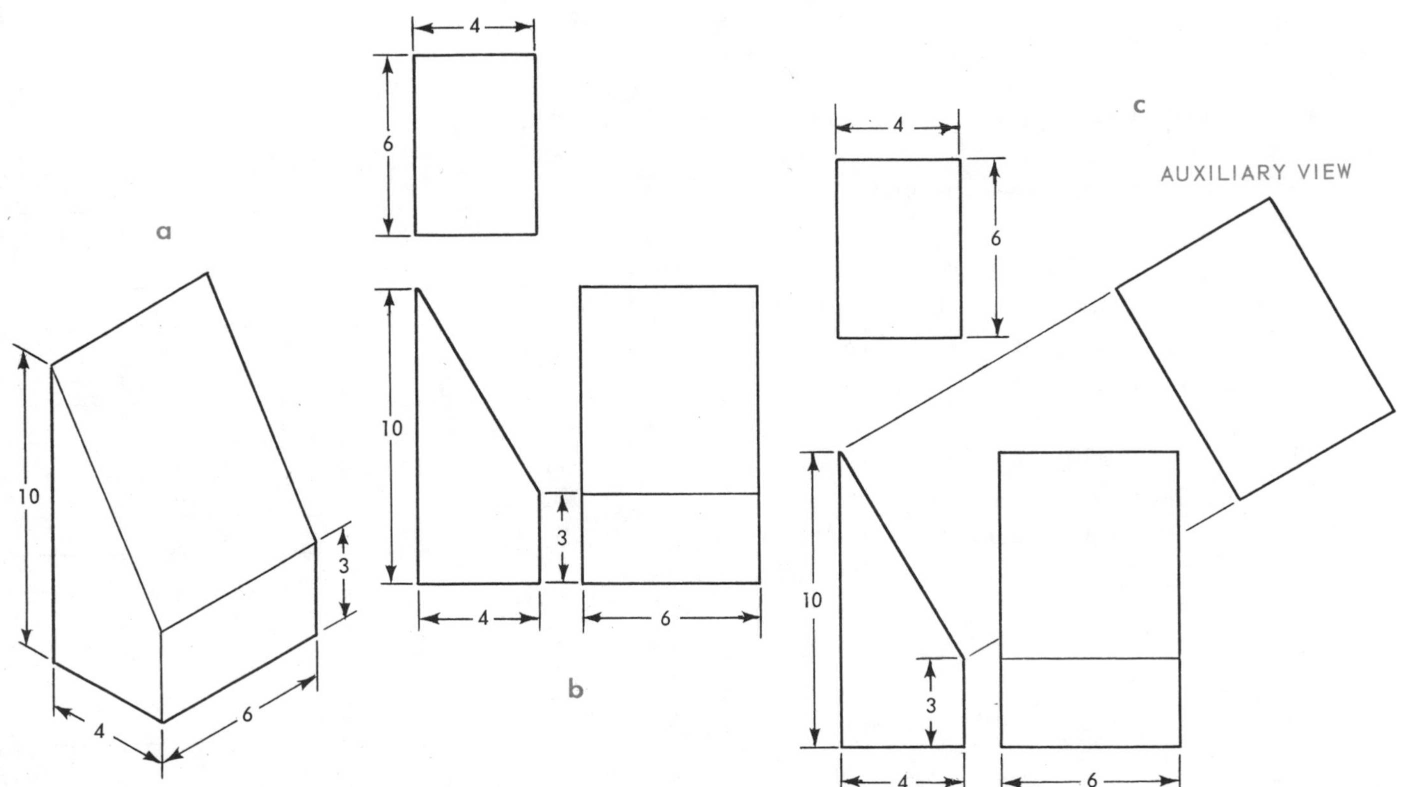

Fig. 7-1. Auxiliary view is added to give size and shape of inclined surface.

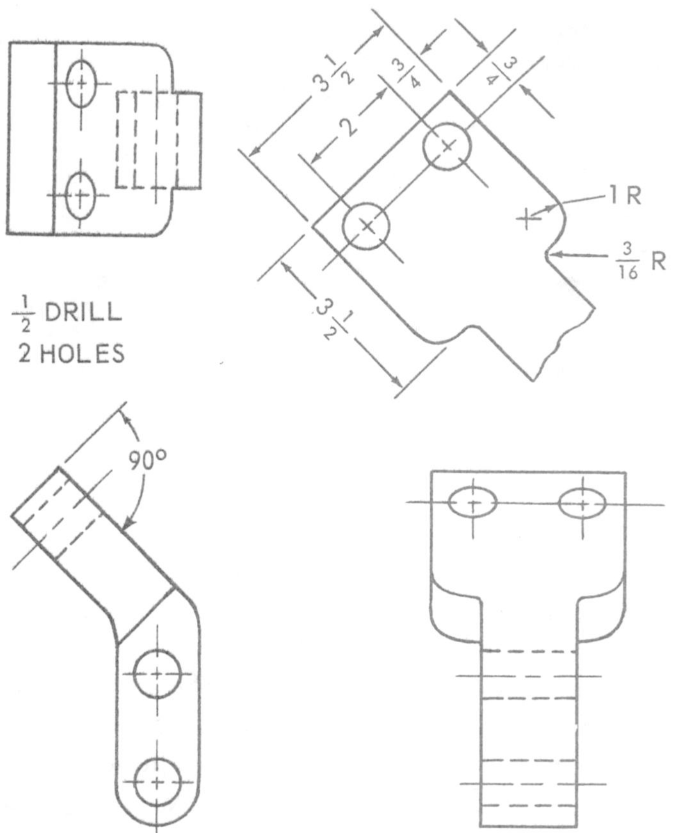

Fig. 7-2 also shows an auxiliary view projected on a plane parallel to the inclined surface, to provide a true representation of the special feature of the object. The distances for this inclined surface are foreshortened in the top and right side views. The surface appears in its true shape and size in the auxiliary view.

When projected at right angles to the surface to be featured, the auxiliary view is obtained in the same manner as all other orthographic projection views. However, since it is projected from an inclined surface onto a parallel plane, the auxiliary view appears on the drawing at an odd angle with the rest of the orthographic projection views.

Fig. 7-2. Auxiliary view showing true shape and size of a feature of an object.

Blueprint Reading Activity 7-BPR-1
AUXILIARY VIEWS

Study blueprint Fig. 7-BPR-1 and answer these questions:

1. What three views are shown?

2-6. Identify, in the other views, the line or surface represented by the letter given. If the feature does not appear in a view, place a dash in the answer space.

7. What is the radius of the auxiliary piece?
8. Give the length of surface:
9. Give the width and length of surface:
10. Give the thickness, width and length of triangle D:
11. What is the material from which the hook body is made?
12. What is the inside radius of the "eye" hook?

1.	______	______	______
	FRONT	TOP	AUXILIARY
2.	A	______	______
3.	______	C	______
4.	______	______	E
5.	B	______	______
6.	______	D	______

7. ______________________
8. A ______________________
9. B ____________ E ____________
10. ______________________
11. ______________________
12. ______________________

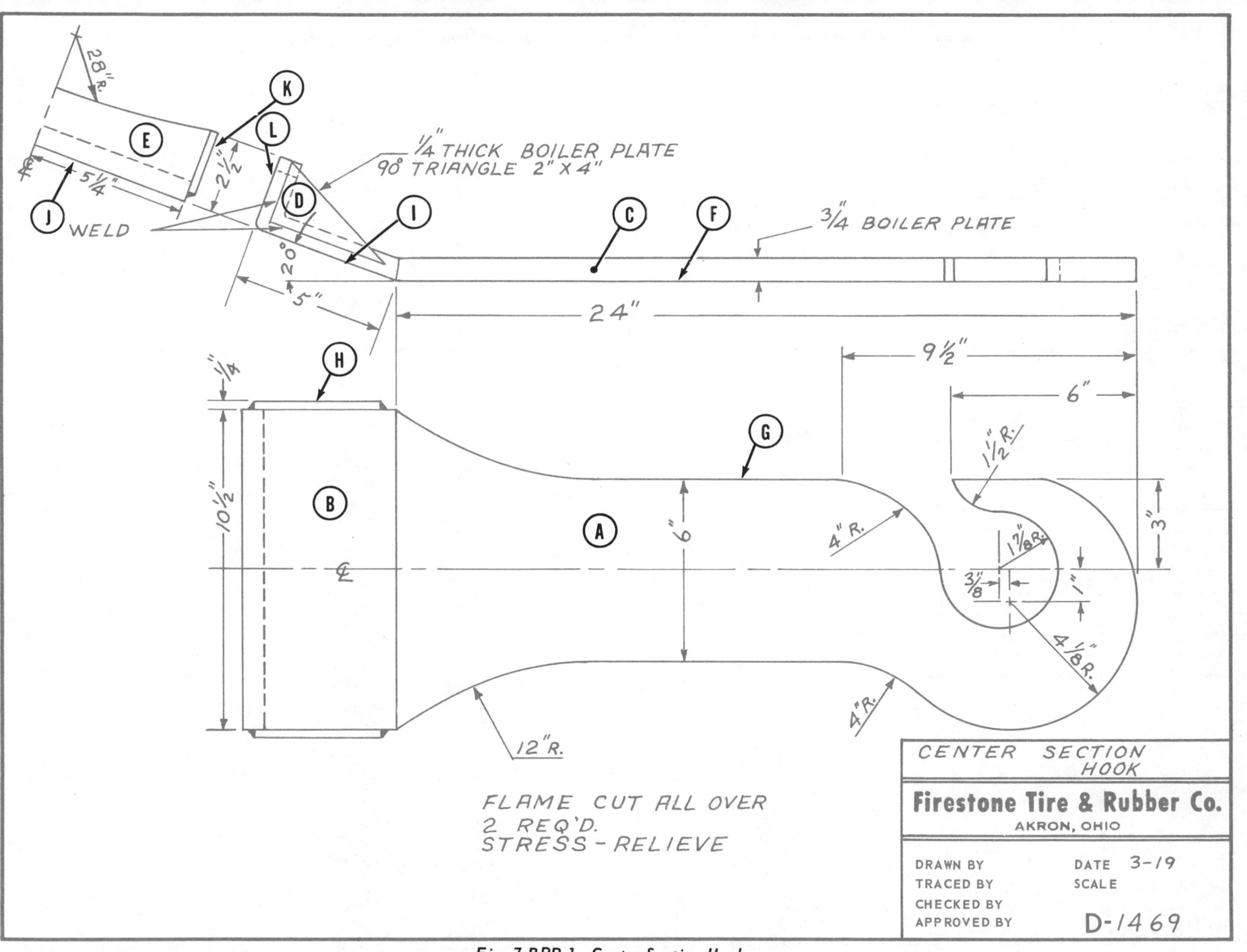

Fig. 7-BPR-1. Center Section Hook.

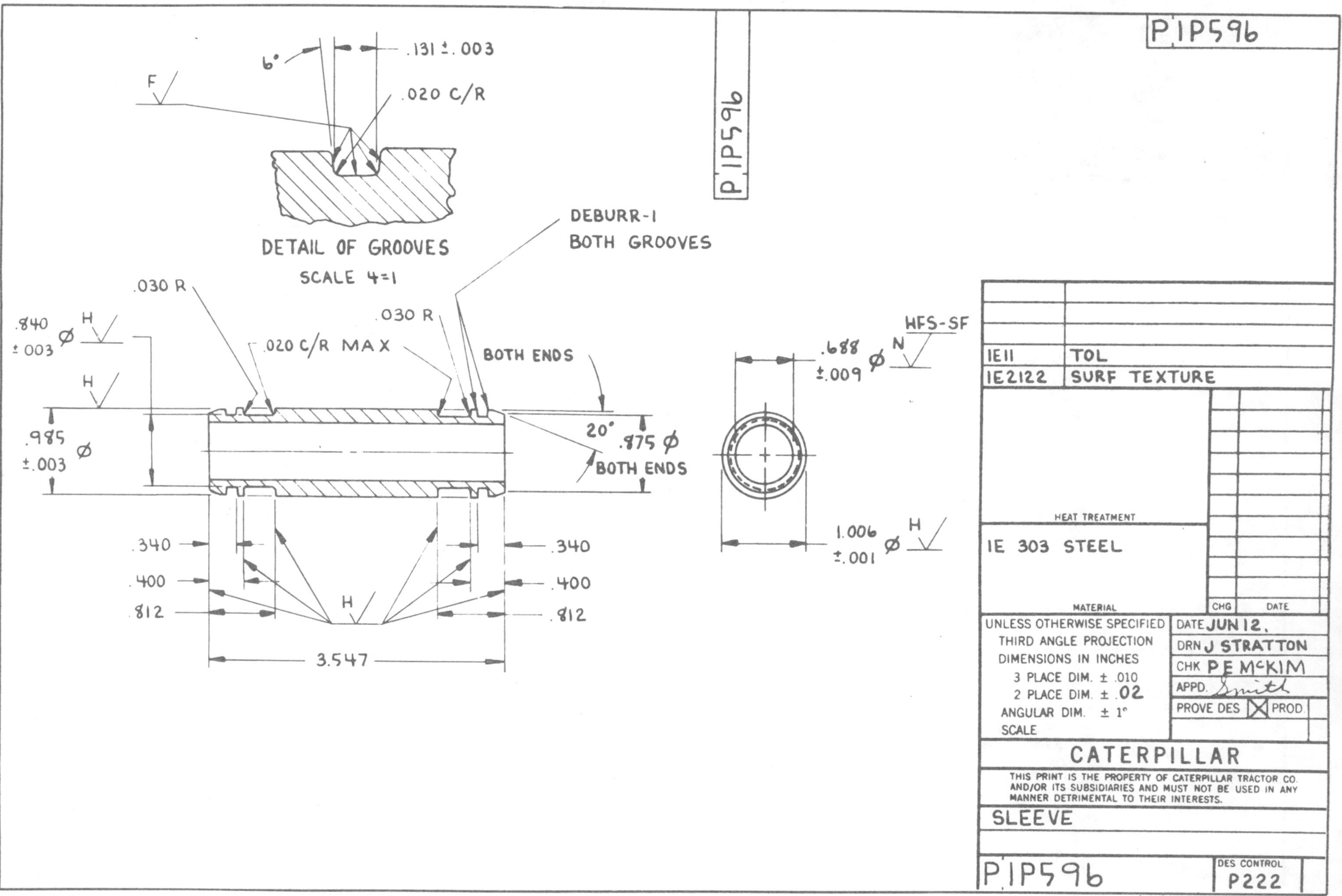

Fig. 8-1. Industry plan. Detail drawing giving constructional details on a metal sleeve. (Caterpillar Tractor Co.)

Unit 8
Detail and Assembly Drawings

In the manufacture of a machine or the building of a structure, a set of drawings is necessary to provide details for the production of each of the parts and information on the correct assembly of the parts. The drawings, called WORKING DRAWINGS, consist of two basic types: DETAIL DRAWINGS for the parts produced, and an ASSEMBLY DRAWING for each unit or sub-unit to be put together.

The nature and purposes of detail and assembly drawings will be discussed in this unit.

Detail Drawings

A DETAIL DRAWING is a drawing made for a single part and provides all the information necessary in the production of that part. It supplies the workman with the:

1. Name of the part.
2. Shape description of the part.
3. Dimensional size of the part and its features.
4. Notes detailing material, special machining, finish, heat-treatment, etc.

Usually only one part is placed on a detail drawing, but some industries detail several related parts on a single sheet. Each part in a mechanism is detailed and a print made available to the mechanic or technician working on that part. Depending on the complexity of the mechanism or machine, several, or several hundred, detail prints may be needed in its manufacture or construction.

Fig. 8-1 is typical of a detail drawing used in industry.

Assembly Drawings

ASSEMBLY DRAWINGS show the working relationship of the various parts of a machine or structure as they fit together. Usually, each part in the assembly is numbered and listed in a table on the drawing. The assembly drawing provides the workman with the following:

1. Name of the assembly mechanism or sub-assembly.
2. Visual relationship of one part to another in order to correctly assemble the various parts.
3. List of parts.
4. Bill of material may be included.
5. Overall size and location dimensions when necessary to check clearance or fitting to another mechanism or foundation.

An assembly drawing may be made for

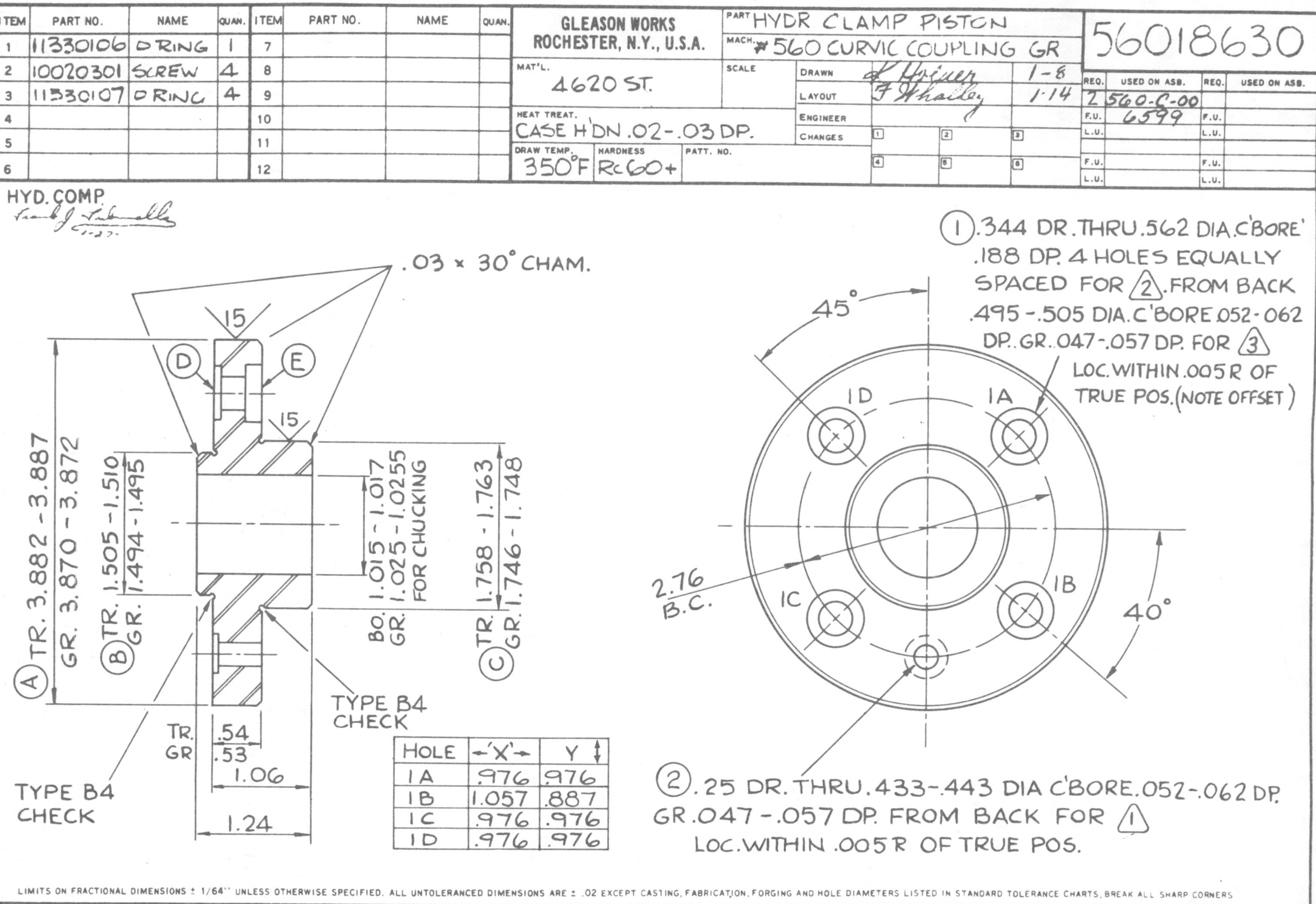

Fig. 8-2. Industry plan. Sub-assembly drawing of hydraulic clamp piston.

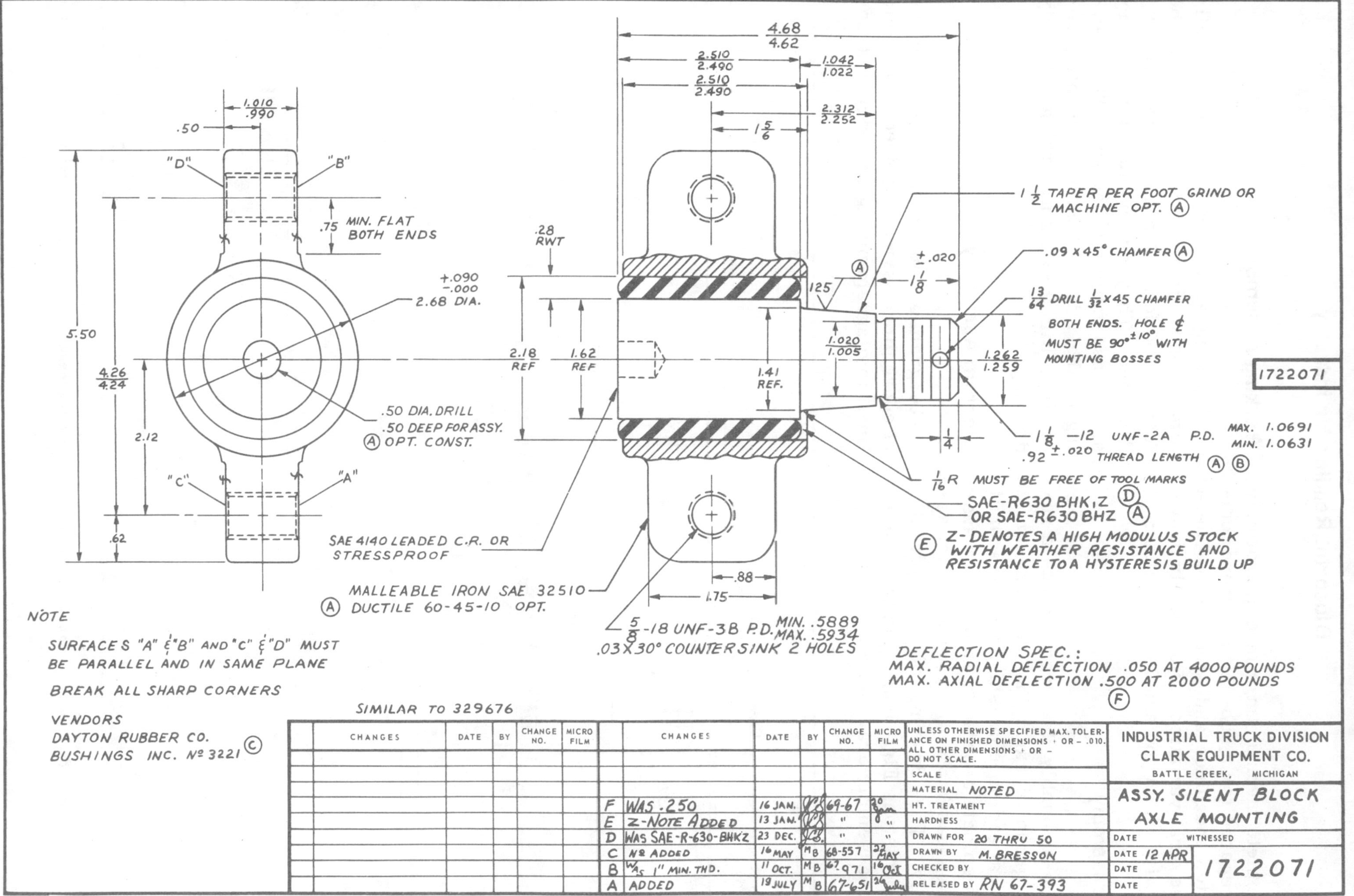

Fig. 8-3. Industry plan. Working assembly of silent block axle mounting.

a complete mechanism or machine, or may be prepared for a sub-unit for more complicated machines and structures. There are several variations of assembly drawings with which you should be familiar:

Sub-assembly

Sub-assembly drawings include a related group of parts that compose a unit in a larger mechanism such as a drill press spindle assembly, automatic transmission assembly or drive sprocket assembly. The drawing of the hydraulic clamp piston, Fig. 8-2, is an example of a sub-assembly drawing.

Working Assembly

Working assembly drawings are fully dimensioned drawings and combine the features of detail drawings with assembly drawings. They are useful only on very simple assembly mechanisms. The drawing of the silent block axle mounting assembly, Fig. 8-3, is a working assembly drawing.

Diagram Assembly

Diagram assembly drawings make use of conventional symbols joined with single lines to diagram piping and wiring struc-

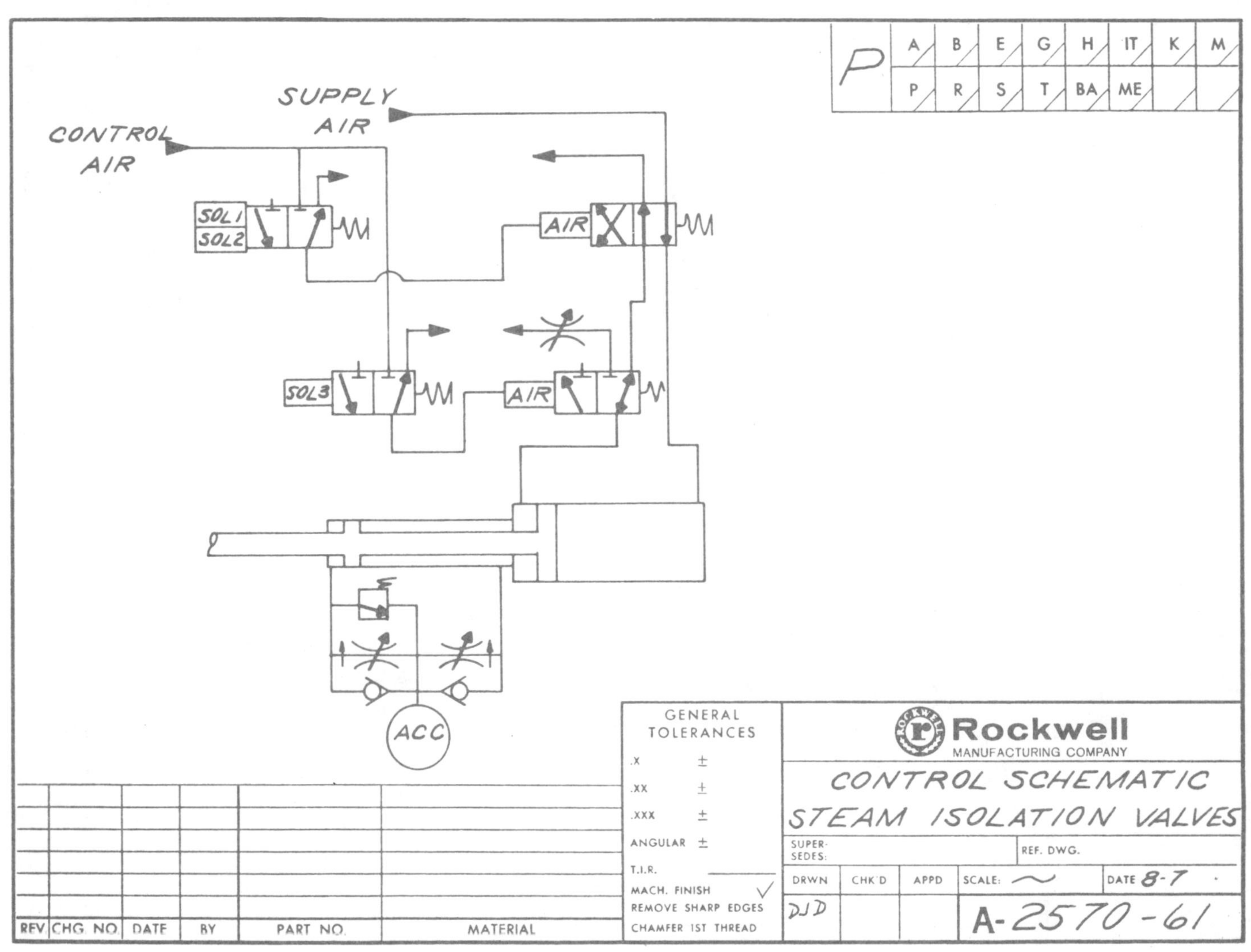

Fig. 8-4. Industry plan. Diagram assembly, control schematic steam isolation valves. (Rockwell Mfg. Co.)

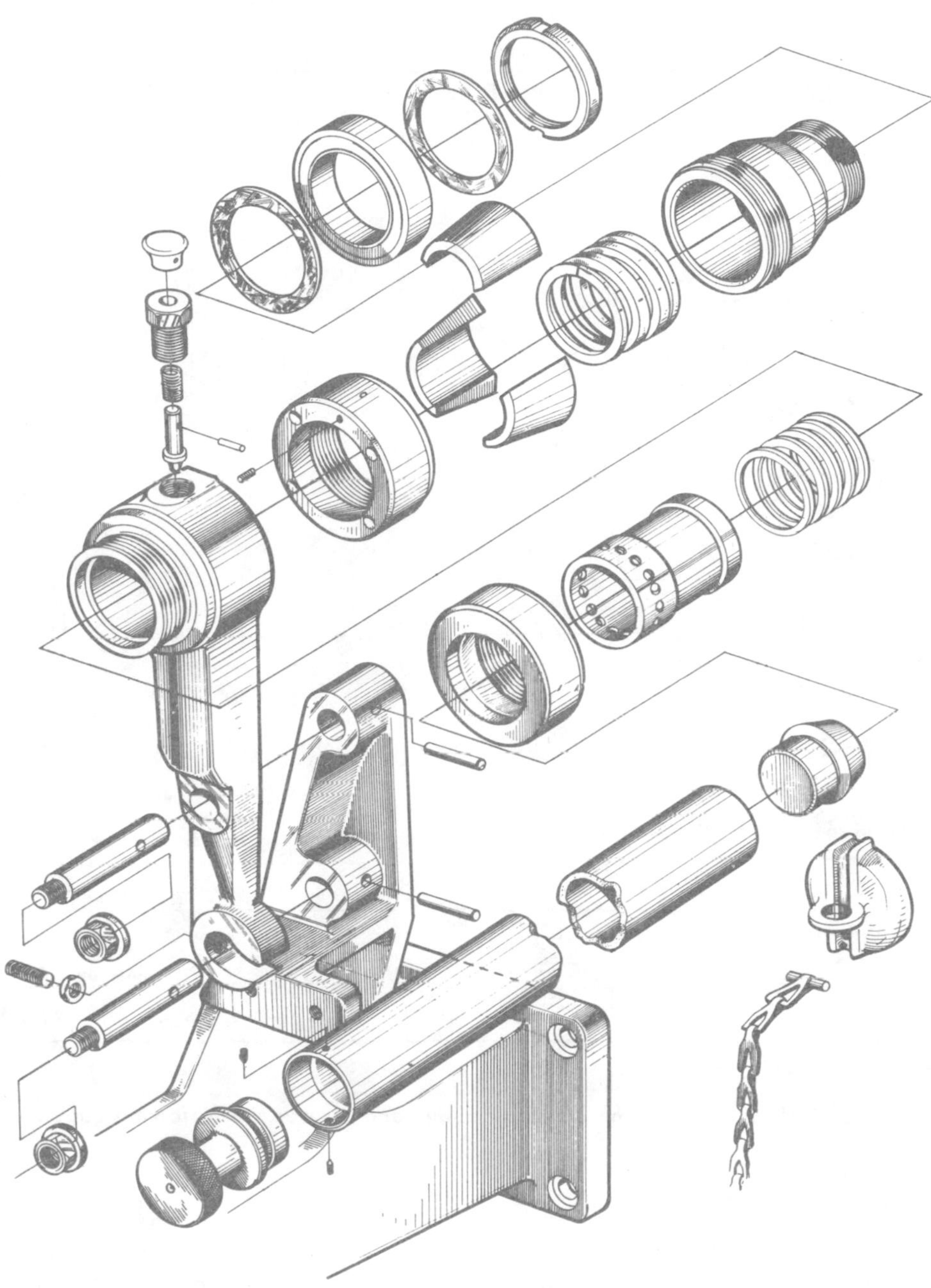

Fig. 8-5. Industry plan. Exploded pictorial assembly showing stock support for screw machine. (Brown and Sharpe)

tures. An example of a diagram assembly drawing is shown in Fig. 8-4.

Installation Assembly

Installation assembly drawings provide the necessary information to install or erect a piece of equipment such as a dust collector system or large piece of equipment shipped "knocked down" and erected at the customer's plant or field location.

Exploded Pictorial Assembly

Exploded pictorial assembly drawings show all the parts of an assembly in an "exploded" view in the correct order of assembly, Fig. 8-5. These are easy to read and understand when assembling a mechanism.

Blueprint Reading Activity 8-BPR-1
MAIN DRIVE SPROCKET

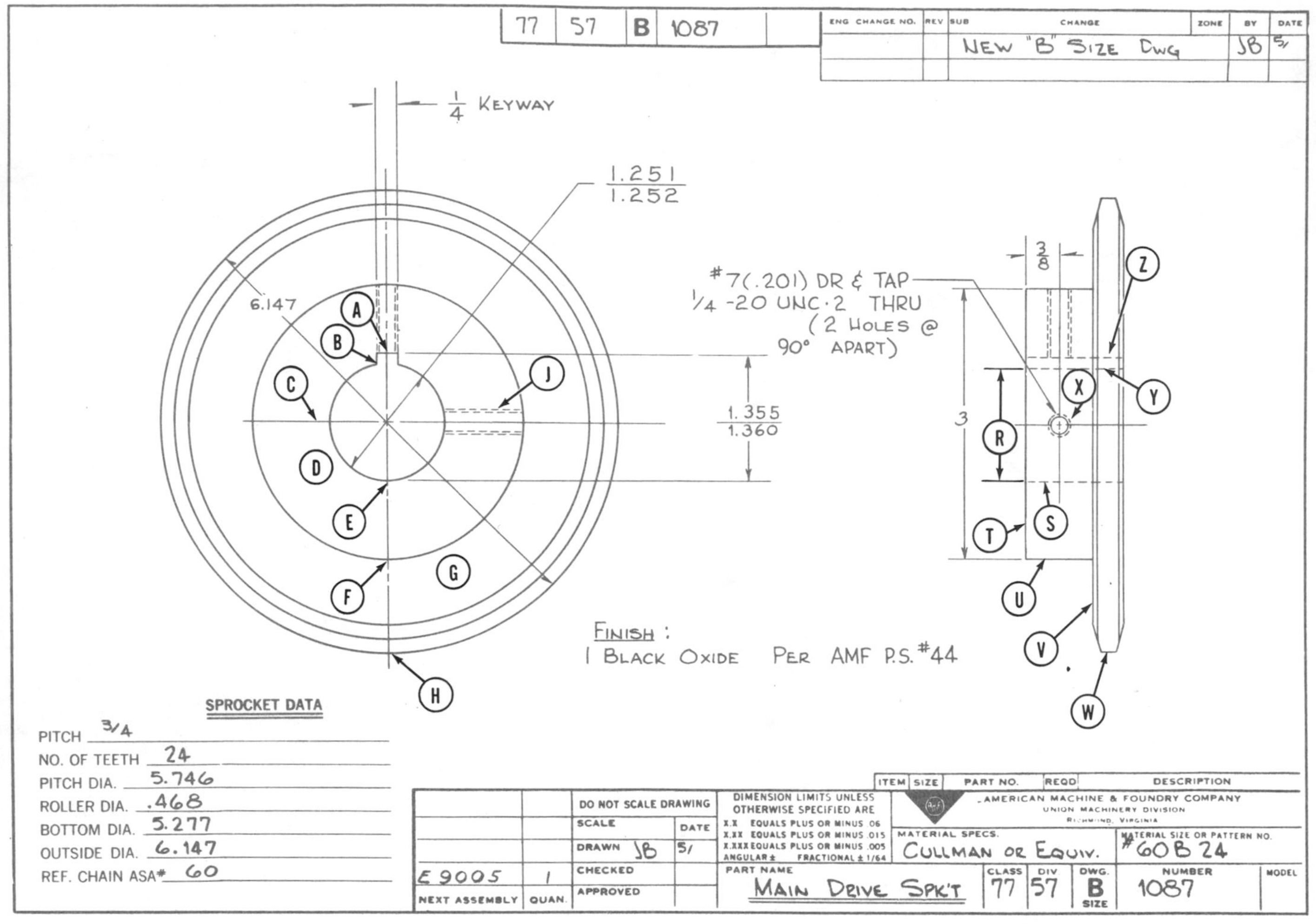

Fig. 8-BPR-1. Refer to the blueprint above and answer the following questions.

1. Circle the type of drawing shown. — 1. detail ______ ______ assembly
2. What views are shown? — 2. ____________
3-10. Identify these features in the other view:
   3. A ____________
   4. B ____________
   5. D ____________
   6. E ____________
   7. F ____________
   8. G ____________
   9. H ____________
   10. J ____________
11. What is the diameter of R? — 11. ____________
12-14. What kinds of lines are:
   12. A ____________
   13. C ____________
   14. S ____________
15. Give the outside diameter of the part. — 15. ____________
16. How many drilled and tapped holes are there? — 16. ____________

For additional Blueprint Activities for Unit 8 see pages 230 and 233.

Blueprint Reading Activity 8-BPR-2
ARM AND HUB ASSEMBLY

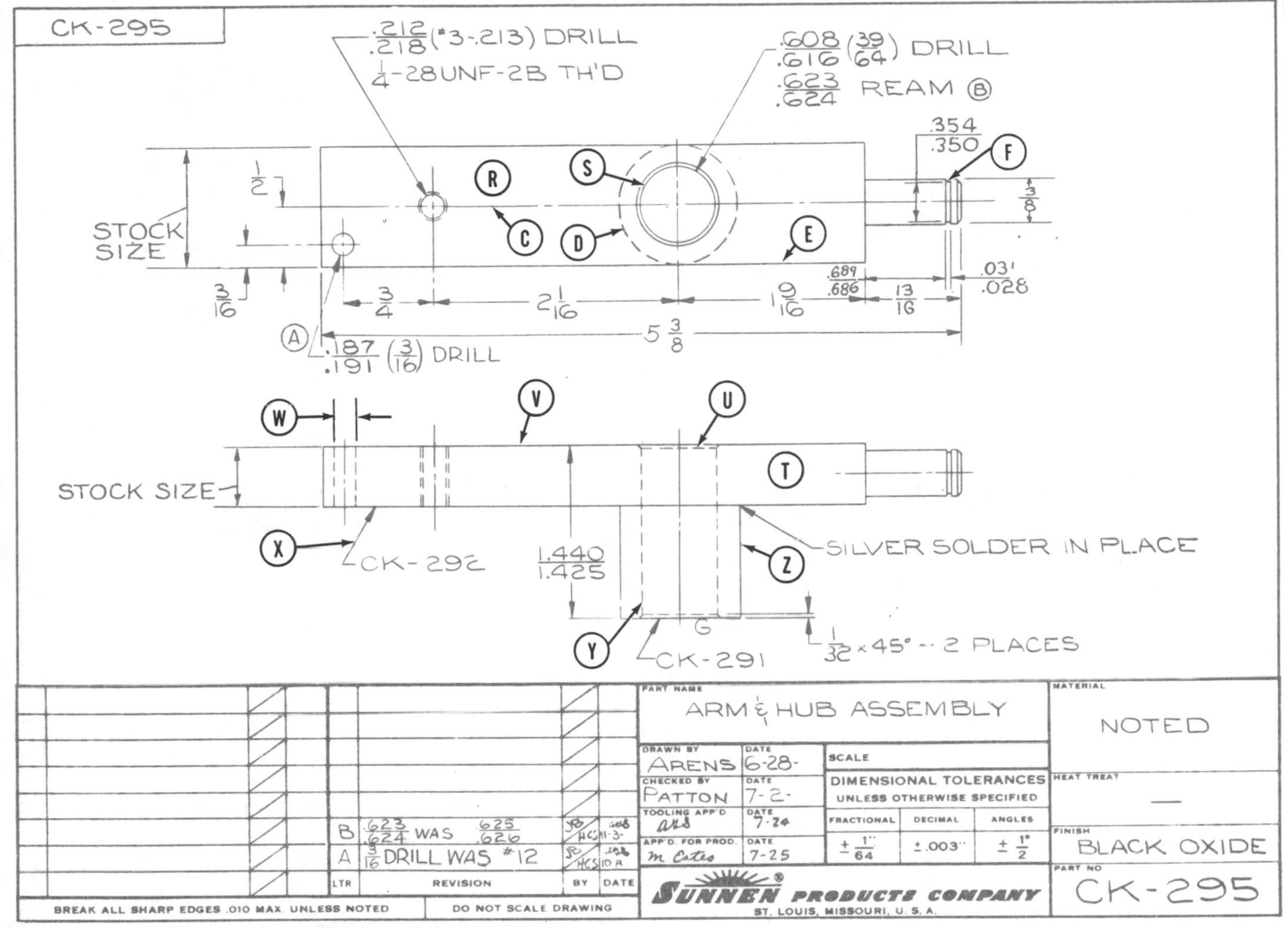

Fig. 8-BPR-2. Refer to the blueprint above and answer the following questions.

1. What kind of an assembly drawing is it? 1. ______
2. How many parts are there? 2. ______
3. How are the parts joined? 3. ______
4. What are the overall dimensions? 4. ______
5. How many holes are drilled in the assembly? 5. ______
6. What is the shape of part CK-292? 6. ______

7-10. Identify the following features in the other view:
7. R ______
8. S ______
9. D ______
10. E ______

11-14. What kind of lines are:
11. S ______
12. C ______
13. D ______
14. X ______

15-16. Give the dimension of:
15. F ______
16. W ______

17. What is the diameter of D? 17. D ______
18. What are the dimensions of the chamfer at U? 18. U ______

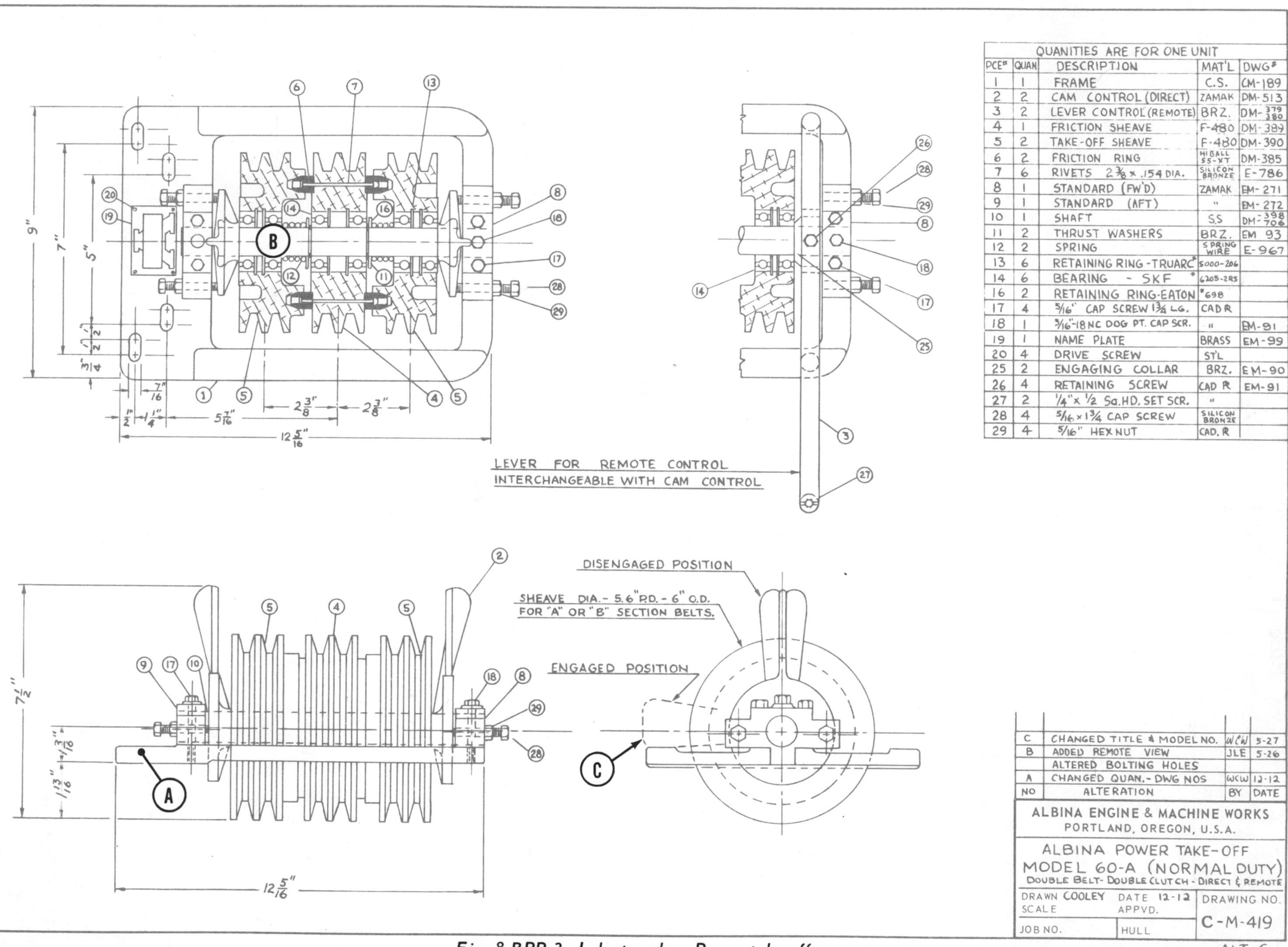

PCE#	QUAN	DESCRIPTION	MAT'L	DWG#
		QUANITIES ARE FOR ONE UNIT		
1	1	FRAME	C.S.	CM-189
2	2	CAM CONTROL (DIRECT)	ZAMAK	DM-513
3	2	LEVER CONTROL (REMOTE)	BRZ.	DM-379 380
4	1	FRICTION SHEAVE	F-480	DM-389
5	2	TAKE-OFF SHEAVE	F-480	DM-390
6	2	FRICTION RING	HI BALL 55-XT	DM-385
7	6	RIVETS 2 3/8 x .154 DIA.	SILICON BRONZE	E-786
8	1	STANDARD (FW'D)	ZAMAK	EM-271
9	1	STANDARD (AFT)	"	EM-272
10	1	SHAFT	S.S	DM-398 706
11	2	THRUST WASHERS	BRZ.	EM 93
12	2	SPRING	SPRING WIRE	E-967
13	6	RETAINING RING-TRUARC*	5000-206	
14	6	BEARING - SKF *	6205-2RS	
16	2	RETAINING RING-EATON	#698	
17	4	5/16" CAP SCREW 1 3/4 LG.	CAD R	
18	1	5/16"-18NC DOG PT. CAP SCR.	"	EM-91
19	1	NAME PLATE	BRASS	EM-99
20	4	DRIVE SCREW	ST'L	
25	2	ENGAGING COLLAR	BRZ.	EM-90
26	4	RETAINING SCREW	CAD R	EM-91
27	2	1/4" x 1/2 Sq.HD. SET SCR.	"	
28	4	5/16 x 1 3/4 CAP SCREW	SILICON BRONZE	
29	4	5/16" HEX NUT	CAD. R	

NO	ALTERATION	BY	DATE
C	CHANGED TITLE & MODEL NO.	WCW	5-27
B	ADDED REMOTE VIEW	JLE	5-26
	ALTERED BOLTING HOLES		
A	CHANGED QUAN.- DWG NOS	WCW	12-12

Fig. 8-BPR-3. Industry plan. Power take-off.

Blueprint Reading Activity 8-BPR-3
POWER TAKE-OFF

Refer to Fig. 8-BPR-3 and answer the following questions.

1. What kind of a drawing is 8-BPR-3? 1. ______

2. What views are shown? 2. ______

3. How many different parts are shown? 3. ______

4. What are the overall dimensions? 4. ______

5. What is the name of the part on which the letter A appears in the front view? 5. ______

6. How many of part No. 11 are required for one unit? What is the name of this part? 6. ______

7. How many drive screws are required to attach the name plate? 7. ______

8. Name the part on which the letter B appears in the top view. 8. ______

9. What kind of a line is C? 9. ______

10. What is part No. 6 and how many are needed for one unit? 10. ______

Unit 9
Review of Shop Mathematics

Craftsmen and technicians frequently need to make calculations involving common fractions and decimals in connection with the reading of blueprints. If you are in need of a quick review of the fundamentals of this area of mathematics, this unit will assist you in your study.

Common Fractions

Common fractions are written with one number over the other, as $\frac{11}{16}$. The number on the bottom, 16, called the DENOMINATOR, indicates the number of equal parts into which a unit is divided and the number on top, 11, called the NUMERATOR, indicates what number of equal parts is taken, Fig. 9-1. In the fraction shown, $\frac{11}{16}$, eleven of the sixteen parts are taken.

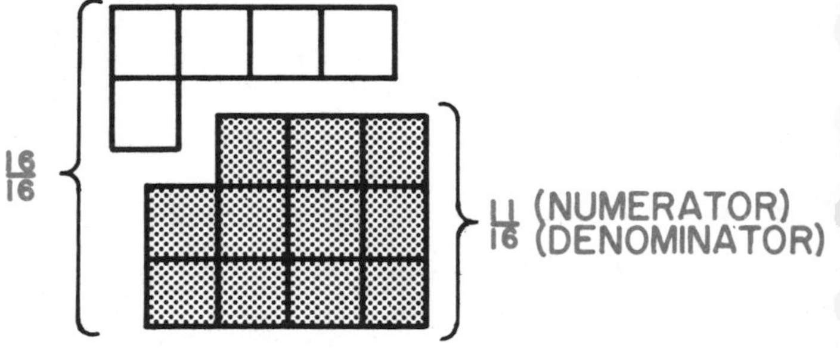

Fig. 9-1. Unit divided into 16 fractional parts.

A PROPER FRACTION is one whose numerator is less than its denominator, as: $\frac{7}{16}$ and $\frac{3}{4}$.

An IMPROPER FRACTION is one whose numerator is greater than its denominator, as: $\frac{5}{4}$ and $\frac{19}{16}$.

A MIXED NUMBER is a number which consists of a whole number and a proper fraction, as: $2\frac{3}{4}$ and $5\frac{1}{8}$.

Fundamental Steps in the Use of Common Fractions

1. Whole numbers may be changed to fractions by multiplying the numerator and denominator by the same number:

Change 6 (whole number) into fourths.

$$\frac{6}{1} \times \frac{4}{4} = \frac{24}{4}$$

Each whole unit contains 4 fourths.

Six units will contain 6 x 4 fourths or 24 fourths.

The value of the number has not changed,

$$\frac{24}{4} = 6.$$

2. Mixed numbers may be changed to fractions by changing the whole number to a fraction with the same denominator as the fractional part of the mixed number and adding the two fractions:

$$3\frac{5}{8} = \frac{3}{1} \times \frac{8}{8} + \frac{5}{8}$$

$$= \frac{24}{8} + \frac{5}{8}$$

$$= \frac{29}{8}$$

Each whole unit contains 8 eights.

Three units contain 3 x 8 eights or 24 eights.

Adding the $\frac{5}{8}$ part of the mixed number to $\frac{24}{8}$ gives us $\frac{29}{8}$.

3. Improper fractions may be reduced to a whole or mixed number by dividing the numerator by the denominator:

$$\frac{17}{4} = 17 \div 4 = 4\frac{1}{4}$$

4. Fractions may be reduced to the lowest form by dividing the numerator and denominator by the same number:

$$\frac{6 \div 2}{8 \div 2} = \frac{3}{4}$$

The value of a fraction is not changed if the numerator and denominator are divided by the same number, since this is the same as dividing by 1.

5. Fractions may be changed to a higher form by multiplying the numerator and denominator by the same number:

$$\frac{5 \times 2}{8 \times 2} = \frac{10}{16}$$

The value of a fraction is not changed by multiplying the numerator and denominator by the same number, since this is the same as multiplying by 1.

Addition of Fractions

To add common fractions, the denominators must all be the same:

Example: $\frac{5}{16} + \frac{3}{8} + \frac{11}{32} = ?$

The LOWEST COMMON DENOMINATOR into which these denominators can be divided is 32.

Change fractions to a higher form (Fundamental Step No. 5)

$$\frac{5}{16} \times \frac{2}{2} = \frac{10}{32}$$

$$\frac{3}{8} \times \frac{4}{4} = \frac{12}{32}$$

Add the fractions with the common denominators

$$\frac{10}{32} + \frac{12}{32} + \frac{11}{32} = \frac{33}{32}$$

Reduce improper fraction (Fundamental Step No. 3)

$$\frac{33}{32} = 1\frac{1}{32}$$

Shop Mathematics Activity 9-BPR-1
ADDITION OF FRACTIONS

Solve the following problems. Reduce answers to lowest form.

1. $\frac{3}{4} + \frac{1}{8} + \frac{1}{2} =$

2. $\frac{7}{8} + \frac{3}{16} =$

3. $\frac{5}{12} + \frac{3}{8} + \frac{3}{4} =$

4. $\frac{3}{10} + \frac{9}{10} + \frac{1}{20} =$

5. $\frac{7}{16} + \frac{3}{32} + \frac{1}{4} =$

6. $1\frac{3}{4} + \frac{7}{8} + 1\frac{1}{16} =$

7. $\frac{5}{32} + \frac{7}{64} + \frac{7}{8} =$

8. $1\frac{3}{8} + \frac{3}{32} + \frac{7}{16} =$

9. $3\frac{1}{16} + \frac{9}{16} + \frac{1}{2} =$

10. $5\frac{1}{5} + 2\frac{3}{10} + 8\frac{1}{2} =$

11. $4\frac{5}{8} + 20\frac{7}{32} =$

12. $\frac{3}{8} + \frac{7}{64} + \frac{9}{16} =$

13. $12\frac{7}{8} + 25\frac{3}{8} =$

14. $\frac{21}{32} + \frac{9}{64} + \frac{1}{4} =$

15. $\frac{3}{8} + 1\frac{1}{2} + \frac{7}{16} + \frac{7}{8} =$

16. $2\frac{1}{4} + \frac{5}{8} + \frac{5}{16} + \frac{17}{32} =$

Subtraction of Fractions

To subtract common fractions, the denominators must all be the same.

Example: $\frac{3}{4} - \frac{5}{16} = ?$

The LOWEST COMMON DENOMINATOR into which these denominators can be divided is 16.

Change fractions to a higher form (Fundamental Step No. 5)

$$\frac{3}{4} \times \frac{4}{4} = \frac{12}{16}$$

Subtract the numerators

$$\frac{12}{16} - \frac{5}{16} = \frac{7}{16}$$

Shop Mathematics Activity 9-BPR-2
SUBTRACTION OF FRACTIONS

Solve the following problems. Reduce answers to lowest form.

1. $\frac{3}{8} - \frac{1}{4} =$

2. $\frac{3}{4} - \frac{5}{16} =$

3. $1\frac{7}{8} - \frac{13}{16} =$

4. $3\frac{1}{2} - \frac{9}{16} =$

(borrow $\frac{16}{16}$ from 3)

5. $10\frac{3}{8} - 7\frac{3}{32} =$

6. $5 - 2\frac{3}{8} =$

7. $12\frac{1}{16} - 8\frac{1}{2} =$

8. $4\frac{1}{4} - 3\frac{1}{16} =$

9. $20\frac{7}{8} - 11\frac{3}{64} =$

10. $15\frac{5}{8} - 5\frac{1}{2} =$

Multiplication of Fractions

Common fractions may be multiplied as follows:

1. Change all mixed numbers to improper fractions.

2. Multiply all numerators to get the numerator part of the answer.

3. Multiply all denominators to get the denominator part of the answer.

4. Reduce the fraction to lowest form.

Example: $\frac{1}{2} \times 3\frac{1}{8} \times 4 = ?$

$$\frac{1}{2} \times \frac{25}{8} \times \frac{4}{1} = \frac{100}{16}$$

$$\frac{100}{16} = 6\frac{4}{16} = 6\frac{1}{4}$$

Shop Mathematics Activity 9-BPR-3
MULTIPLICATION OF FRACTIONS

Solve the following problems. Reduce answers to lowest form.

1. $\frac{3}{4} \times \frac{1}{2} =$

2. $2\frac{5}{8} \times \frac{1}{4} =$

3. $\frac{7}{8} \times 5 =$

4. $6\frac{3}{4} \times \frac{1}{3} =$

5. $12\frac{1}{2} \times \frac{1}{2} =$

6. $4\frac{3}{4} \times \frac{1}{2} \times \frac{1}{8} =$

7. $16 \times \frac{3}{4} =$

8. $9\frac{5}{8} \times \frac{1}{2} =$

9. $10 \times \frac{4}{5} =$

10. $\frac{14}{3} \times 6 =$

Division of Fractions

Common fractions may be divided as follows:

1. Change all mixed numbers to improper fractions.

2. Invert (turn upside down) the divisor and proceed as in multiplication.

Example: $5\frac{1}{4} \div 1\frac{1}{2} = ?$

$$\frac{21}{4} \div \frac{3}{2} =$$

$$\frac{21}{4} \times \frac{2}{3} = \frac{42}{12}$$

$$\frac{42}{12} = 3\frac{6}{12} = 3\frac{1}{2}$$

Shop Mathematics Activity 9-BPR-4
DIVISION OF FRACTIONS

Solve the following problems. Reduce answers to lowest form.

1. $2\frac{3}{4} \div 6 =$

2. $12 \div \frac{3}{4} =$

3. $16\frac{1}{8} \div 2 =$

4. $8\frac{2}{3} \div \frac{1}{3} =$

5. $16\frac{1}{4} \div 20 =$

6. $\frac{7}{8} \div \frac{7}{16} =$

7. $15 \div 1\frac{1}{4} =$

8. $21\frac{3}{8} \div 3\frac{1}{8} =$

9. $5\frac{1}{4} \div \frac{3}{8} =$

10. $3\frac{5}{8} \div 2 =$

Decimal Fractions

The denominator in decimal fractions is 10 or a multiple of 10 (that is 100, 1000, etc.) When writing decimal fractions, we omit the denominator and place a decimal point in front of the numerator:

$\frac{3}{10}$ is written .3 (three tenths)

$\frac{87}{100}$ is written .87 (eighty seven hundredths)

$\frac{375}{1000}$ is written .375 (three hundred seventy five thousandths)

$\frac{4375}{10000}$ is written .4375 (four thousand three hundred seventy five ten thousandths)

Whole numbers are written to the left of the decimal point and fractional parts are to the right.

$5\frac{253}{1000}$ is written 5.253 (five and two hundred fifty three thousandths)

Addition and Subtraction of Decimals

Decimals are added and subtracted in the same manner as whole numbers, except in decimals we write the figures so that the decimal points line up vertically.

Example:

Add		Subtract	
	7.3125		8.625
	1.25		2.25
	.625		6.375
	3.375		
	12.5625		

The decimal point in the answer is directly below the decimal points in the problem.

Shop Mathematics Activity 9-BPR-5
ADDITION AND SUBTRACTION OF DECIMALS

Solve the following problems:

Add: 1.
```
 4.5625
  .875
 2.75
 5.8137
-------
```

2. 7.0625 + .125 + 8.0

3. .832 + .4375 + .27

Subtract: 4.
```
 27.9375
 16.937
--------
```

5. 4.0 - .0625

6. 2.25 - 1.125

Multiplication of Decimals

Decimals are multiplied in the same manner as whole numbers and the decimal points are disregarded until the multiplication is completed. To find the position of the decimal point in the answer, count the total number of decimal places to the right of the decimal point in the numbers being multiplied and set off this number of decimal places in the answer starting at the right.

Example:

```
  6.25 }
   1.5 }  3 decimal places
  ----
  3125
  625
 -----
 9.375 }  3 decimal places
```

Shop Mathematics Activity 9-BPR-6
MULTIPLICATION OF DECIMALS

Solve the following problems:

1.
```
 4.825
  1.75
------
```

2.
```
 167
 .25
----
```

3.
```
 65.96
   .37
------
```

4. 4.95 x 1.35

5. 93.18 x .07

Division of Decimals

In dividing decimals we proceed as in the division of whole numbers except that the decimal point must be properly placed in the quotient (answer).

To place the decimal point in the quotient, count the number of places to the right of the decimal point in the divisor. Add this number of places to the right of the decimal point in the dividend and place the decimal point directly above in the quotient.

Example: 36.5032 ÷ 4.12 = ?

```
                   8.86
Divisor  4.12^ |36.50^32   Dividend
                32 96
                 3 543
                 3 296
                   2472
                   2472
```

Shop Mathematics Activity 9-BPR-7 DIVISION OF DECIMALS

Solve the following problems:

1. 2.7 ⟌9.45

2. 35 ⟌654.5

3. 1.65 ⟌1386.0

4. 331.266 ÷ 80.6

5. 4401.25 ÷ 503

Metric System

The metric system of numbers works in the same way as the decimal-inch system. Only the size of the units and terms differ. Both are on the base-ten number system which makes it easy to shift from one multiple or submultiple to another.

Addition and Subtraction in Metric System

Numbers in the metric system are added and subtracted in the same manner as they are in the decimal system. Write the figures so the decimal points line up vertically.

Example:

```
Add   38.35        Subtract  118.06
      20.666                 107.902
     116.59                   10.158
     175.606
```

Shop Mathematics Activity 9-BPR-8

Solve the following problems:

Add:

```
1.  66.67
     1.42
     3.76
     1.24
```

2. 41.88 + 89.112 + 8.38

3. 4.19 + 49.25 + 2.6

Subtract:

4.
```
  66.68
  41.88
  -----
```

5. 26.97 - 7.1

6. 102.85 - 16.302

Multiplication and Division of Metric Decimal Numbers

Metric decimal numbers are multiplied and divided in the same manner as decimal numbers in the customary system.

Example:

81.6 millimetres multiplied by three

Multiply
```
   81.6 mm
      3
  --------
  244.8 mm
```

Divide 103.42 millimetres into four equal parts

Divide
```
     25.85 mm
  4 |103.42 mm
     8
     --
     23
     20
     --
      34
      32
      --
       22
       20
       --
        20
        20
        --
```

Shop Mathematics Activity 9-BPR-9

1. Find the total length of six sections when each section is 72.5 mm in length.

2. What is the total thickness of five spacers, each of which is 1.22 mm thick?

3. Find the length in millimetres of each part when a rod 108 millimetres in length is divided into 7 equal parts.

Unit 10
Measurement Tools

The ability to make accurate measurements is a fundamental skill needed by all who read and use blueprints. See Fig. 10-1. This unit is intended as a review of the basic principles of reading the Draftsman's Scale, the Micrometer and the Vernier Scale.

Fig. 10-1. Machinist checks a blueprint. (McDonnell Douglas)

How to Read the Draftsmen's Scale

You have been using the full-size scale in the form of a 6-inch steel rule. Although the skilled craftsman or technician NEVER scales a blueprint for an actual measurement (most prints shrink or stretch in the development and drying process) it is helpful for reference and sketching purposes to know and understand the scale of a print.

Principle of Scale Measurements

The principle of scale measurements on a drawing is simply letting a unit, such as 1/2 inch on the scale, represent some larger unit such as 1 inch on the actual object. This scale would appear on the print as "HALF SIZE" or 1/2" = 1". See Fig. 10-2.

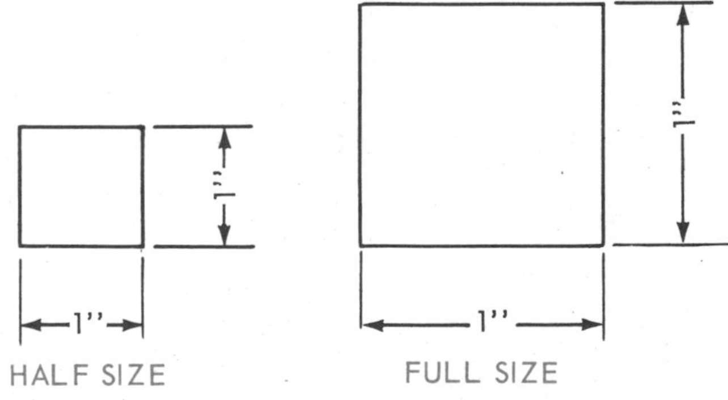

Fig. 10-2. Half-size scale.

Other scales most frequently used are "QUARTER SIZE" or 1/4" = 1"; and "FULL SIZE", or 1" = 1". Scales of this type are found on the mechanical engineer's scale, Fig. 10-3. Some smaller parts of an assembly may need to be enlarged on the drawing for clarity in shape and size. Typical scales would be: TWICE SIZE, or 2" = 1"; or 5X, meaning 5 TIMES or 5 inches on the drawing equals 1 inch on the actual object.

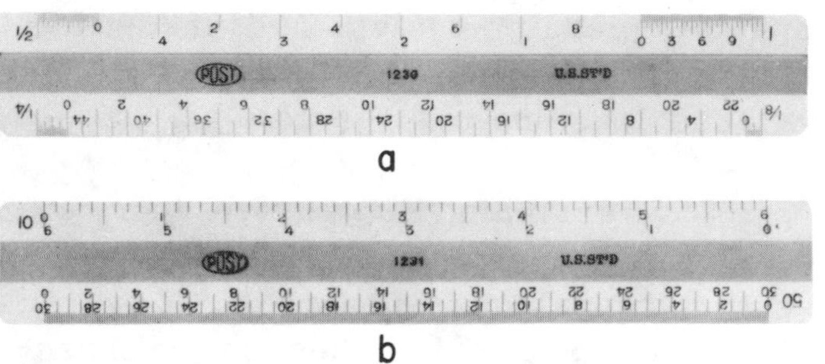

Fig. 10-3. Drafting scales. (Frederick Post Co.)

Scales on drawings of large structures would have much greater reductions and the inch or fractional part of an inch on the scale would represent feet on the actual object, such as 1/4" = 1' - 0". Scales used for drawings of this type reduction are called architect's scales, Fig. 10-3(a).

Some scales are marked off in divisions of 10 (or multiples of 10 as: 20, 30, 40, 50, or 60) parts to the inch and are used in laying out drawings based on decimal dimensions. These scales are called civil engineer's, engineer's or decimal scales, Fig. 10-3(b).

Reading the Fractional-Inch Scale

Observe Fig. 10-4 which illustrates the 1/2 size scale on the mechanical engineer's scale. If a mechanical engineer's scale is not available, an architect's scale may be used by letting the 1/2" units represent inches instead of feet. This 1/2 inch unit represents 1 inch on the actual object. To measure with the half-size scale, think of each dimension in its full size. Lay off measurements in the following manner:

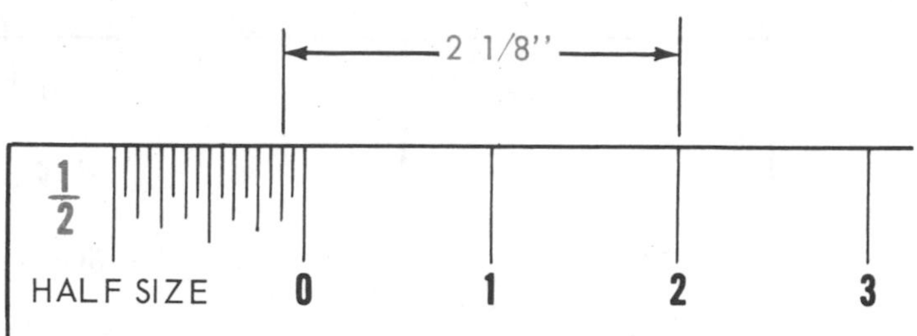

Fig. 10-4. Mechanical engineer's half-size scale.

1. Place the scale along the distance to be measured.

2. Mark a short vertical dash at the whole number (2 inches in our example), Fig. 10-4.

3. The fractional part of an inch is marked in the subdivided end unit, Fig. 10-4.

4. The distance marked off is 2 1/8 inches on the half-size scale.

The scale has done the calculations and assures you of the correct distance if you have read the scale correctly.

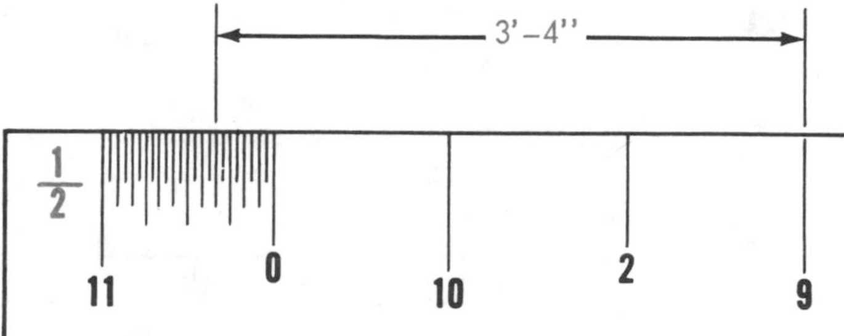

Fig. 10-5. Architect's scale 1/2" = 1' – 0".

Now, using the scale 1/2" = 1' - 0" (1/24 size) on the architect's scale, let us see how this scale is used. You should think of each dimension in terms of its full size, so think of the 1/2 inch division representing one foot, Fig. 10-5. Lay off

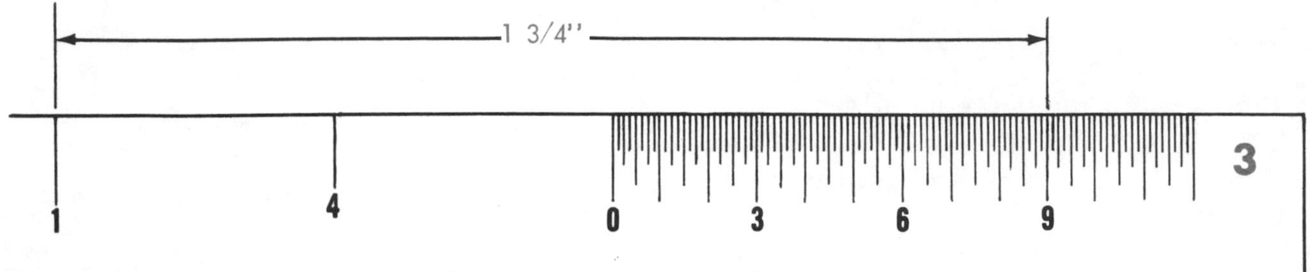

Fig. 10-6. Reading the scale: Three-times size.

measurements in the following manner:

1. Place the scale along the distance to be measured.
2. Mark a short vertical dash at the whole number (3 feet in our example), Fig. 10-5.
3. The inches are marked in the subdivided end unit, (4 inches in our example), Fig. 10-5.
4. The distance marked off is 3 feet and 4 inches on the scale: 1/2" = 1' - 0".

Small machine parts are frequently drawn three-times size and on very small parts of a watch or instrument, the parts may be drawn 10 to 1 or 20 to 1, using special enlarging scales. The three-times size scale is illustrated in Fig. 10-6 and is read in same manner as other scales.

In our example, Fig. 10-6, 0 to 1 represents 1 inch and 0 to 9 in the subdivided section represents 3/4 inch (9/12 = 3/4), a distance of 1 3/4 inches on the three-times size scale.

Measurement Activity 10-BPR-1 Fractional-Inch SCALE MEASUREMENT

Measure the length of the following lines, using the scales indicated. Place your readings on the line above the scale.

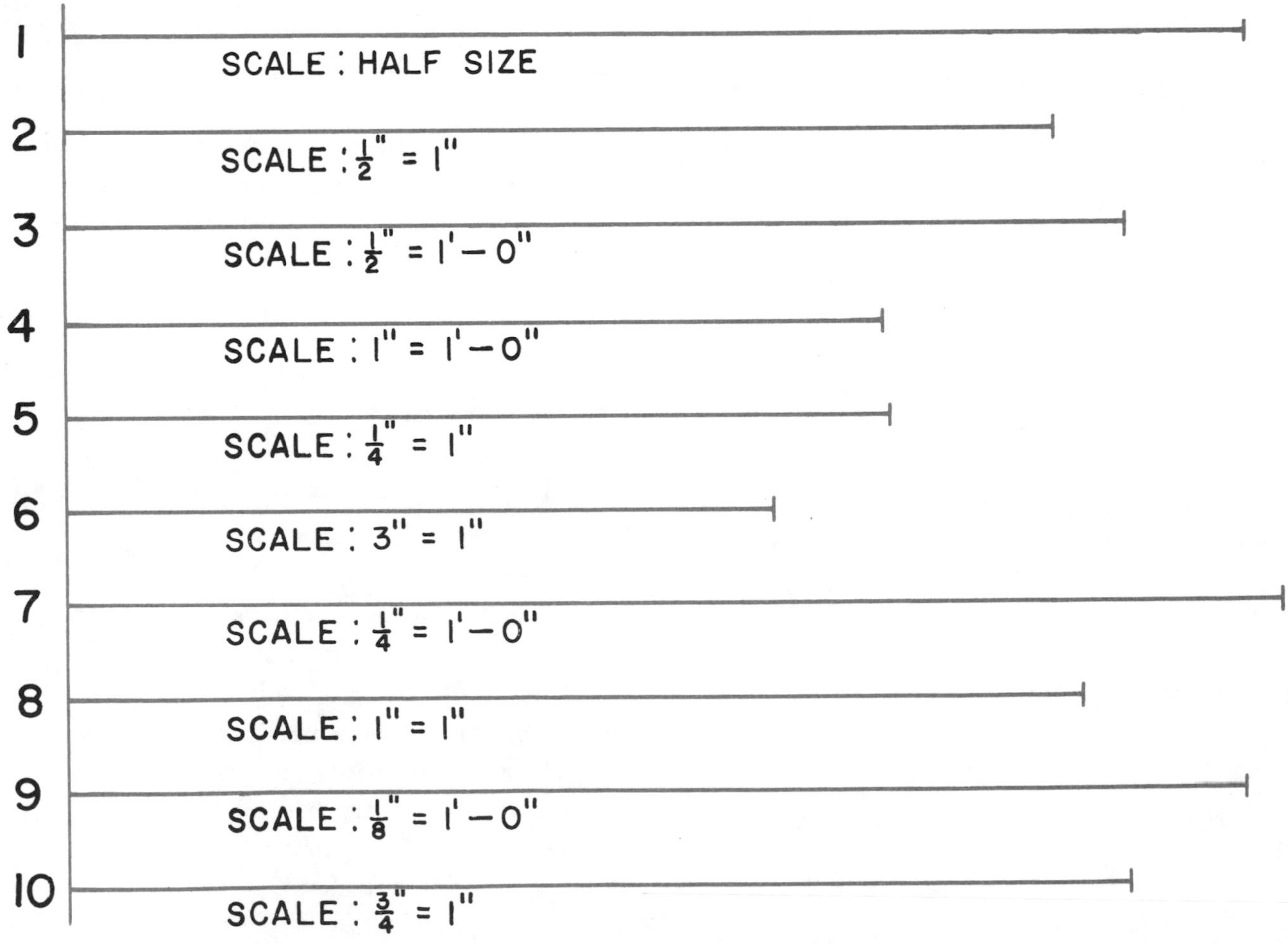

1 SCALE: HALF SIZE

2 SCALE: $\frac{1}{2}$" = 1"

3 SCALE: $\frac{1}{2}$" = 1'—0"

4 SCALE: 1" = 1'—0"

5 SCALE: $\frac{1}{4}$" = 1"

6 SCALE: 3" = 1"

7 SCALE: $\frac{1}{4}$" = 1'—0"

8 SCALE: 1" = 1"

9 SCALE: $\frac{1}{8}$" = 1'—0"

10 SCALE: $\frac{3}{4}$" = 1"

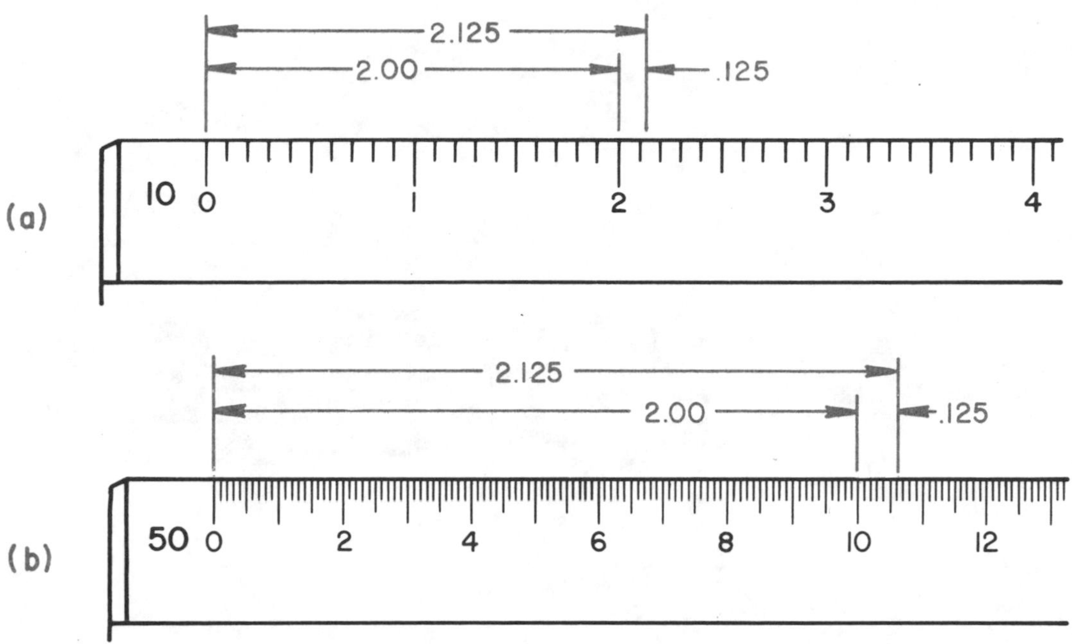

Fig. 10-7. Measuring with the civil engineer's or decimal scale.

Reading the Decimal-Inch Scale

To read a decimal dimension of 2.125 inches on the civil engineer's fully divided 10-scale, Fig. 10-7(a), start from the 0 and move past the 2. Continue past the first tenth (.10) to one-fourth (.025) of the next tenth (judged by eye). This represents the decimal .125, Fig. 10-7(a).

The same reading can be made with greater accuracy on the 50-scale, where each subdivision equals 1/50 or .02 of an inch, Fig. 10-7(b). Move from 0 to 10 (five major divisions to the inch, 10 = 2 inches) and on to six subdivisions (6 x .02 = .12). Continue to one-fourth of the next subdivision (1.4 of .02 = .005) for a measurement of 2.125 inches. The 50-scale is most commonly used in the machine-parts manufacturing industries where decimal dimensioning is standard practice.

Measurement Activity 10-BPR-2
Decimal-Inch SCALE MEASUREMENT

Measure the length of the following lines, using the decimal-inch scales indicated. Place your readings on the line above the scale.

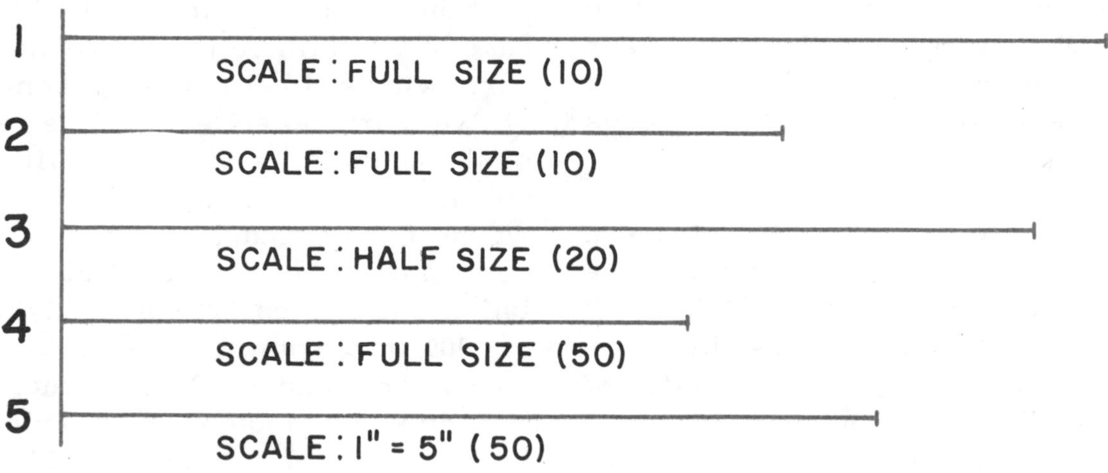

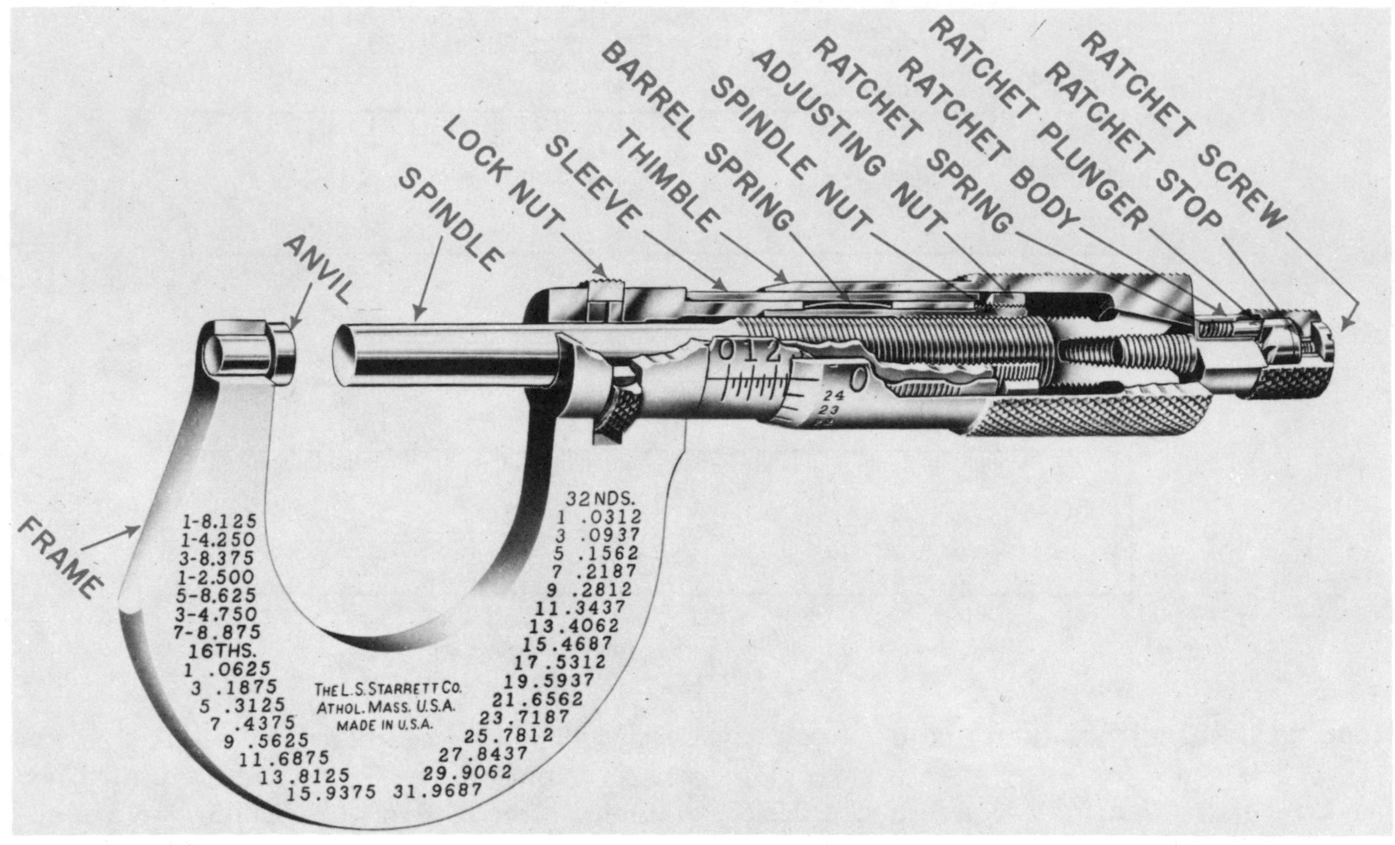

Fig. 10-8. Micrometer with parts identified. (L. S. Starrett Co.)

How to Read the Inch Micrometer

The precision measurement capability of the micrometer is based upon a very accurate screw of 40 threads per inch on the spindle which rotates in a fixed nut, thus opening or closing the distance beteween the measuring surfaces of the anvil and spindle, Fig. 10-8. The spindle is attached to the thimble and as the thimble is turned ONE revolution, the spindle advances toward or withdraws from the anvil face precisely 1/40 (screw of 40 threads per inch) or .025 of an inch.

Refer again to Fig. 10-8 and note there is a line that extends the length of the sleeve. This line is called the INDEX LINE and is divided into 40 equal parts by vertical lines which correspond to the pitch of the thread (distance from one thread to another) on the spindle. Therefore, each vertical line on the sleeve designates 1/40 or .025 of an inch; two of these lines or spaces equal .050 (.025 + .025) of an inch. Every fourth line marked with a number (1, 2, 3, etc.) designates hundreds of thousandths (4 x .025 = .100). The line marked "1" represents .100 inch, line marked "2" represents .200 inch, etc.

Note again in Fig. 10-8, the beveled edge of the thimble is divided into 25 equal parts with the lines numbered consecutively. We learned earlier that one revolution of the thimble, advanced or withdrew the spindle .025 inch. Moving from one line on the beveled edge of the thimble to the next moves the thimble 1/25 of a revolution and moves the spindle 1/25 of .025 or .001 inch; rotating two divisions or lines moves the spindle .002 inch, etc. Twenty-five divisions indicate a complete revolution of the thimble or the .025 inch.

To read the micrometer in thousandths, refer to Example 1 for a sample reading. The reading is taken at the edge of the thimble at the index line.

Example 1:

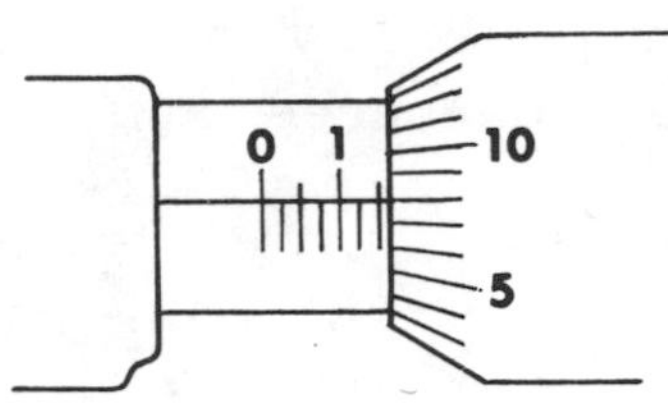

Line "1" on sleeve is visible, representing .100
Two additional lines are visible representing .025 each (2 x .025 = .050) .050
Line 8 on the thimble is at the index line, representing .008

The micrometer reading is .158 inch

Let's try another sample reading.

Example 2:

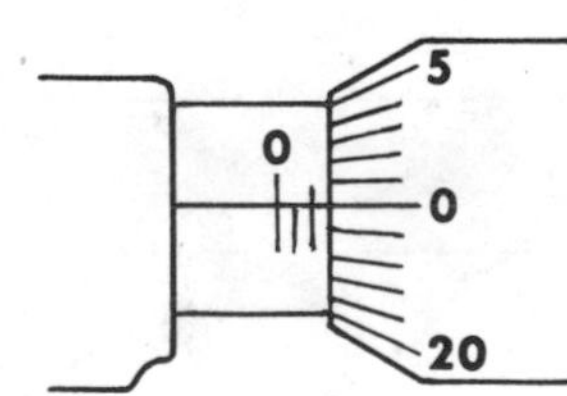

Three lines are visible representing .025 each (3 x .025 = .075) .075
The thimble reading is at 0 so there is nothing to add

The micrometer reading is .075 inch

A Help in Remembering

An easy way to remember the values of the various divisions in reading a micrometer is to think of them as U.S. money. Count the figures on the sleeve . . . 1, 2, etc. as dollars, the extra vertical lines as quarters, and the divisions on the thimble as pennies. Add up your money and put a decimal point in front of the figure instead of a dollar sign. This is your "mike" reading.

Measurement Activity 10-BPR-3
INCH MICROMETER READING
(Paper Exercise)

Record the readings in the space provided for the following micrometer settings:

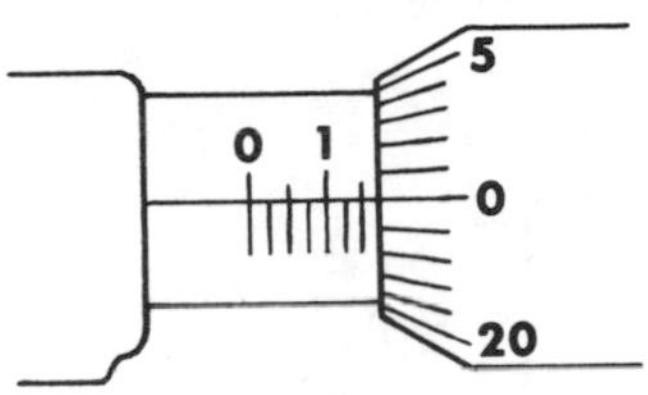

1. Reading __________

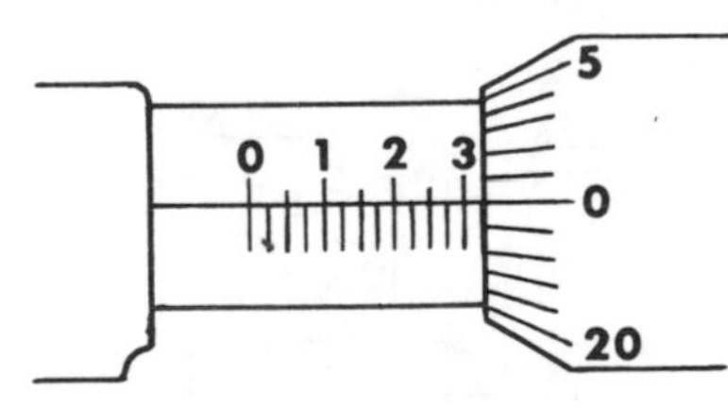

2. Reading __________

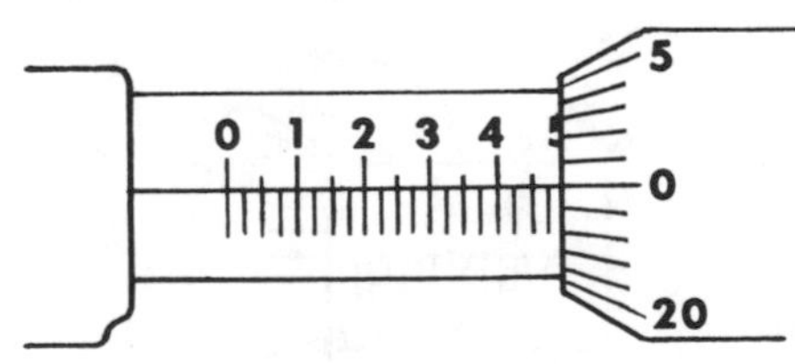

3. Reading __________

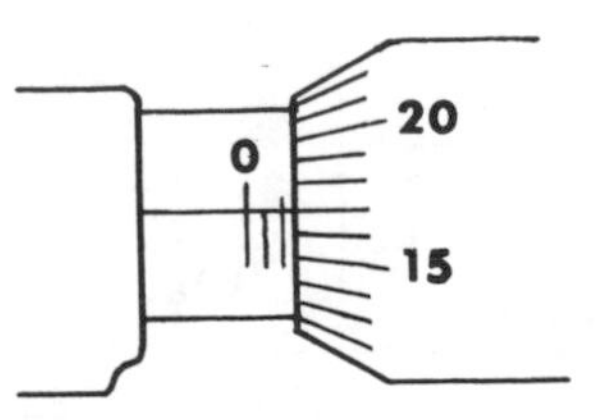

4. Reading __________

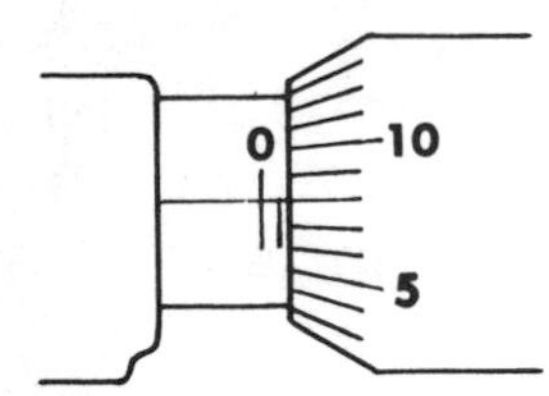

5. *Reading* ____________

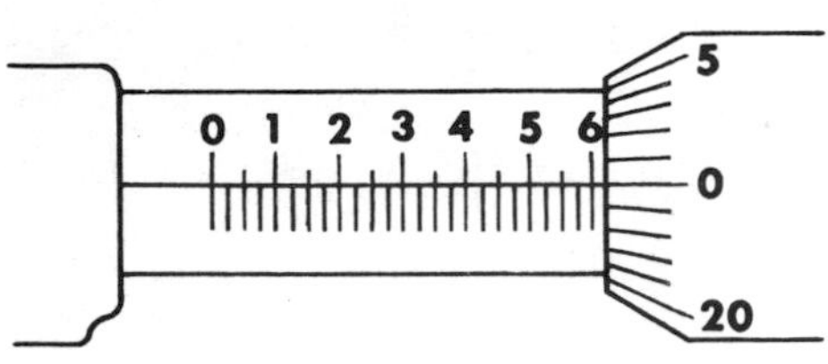

6. *Reading* ____________

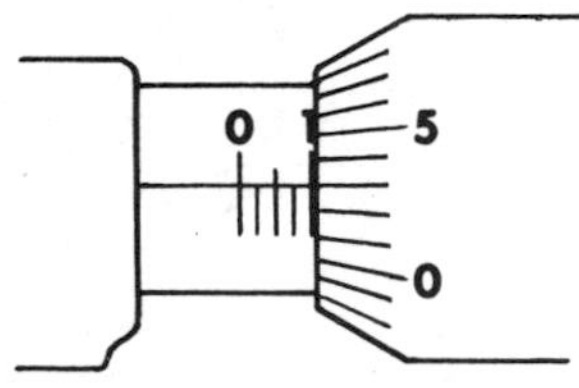

7. *Reading* ____________

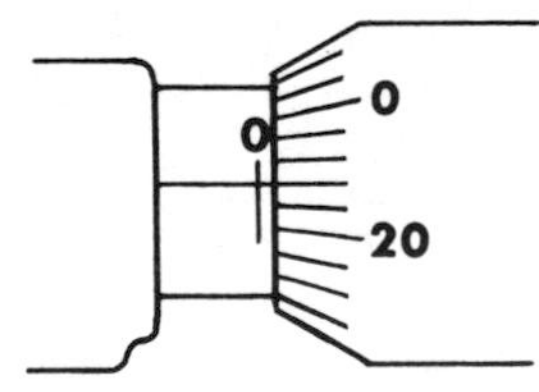

8. *Reading* ____________

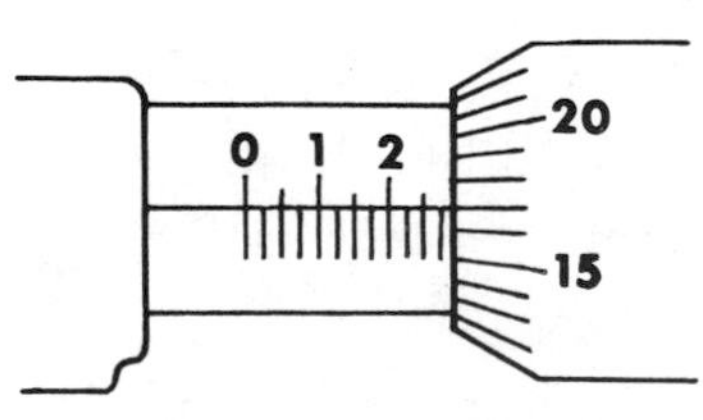

9. *Reading* ____________

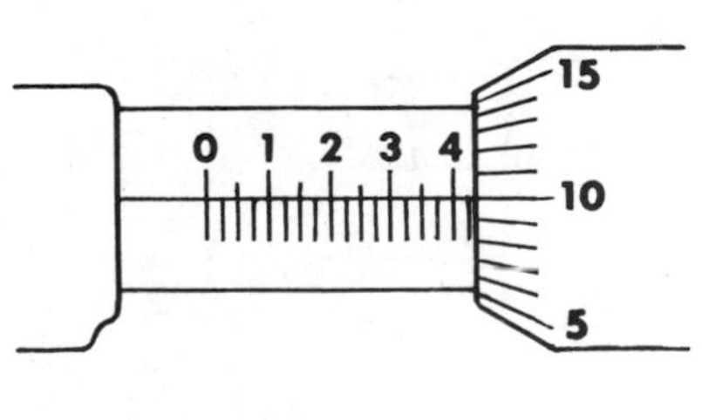

10. *Reading* ____________

Measurement Activity 10-BPR-4
INCH MICROMETER READING
(With Micrometer)

With a micrometer, make actual measurements on machine parts, such as a spindle, pin, plate, key for shaft, etc. Record items measured and micrometer readings in spaces provided.

	Item	Reading
1.	____________	____________
2.	____________	____________
3.	____________	____________
4.	____________	____________
5.	____________	____________
6.	____________	____________
7.	____________	____________
8.	____________	____________
9.	____________	____________
10.	____________	____________
11.	____________	____________
12.	____________	____________
13.	____________	____________
14.	____________	____________

How to Read a 25-Division Vernier Scale

Slide the Vernier plate along the bar until the jaws nearly contact the work. Lock the clamping screw on the Vernier and make the final adjustment with the fine adjusting screw on the Vernier, Fig. 10-9.

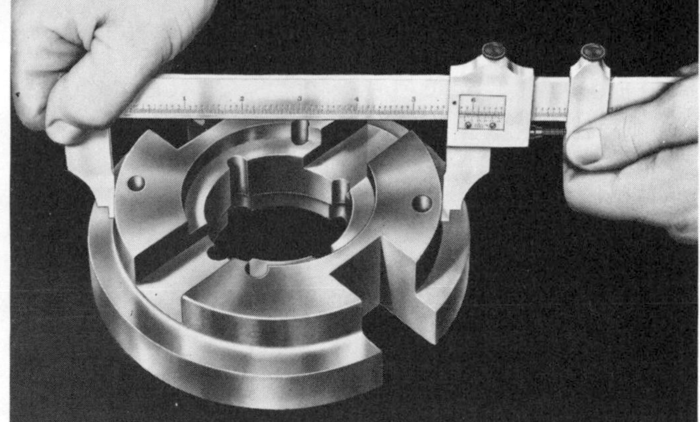

Fig. 10-9. Measuring with the Vernier calipers. (L. S. Starrett Co.)

The BAR (top row of numbers) of the Vernier is graduated into 40ths (.025) of an inch, Fig. 10-10, with every fourth division, representing a tenth of an inch, being numbered. The Vernier PLATE (bottom row of numbers) is divided into twenty-five divisions numbered 0, 5, 10, 15, 20 and 25. The twenty-five divisions on the plate occupy the same space as twenty-four divisions on the bar. This slight differencc, equal to 1/1000 of an inch per division, is the principle behind Vernier measuring.

To read Vernier, note number of full inches (1, 2, etc.), tenths (.100, .200, etc.), and fortieths (.025, .050, etc.) the 0 mark on the plate is from the 0 mark on the bar. Add to this the number of thousandths indicated by THE LINE ON THE PLATE that exactly coincides with a line on the bar (11 as indicated by stars, Fig. 10-10).

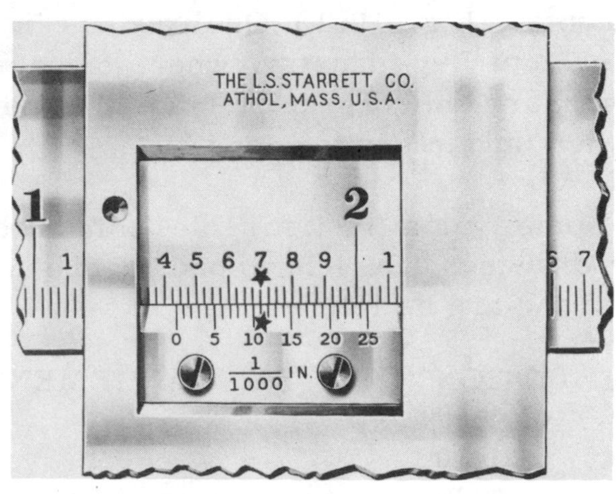

Fig. 10-10. Photo showing Vernier scale. (L. S. Starrett Co.)

Example: Refer to Fig. 10-10.

Full inches (0 mark is to right of 1)	1.000
Tenths (4)	.400
Fortieths (1/40 = .025)	.025
Thousandths (11)	.011
Vernier reading is	1.436 inches

Measurement Activity 10-BPR-5
VERNIER SCALE READING

Using a Vernier tool, make actual measurements on machine parts, such as slots on angle plate, diameters of shafts, diameters of holes, heights of offsets, etc. Record items measured and vernier readings in spaces provided.

	Item	Reading
1.	______________	________________
2.	______________	________________
3.	______________	________________
4.	______________	________________
5.	______________	________________
6.	______________	________________
7.	______________	________________
8.	______________	________________
9.	______________	________________
10.	______________	________________
11.	______________	________________
12.	______________	________________
13.	______________	________________

SI METRIC BASE UNITS

QUANTITY	NAME OF UNIT	SYMBOL
Length	metre	m
Mass (weight)	kilogram	kg
Time	second	s
Electric current	ampere	A
Temperature	kelvin	K
Luminous intensity	candela	cd
Amount of substance	mole	mol

SUPPLEMENTARY UNITS

UNIT	NAME OF UNIT	SYMBOL
Plane Angle	radian	rad
Solid angle	steradian	sr

Fig. 10-11. The seven base and two supplementary units of the International System of Units.

Metric Scales

The United States is converting to the metric system. In the transition period of the next few years, both sets of measurements, the metric and English, are likely to be used. Many industries today are using a system of "dual dimensioning" (decimal inch and metric) as shown on the blueprint on page 236.

The International System (SI) units of metric measure conform to reason, are consistent and fit together. The seven base and two supplementary units of the International System are shown in Fig. 10-11.

The metric system of numbers works in the same way as the decimal-inch system. Only the size of the units and terms vary. Both are on the base-ten number system which makes it easy to shift from one multiple or submultiple to another.

A scale of 1:20 would be referenced on the drawing as 5 cm = 1 m, Fig. 10-12(a). A scale of 1:100 means the drawing is 1/100 the size of the actual object and would be referenced 1 cm = 1 m, as in Fig. 10-12(b). The same scale (1:100) could be used to represent 1 cm = 1 km, as in Fig. 10-12(c). The reduction would be 1/100,000 (multiple factor of 10^{-2} to 10^{3} = 10^{5}). The 1:100 scale may also be used as a full-size scale when

SCALE 1:20	SCALE 1:100	SCALE 1:100,000
5 cm = 1 m	1 cm = 1 m	1 cm = 1 km
(a)	(b)	(c)

Fig. 10-12. Metric scales are referenced on a drawing to indicate the units of the ratio.

dimensions are in millimetres since the smallest divisions are millimetres.

Any craftsman using the metric system of linear measure should learn all common units and their relation to other factors, Fig. 10-13. Actually, changing from kilometres to metres is a simple shift of the decimal point, Fig. 10-14.

Kilometre	=	1 000	metres	(thousands)
Hectometre	=	100	metres	(hundreds)
Dekametre	=	10	metres	(tens)
Metre	=	1	metre	(unit of linear measure)
Decimetre	=	0. 1	metre	(tenths)
Centimetre	=	0. 01	metre	(hundredths)
Millimetre	=	0. 001	metre	(thousandths)
Micrometre	=	0. 000 001	metre	(millionths)

Fig. 10-13. Common linear units in the metric system.

Metric Units used on Industrial Blueprints

The base unit of the metric system is the metre. However, derived-decimal units are used in various industrial fields as the official unit of measure on metric drawings, Fig. 10-15. On drawings in the manufacturing industry, where measurements are more precise, the millimetre is the standard unit of length.

Some typical metric scale designations for drawings which have been reduced or enlarged are as follows:

REDUCTIONS		ENLARGEMENTS
1:2	1:50	2:1
1:2.5	1:100	5:1
1:5	1:200	10:1
1:10	1:500	20:1
1:20	1:1000	50:1

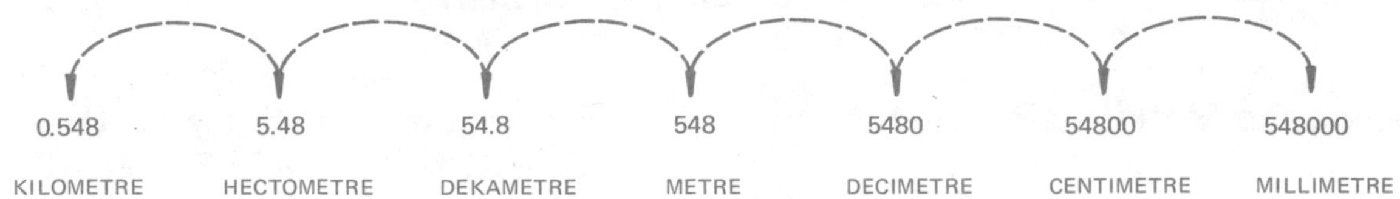

Fig. 10-14. Numbers in the metric system are added, subtracted, multiplied, and divided in the same way as decimal-inch numbers.

Measuring with the metric scale is shown in Fig. 10-16, using the .01 scale as a full size scale. The measurement reads 43.5 mm. To use the same scale as a reducing scale of 1:100, let each numbered unit (actually a centimetre) represent a metre - a reduction of 1 to 100, Fig. 10-16. That is, one unit on the drawing represents 100 on the object. The measurement in Fig. 10-16 would then be 4.35 metres.

INDUSTRY	UNIT OF MEASURE	INTERNATIONAL SYMBOL	MULTIPLE FACTOR
Topographical	kilometre	km	10^3 = 1 000 m
Building, Construction	metre	m	10^0 = 1 m
Lumber, Cabinet	centimetre	cm	10^{-2} = 0. 01 m
Mechanical Design, Manufacturing	millimetre	mm	10^{-3} = 0. 001 m

Fig. 10-15. Metric units of measure on industrial drawings by type of industry.

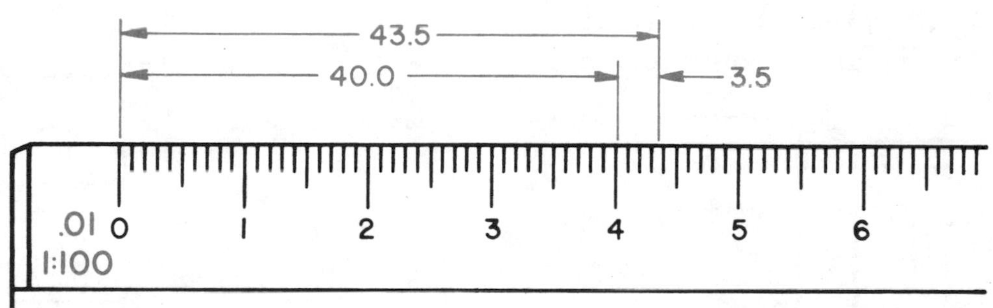

Fig. 10-16. Measuring with the metric scale. Scale enlarged for easier reading.

Measurement Activity 10-BPR-6
METRIC SCALE MEASUREMENT

Lines below have been drawn to a certain metric scale. Measure the length of each using the metric scales called for under the line. Before you write in the answer it is suggested you check it against the example given in Fig. 10-16. Place your answer above each line measured.

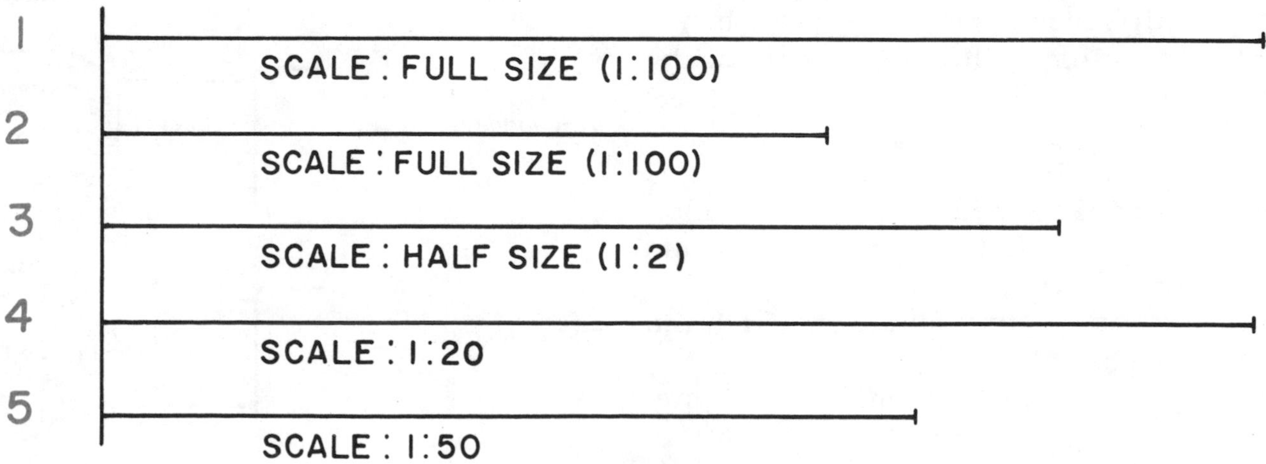

How to Read the Metric Micrometer

Reading a metric micrometer which measures in hundredths of a millimetre (0.01 mm) is quite similar to reading the inch micrometer - only the units are different.

The screw on the metric micrometer advances 1/2 millimetre per revolution of the thimble so two revolutions move the spindle 1 mm.

The sleeve of the micrometer is graduated in millimetres below the datum line (see below) and in half millimetres above the line. The thimble is marked in fifty divisions so that each small division represents 1/50 of 1/2 mm (one revolution of thimble) which equals 1/100 (0.01) mm.

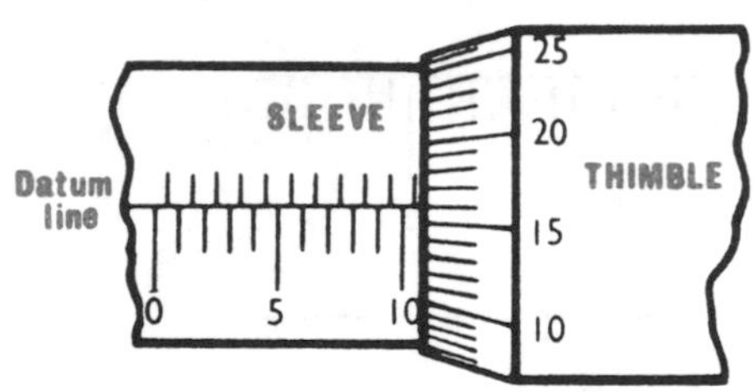

To read the metric micrometer, follow these steps:

1. Note the number of whole millimetre divisions (below datum line) on the sleeve.
2. Note whether there is a half millimetre division (above datum line) visible between the whole millimetre division and the thimble.
3. Finally, read the thimble for hundredths (division on thimble aligning with datum line).

In our example above:

Whole millimetre Divisions =
10 x 1 mm = 10.00 mm
Half millimetre Divisions =
1 x 0.50 mm = 0.50 mm
Reading on Thimble =
16 x 0.01 mm = 0.16 mm
Reading = 10.66 mm

Measurement Activity 10-BPR-7
METRIC MICROMETER READING

Record the readings in the spaces provided for the following micrometer settings.

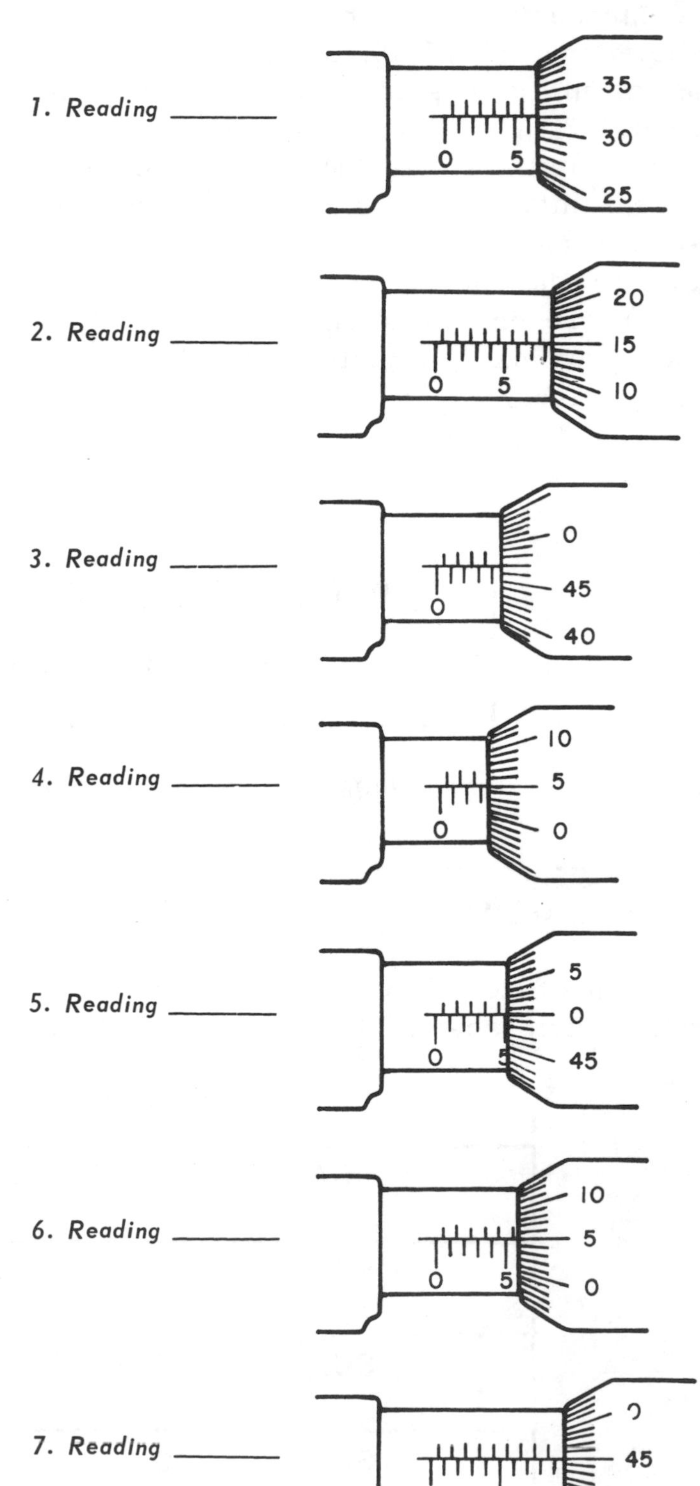

Unit 11
Dimensions and Tolerances

Most major industries do not manufacture all of the parts and sub-assemblies required in their products. Frequently the parts are manufactured by specialty industries, to standard specifications or to specifications provided by the major industry. The key to the successful operation of the various parts and sub-assemblies in the major product is the ability of two or more nearly identical duplicate parts to be used individually in an assembly and function satisfactorily.

The meaning of various terms and symbols as well as procedures and techniques relating to dimensioning and tolerancing will be studied in this Unit to assist you in accurately interpreting industrial blueprints.

Definition of Terms

The following terms are used frequently in industry and it is necessary to understand their true meaning and application in order to satisfactorily read and interpret blueprints. Other terms relating to conditions and applications of dimensioning and tolerancing which are of less importance, will be found in the Glossary of this text, page 316.

NOMINAL SIZE is a general classification term used to designate size for a commercial product, Fig. 11-1. It may or may not express the true numerical size of the part or material. For example, a seamless wrought-steel pipe of 3/4 inch nominal size has an actual inside diameter of 0.824 and

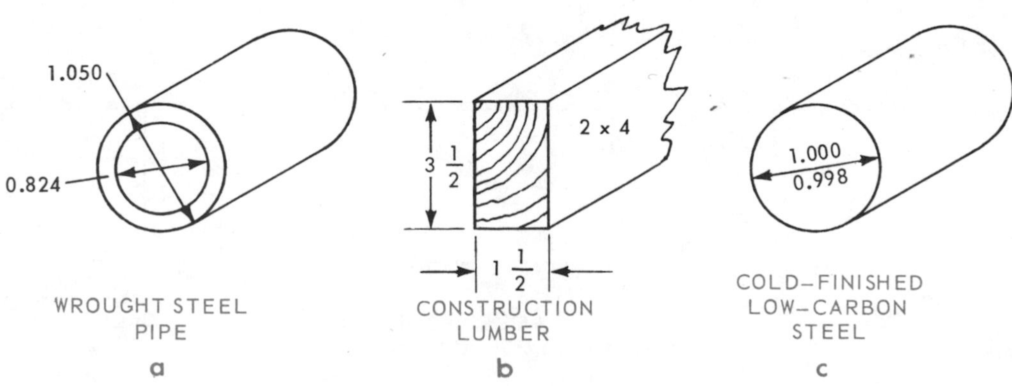

Fig. 11-1. Nominal sizes of objects.

an actual outside diameter of 1.050, Fig. 11-1(a).

Another example of nominal size is the 2 x 4 piece of lumber used in building construction which has an actual size of 1-1/2 x 3-1/2 inches, Fig. 11-1(b). However, in the case of cold-finished low-carbon steel rounds, the nominal 1-inch size comes within two thousandths (.002) tolerance of actual size, Fig. 11-1(c). So it may be seen that the nominal size may or may not be the true numerical size of a material. When an industry uses the term nominal size synonymous with basic size (uses both terms as being alike in meaning), they are assuming the exact or theoretical size from which all limiting variations are made.

BASIC SIZE is the size of a part determined by engineering and design requirements. It is from this size that allowances and tolerances are applied. For example, strength and stiffness may require a 1-inch diameter shaft. The basic 1-inch size (with tolerance) will most likely be applied to the hole size since allowance is usually applied to the shaft, Fig. 11-2(a).

ALLOWANCE is the intentional difference in the dimensions of mating parts to provide for different classes of fits. It is the minimum clearance space or maximum interference, whichever is intended, between mating parts. In our sample, we have allowed .002 on the shaft for clearance (1.000 - .002 = .998), Fig. 11-2(b).

DESIGN SIZE is the size of a part after an allowance for clearance has been applied and tolerances have been assigned. The design size of our shaft, Fig. 11-2(b), is .998 after the allowance of .002 has been made. A tolerance of ± .002 is assigned after the allowance is applied. See Fig. 11-2(c).

ACTUAL SIZE is a measured size.

LIMITS are the extreme permissible dimensions of a part resulting from the application of a tolerance. Two limit dimensions are always involved, a maximum size and a minimum size. For example, the design size of a part may be 1.375. If a tolerance of plus or minus two thousandths (±.002) is applied, then the two limit dimensions are maximum limit 1.377 and minimum limit 1.373, Fig. 11-3(a).

TOLERANCE is the total amount of variation permitted from the design size of a part. Tolerance may be expressed as variation between limits, Fig. 11-3(a); as the design size followed by the tolerance, Fig. 11-3(b); or when only one tolerance value is given, the other value is assumed to be zero, Fig. 11-3(c). Tolerance is also applied to location dimensions for other features (holes, slots, surfaces, etc.) of a part, Fig. 11-3(d).

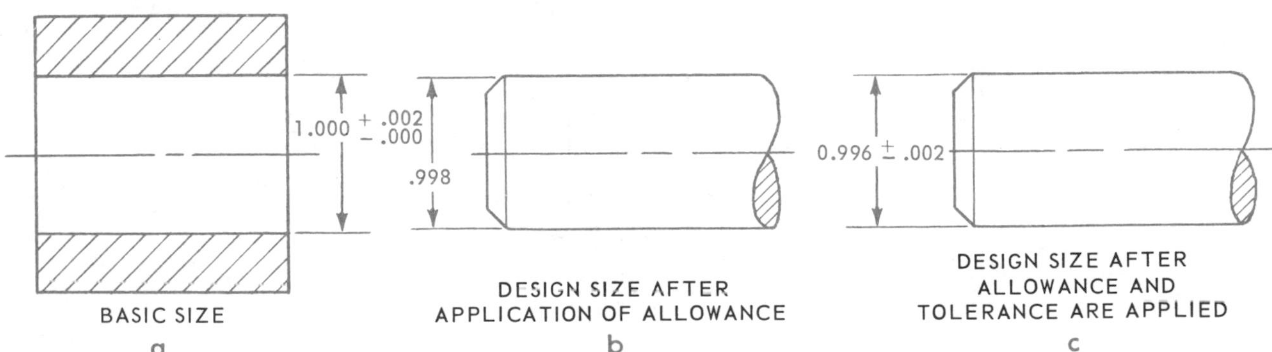

Fig. 11-2. Size terminology used in dimensioning and tolerancing.

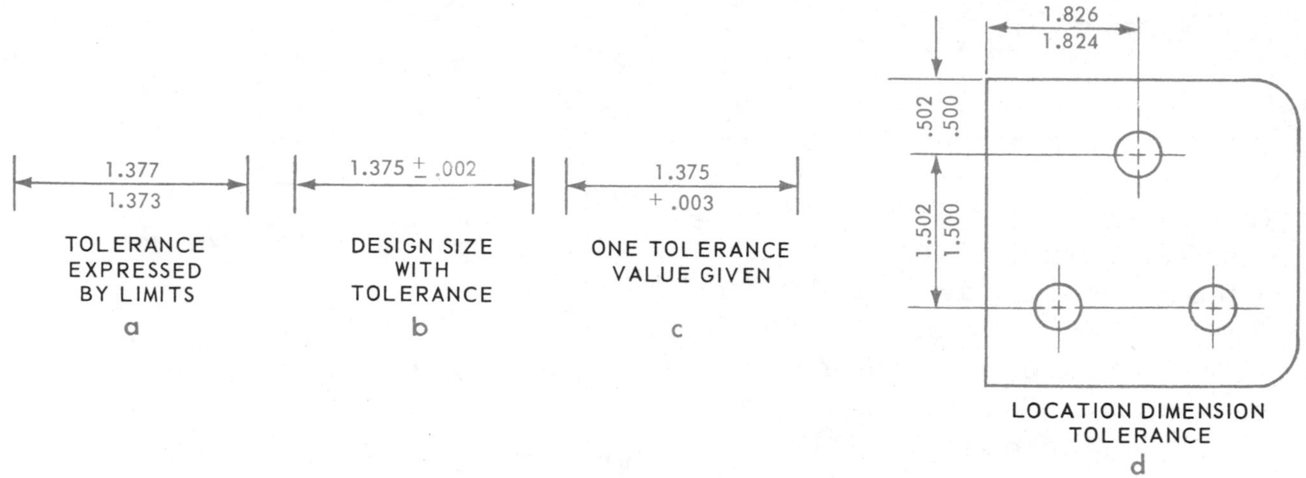

Fig. 11-3. Tolerances indicated in dimensioning.

Since the size of our shaft in Fig. 11-2(b) is .998 after the allowance has been applied, our tolerance must be applied below this size, in order to assure the minimum clearance (allowance) of .002. If a tolerance of ± .002 is permitted on the shaft, the total variation of .004 (+.002 and -.002) must occur between .998 and .994. Then, the design size for our shaft is given a bilateral tolerance (variation is permitted in both directions from the design size) of .996 ± .002 after the application of allowance for minimum clearance and tolerance for variance in manufacture, Fig. 11-2(c).

Tolerance should always be as large as possible, other factors considered, to reduce manufacturing costs. Tolerances may be specific and given with the dimension value or, general and given as a note or included in the title block of the drawing and apply to all dimensions on the blueprint, unless otherwise noted, Fig. 11-4.

TOLERANCE UNLESS OTHERWISE NOTED			
FRAC. TOL. ± 1/64		ANGLES ± 0° 30'	
DEC. TOL.	1 PLACE .X ± .1	2 PLACE .XX ± .03	3 PLACE .XXX ± .005

Fig. 11-4. Tolerances given in the title block.

UNILATERAL TOLERANCE is a tolerance in which variation is permitted only in one direction from the design size, Fig. 11-3(c).

BILATERAL TOLERANCE is a tolerance in which variation is permitted in both directions from the design size, Fig. 11-3(b).

DATUMS are points, lines, surfaces or planes of a part that are assumed to be exact for purpose of reference and from which the location of other features are established. The datum plane is indicated by the assigned letter preceded and followed by a dash, enclosed in a small rectangle or box, Fig. 11-5.

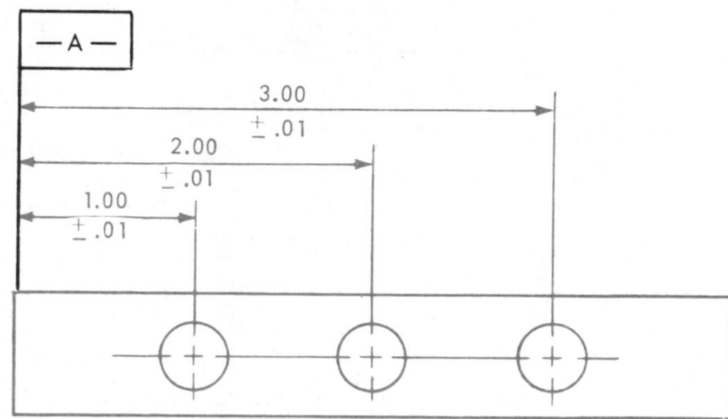

Fig. 11-5. Datum surface used in dimensioning.

Types of Dimensions

LINEAR DIMENSIONS are usually given in inches for measurements of 72 inches and under, and in feet and inches if greater than 72 inches. Where the blueprint calls for accurate machining with close tolerances, the dimensions are usually given in inches and decimal fractions. This is the general practice in aerospace, automotive, electrical and electronic, machine tool, sheet metal, and similar industries. In the cabinetmaking, construction and structural industries, dimensions are given in feet, inches and common fractional parts of an inch.

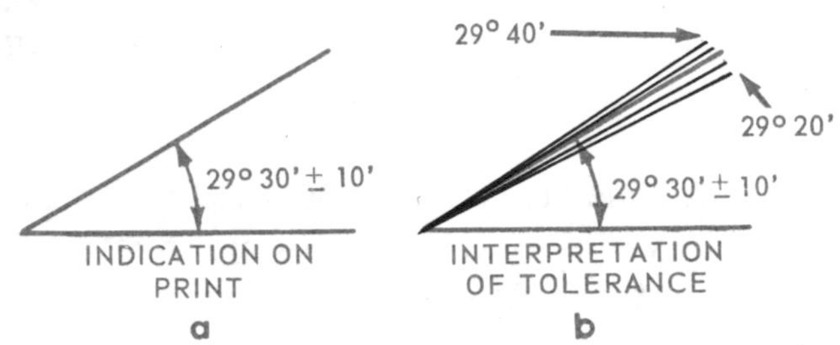

Fig. 11-6. Angular dimension and tolerance.

ANGULAR DIMENSIONS are used on blueprints to indicate the size of angles in degrees (°) and fractional parts of a degree, minutes (') and seconds ("). A complete circle contains 360°, one degree contains 60' (minutes) and one minute contains 60" (seconds). The size of an angle with the tolerance indicated is shown in Fig. 11-6(a) and the interpretation of the tolerance is shown in Fig. 11-6(b).

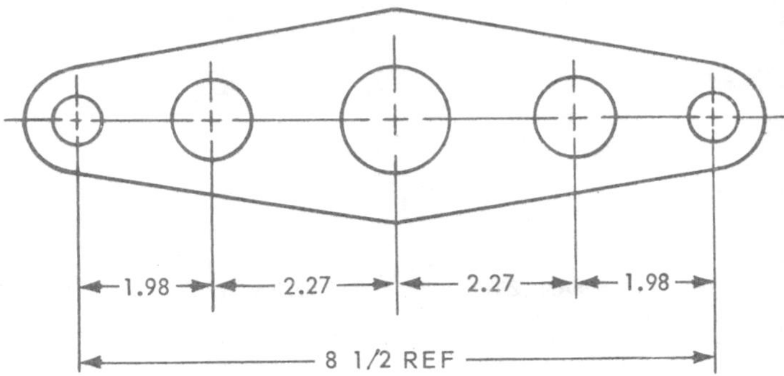

Fig. 11-7. Dimension for reference only.

REFERENCE DIMENSIONS are occasionally given on drawings for reference and checking purposes. These dimensions will be marked REF, Fig. 11-7. They will be without tolerance and are not to be used for layout, machining or inspection operations.

TABULAR DIMENSIONS are used when a company manufactures a series of sizes of an assembly or part. Dimensions on the

PART NO.	A	B	C	D	E	F
41-8706	.625	1.00	1.312	.156	.875	.250
41-8707	.750	1.250	1.812	.484	1.00	.312
41-8708	.875	1.437	2.062	.562	1.125	.375

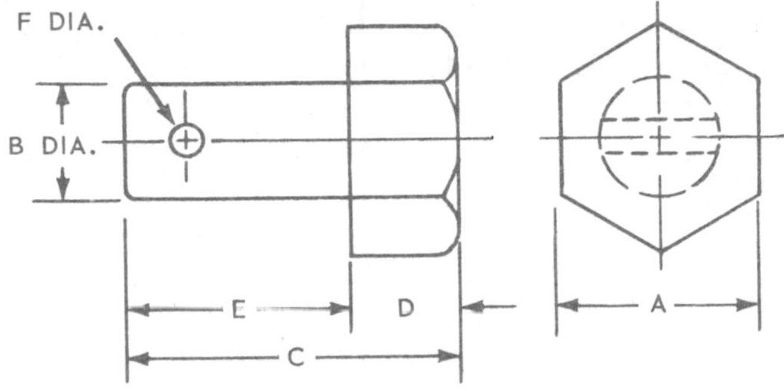

Fig. 11-8. Dimensioning a series of sizes.

HOLE NO.	POSITION			
	+X	+Y	−X	−Y
1	1.730	2.460		
2		.830	4.365	
3	2.900			.475
4			3.150	.545
5	4.220			2.340
6			.430	3.495
7			3.150	4.135
8	2.900			4.205
9			4.365	5.510
10	4.385			5.550

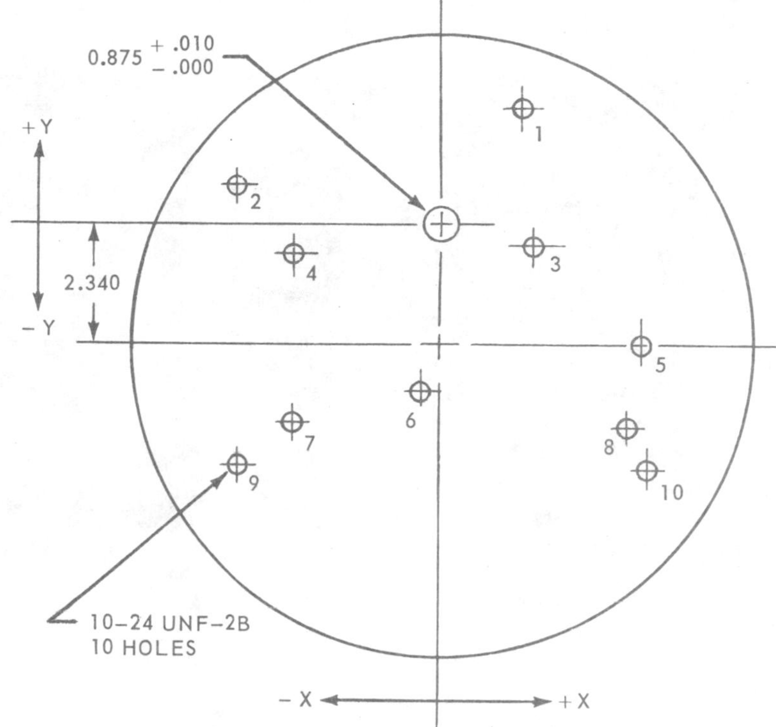

Fig. 11-9. Tabular dimensions.

drawing are replaced by reference letters and a table on the drawing lists the corresponding dimensions, Fig. 11-8.

Another use made of tabular dimensions is in the dimensioning of a part with a large number of repetitive features such as holes, Fig. 11-9. Running extension and dimension lines to each hole would make the drawing difficult to read. Instead, each hole or feature is assigned a letter or numeral (or letter with numeral subscript) and the dimensions of the feature and the feature location along the X and Y axes are given in a table on the drawing, Fig. 11-9.

ARROWLESS DIMENSIONING is frequently used on drawings which contain datum lines or planes, Fig. 11-10. This practice eliminates numerous dimension and extension lines and improves the clarity of the drawing. Arrowless dimensioning is especially useful on drawings of parts which are to be machined on numerically controlled equipment (refer to blueprint on page 188). Arrowless dimensioning is also referred to as coordinate dimensioning or datum dimensioning.

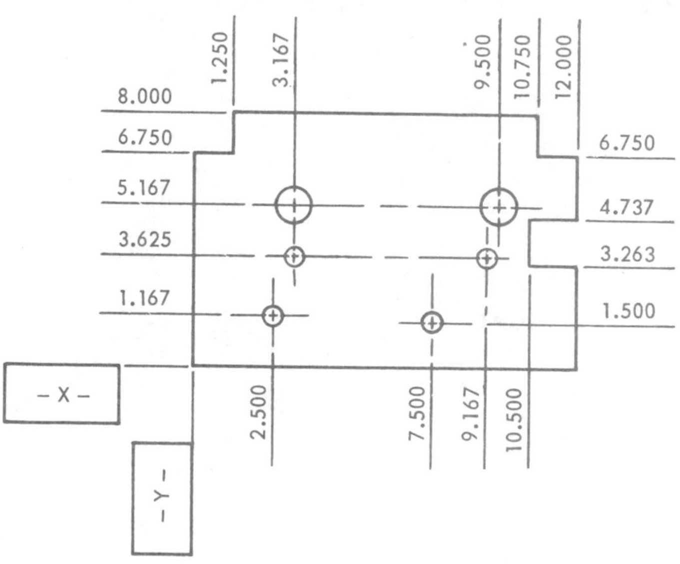

Fig. 11-10. Application of arrowless dimensioning.

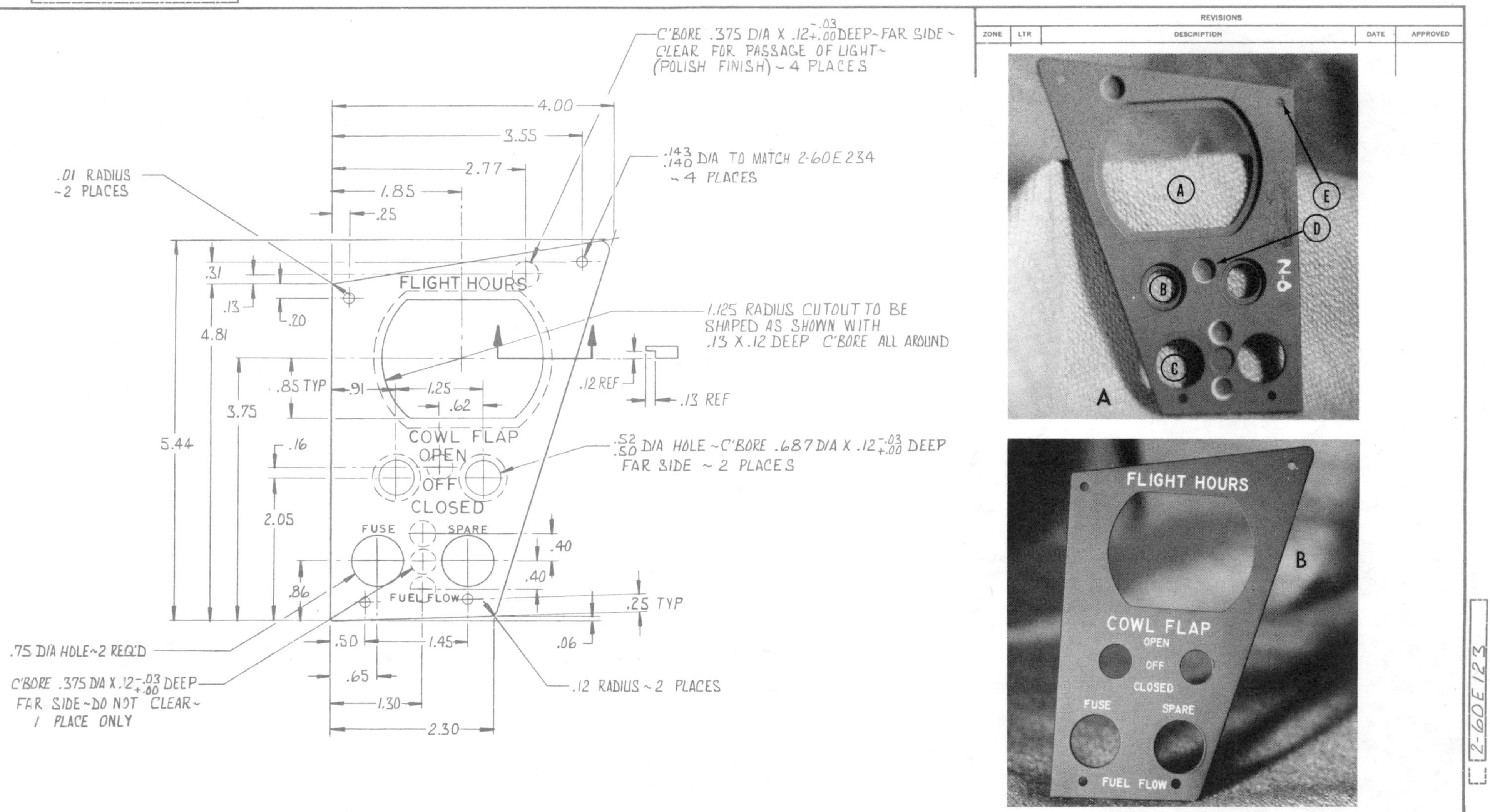

Fig. 11-BPR-1. Industry blueprint. (Beech Aircraft Corporation)

Blueprint Reading Activity 11-BPR-1
DIMENSIONS AND TOLERANCES

Refer to the blueprint in Fig. 11-BPR-1 and answer the following questions.

1. Give the name of the part. 1. ______________________________

2. What is the drawing number? 2. ______________________________

3. What material is required? 3. ______________________________

4. List the material size when issued? 4. ______________________________

5. What is the finished overall size (include upper corner point extension)? 5. ______________________________

6. Give the overall size of the cutout for A (see photos). 6. ______________________________

7. What is the size of the counterbore around the above cutout? 7. ______________________________

8. Give the diameter of hole B and the counterbore. 8. ______________________________

9. What is the diameter of C? 9. ______________________________

10. Give the dimensions for the counterbore at D. 10. ______________________________

11. Give the diameter of E. How many holes are of this diameter? 11. ______________________________

12. What is the distance between centers of the two .75 diameter holes? 12. ______________________________

13. What is the radius for the upper corner of the piece? 13. ______________________________

14. From the far side, give the locating dimensions for the feature at D. 14. ______________________________

For additional Blueprint Activities for Unit 11 see pages 234 and 237.

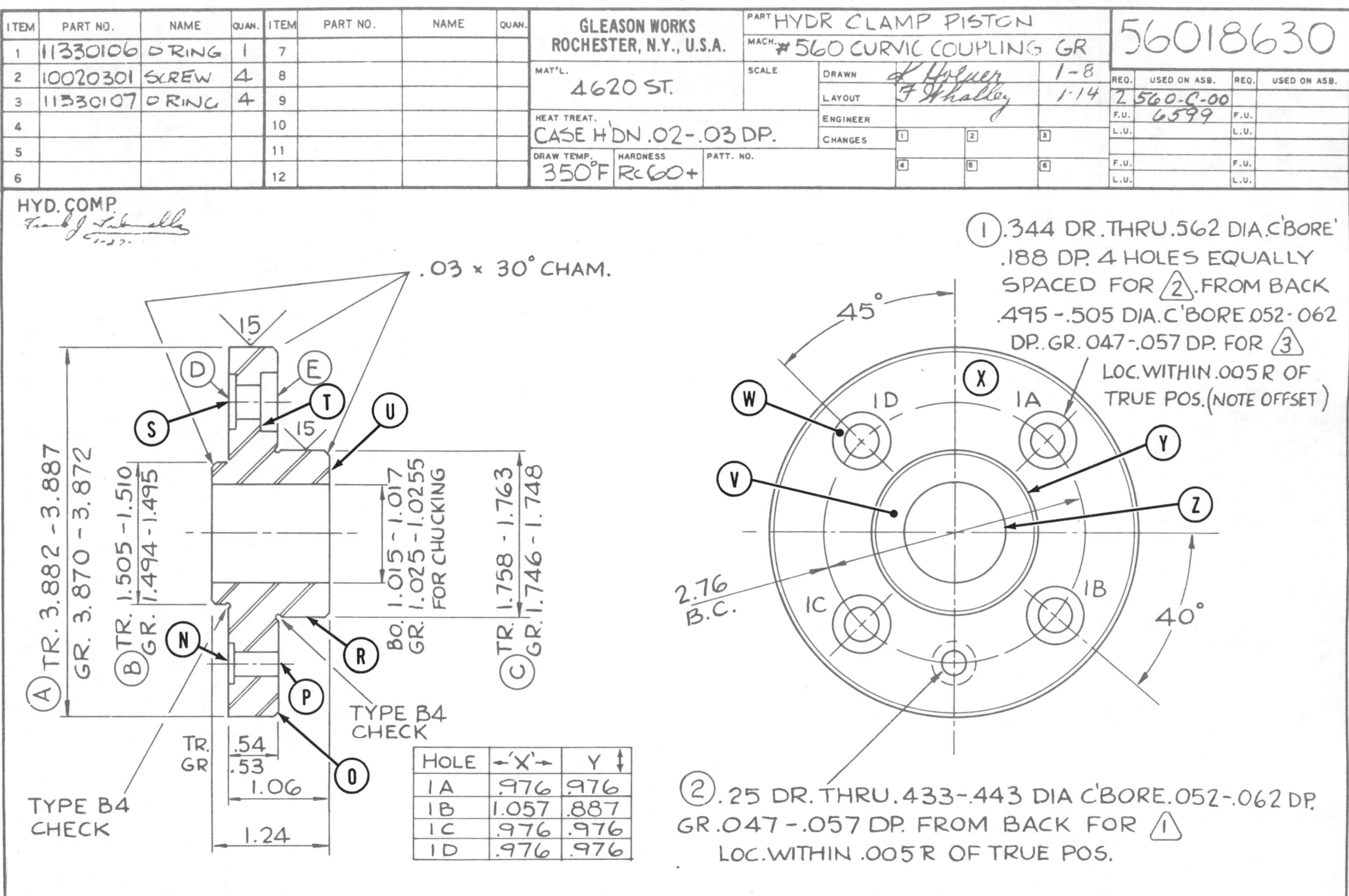

Fig. 11-BPR-2. Industry blueprint.
(Gleason Works)

Blueprint Reading Activity 11-BPR-2
DIMENSIONS AND TOLERANCES

Refer to the blueprint in Fig. 11-BPR-2, and answer the following questions.

1. What is the name of the part? 1. ______________________

2. Give the drawing number. 2. ______________________

3. What material is used? 3. ______________________

4. Give the overall finished dimensions of the part. 4. ______________________

5. Give the diameter of the bolt circle. 5. ______________________

6-9. Identify in the left side view the following lines and surfaces in the front views:

	Front	Left Side
6.	V	______________
7.	W	______________
8.	X	______________
9.	Y	______________

10. Give the limit dimensions for the thickness of the part at X. 10. ______________________

11. Give the diameter and the depth of the counterbore at T. 11. ______________________
____

12. What is the diameter and the depth after grinding of the counterbore at S? 12. ______________________

13. Give the diameter of the hole at P. 13. ______________________

14. What is the diameter and the depth before grinding of the counterbore at N? 14. ______________________

15. How far above the horizontal center line is hole: 15. 1A ______________________
1D ______________________

16. How far to the right of the vertical center line is hole 1B? Below the horizontal center line? 16. ______________________
____

17. What is the diameter at Z after grinding? 17. ______________________

Unit 12
Sectional Views

Sometimes it may be necessary to show more clearly the interior construction of an object than is revealed in the regular views of a drawing. This is done by passing an imaginary cutting plane through the object, removing the part of the object nearest the observer and permitting a direct view of the interior details, Fig. 12-1.

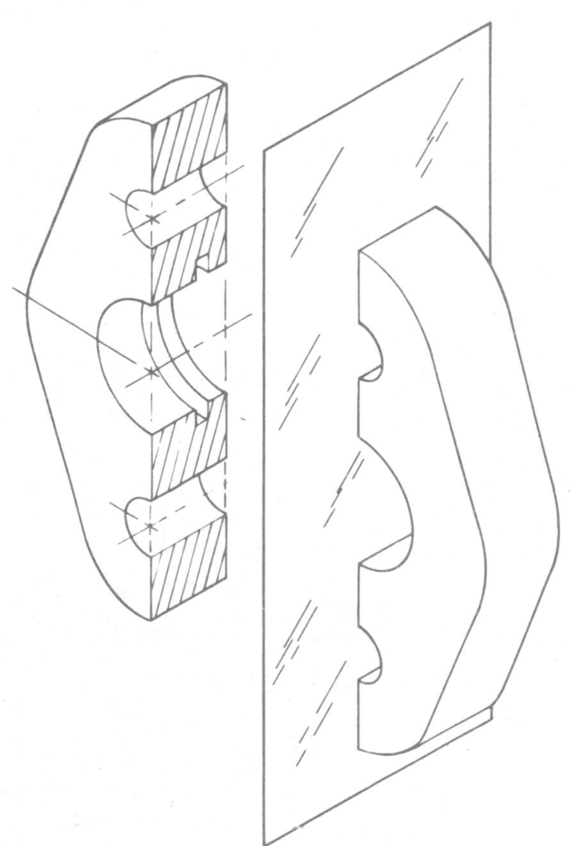

Fig. 12-1. Imaginary cutting plane showing a sectional view.

When the sectional view is drawn, the material that is cut by the imaginary cutting plane is shown in "section." That is, crosshatch lines are drawn at an angle to further contrast and clarify the detail. Crosshatch lines like those shown in Fig. 12-1 are used to indicate cast iron and malleable iron and are also used generally for all materials in section. Other materials in section are shown on page 329.

The material symbols shown in sectional views are used to clarify the drawing for the reader, not to designate a specific composition of material. The material specification is listed in the Title Block, Materials Block or in a note on the drawing.

Crosshatch lines are usually drawn in at 45 deg. but may be at any angle as long as they are not parallel or perpendicular to one of the major visible lines outlining the sectional part. A sectional view may serve as one of the regular views on the drawing or it may appear as an additional view.

In addition to showing internal construction and details, sectional views are also used to show the exact shape or contour of exterior parts which are not shown in the regular views such as wheel spokes, handles, airplane propellers and wings.

Types of Sections

There are several types of sections and each is designed for a specific purpose. These are discussed and illustrated here to familiarize you with their nature and purpose.

Full Section

When the cutting plane passes entirely through the object, the view is called a FULL SECTION, Fig. 12-2.

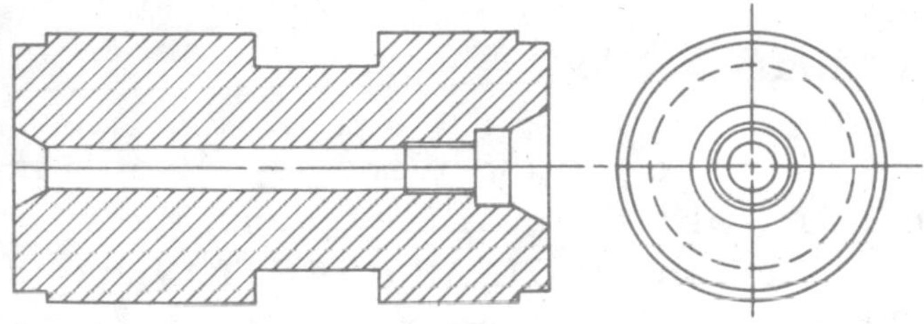

Fig. 12-2. A full section.

Note how much clearer the internal details of the sectional view appear than the right view with its invisible lines. Lines which are visible in the sectional view are shown, but invisible lines usually are omitted.

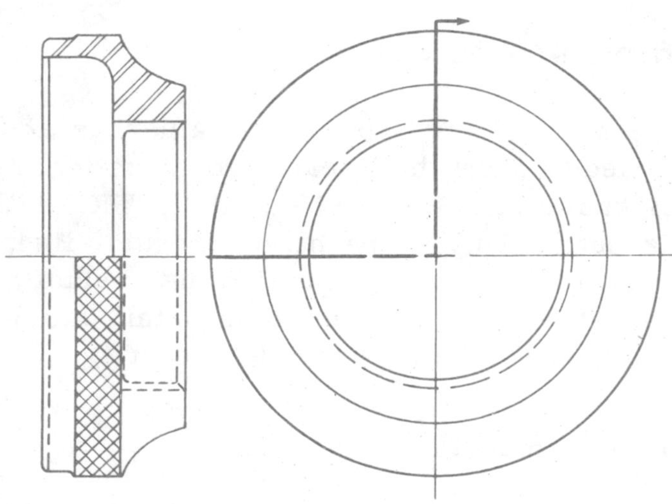

Fig. 12-3. Half section.

Half Section

Objects which are symmetrical (both halves are similar) may be drawn with one view as a half section and the remaining half as an external view, Fig. 12-3. The cutting plane line is shown on the other view indicating the part which has been "removed" and the direction in which the section is viewed.

Offset Section

There are objects in which the essential internal details do not appear on one straight line through the object such as in

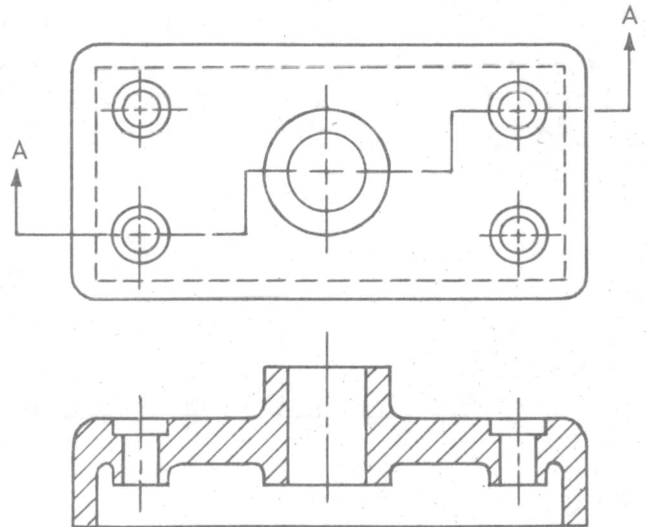

Fig. 12-4. Offset section.

the full section. When this occurs, an offset cutting plane line is drawn through the object to include the desired features which are shown on one plane in the sectional view, Fig. 12-4.

Aligned Section

The cutting plane line on an ALIGNED SECTION indicates the features of a part which are rotated to the path of the principal plane and projected to the sectional

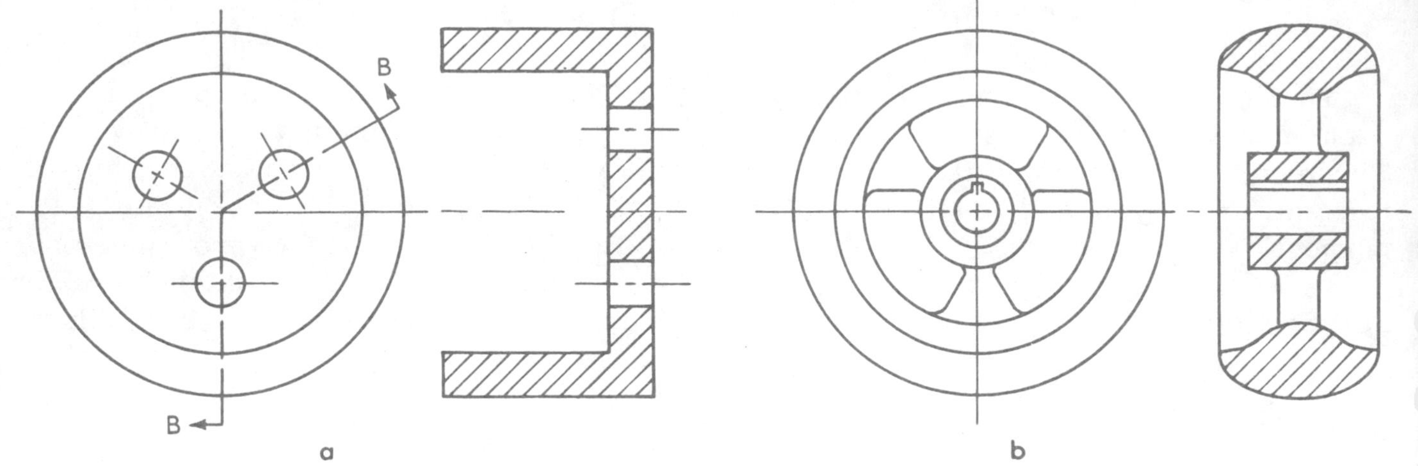

Fig. 12-5. Aligned sections.

view, Fig. 12-5(a). The aligned section is also used to show spokes of a wheel and other unsymmetrical objects in a balanced or aligned position, Fig. 12-5(b). Spokes and arms in a sectional view are not cross-hatched to distinguish between these and a solid web.

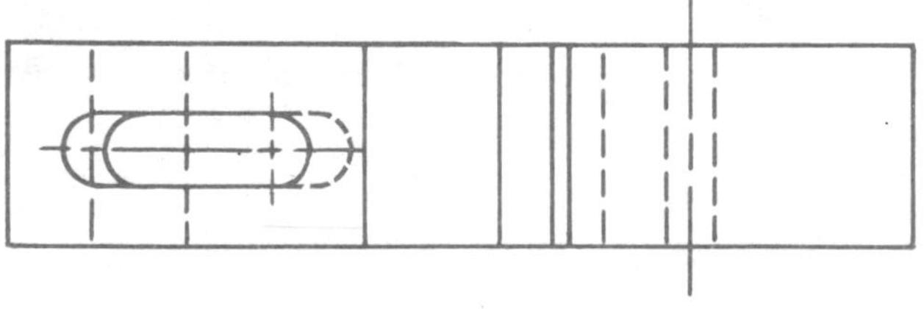

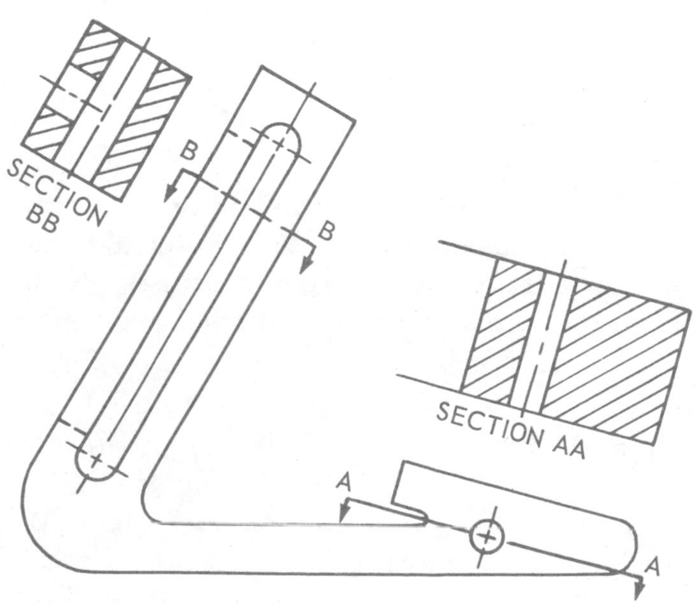

Fig. 12-6. Auxiliary sections.

Auxiliary Section

When the cutting plane is selected in parallel with the auxiliary plane, the view is called an AUXILIARY SECTION and is shown on the blueprint parallel to the auxiliary view, Fig. 12-6. Any type of section found on regular views--full, half, broken-out, etc.--may also be found on auxiliary views. When a section is made of an auxiliary and the portion extending beyond the section is not fully shown, the view is sometimes called a partial section, for example, Section AA, Fig. 12-6. Section BB in Fig. 12-6 could be called a removed section in auxiliary position.

Broken-out Section

When a small portion of a part is exposed to show the interior construction, it is known as a BROKEN-OUT SECTION. It is outlined by a freehand irregular line, Fig. 12-7. This type section also permits the showing of the exterior detail which may need to be preserved for clarity.

Revolved Section

Sectional views of objects which have a more or less uniform exterior shape or

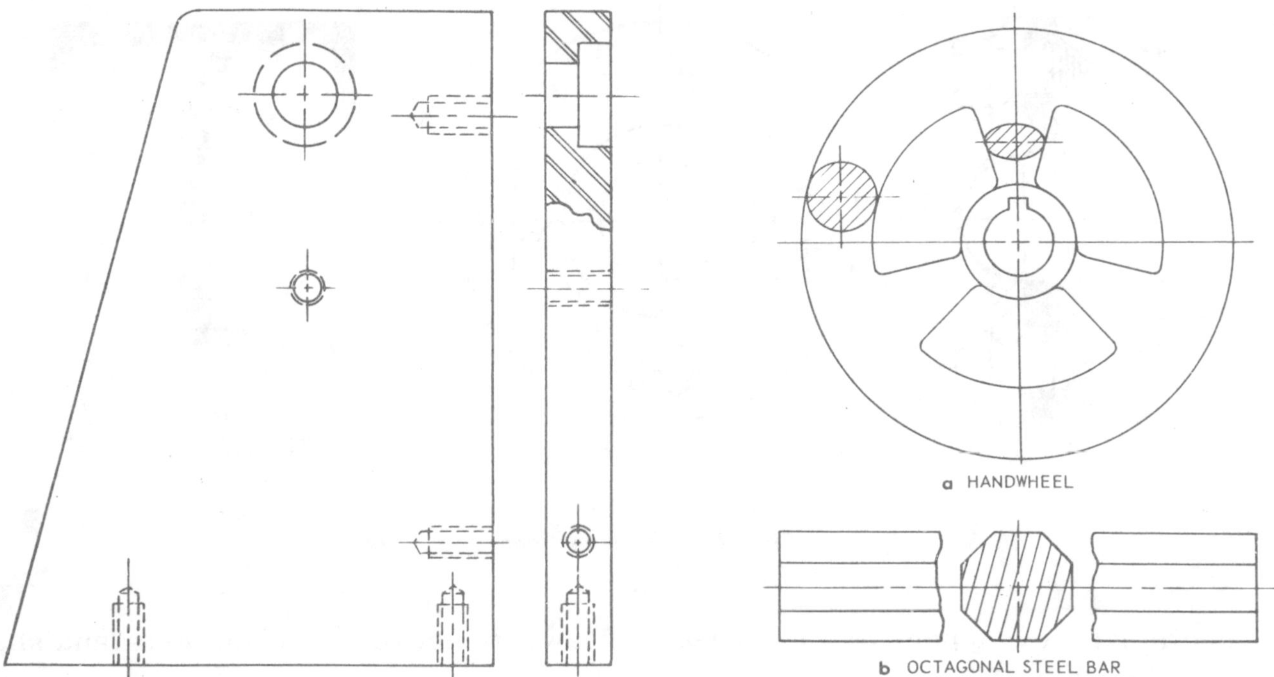

Fig. 12-7. Left. Broken-out section. Fig. 12-8. Right. Revolved sections.

contour, such as wheel spokes and steel bars, are shown in place with the axes revolved 90 deg., Fig. 12-8(a). To further clarify the view, the draftsman may break out the part in section, Fig. 12-8(b).

Removed Section

When a revolved section is removed from its place on the object and is shown in another place on the drawing, it is known as a REMOVED SECTION. See Section BB, Fig. 12-6. Removed sections are frequently used as DETAIL SECTIONS and shown in an enlarged view to clarify detail, Fig. 12-9.

Thin Sections

Where thin parts, such as plates, gaskets, angles and channels are shown in section, the shape and thickness of the part

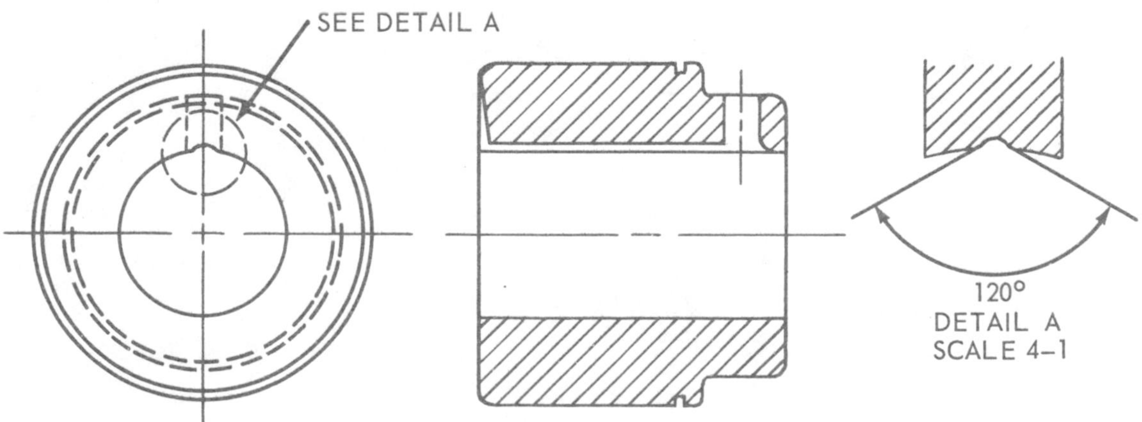

Fig. 12-9. Removed detail section.

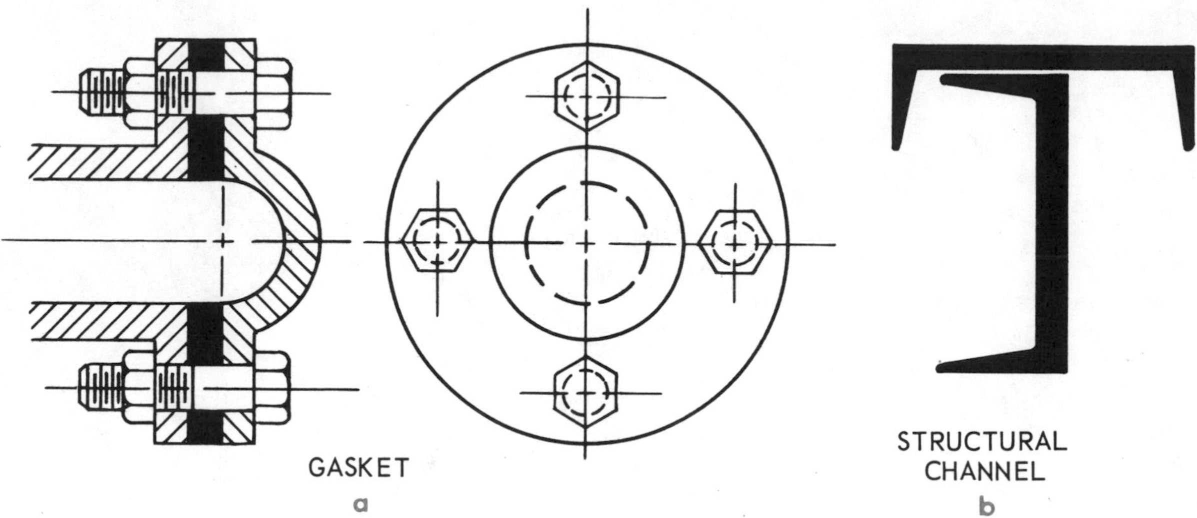

Fig. 12-10. Thin sections.

in the sectional view appears as a shaded solid, Fig. 12-10.

Parts not Shown in Section

It is standard practice in industrial drawings not to show fasteners and shafts in section when they occur in the cutting plane. They are usually more recognizable by their exterior features, as illustrated in Fig. 12-10(a).

Blueprint Reading Activity 12-BPR-1
SECTIONAL VIEWS

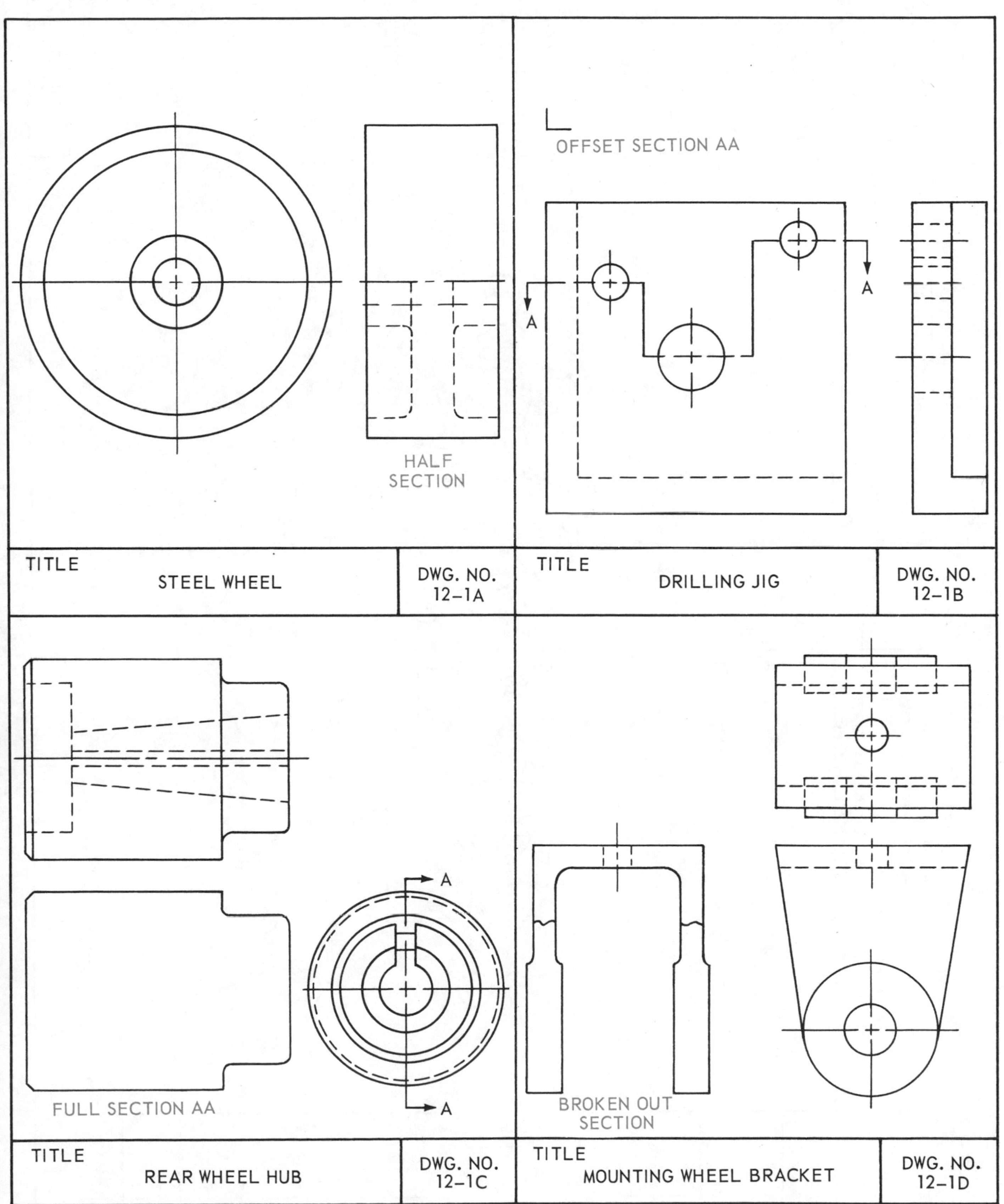

Fig. 12-BPR-1. Complete the unfinished view in each problem above by sketching the type section required.

MATERIAL		NAME	DRAWN BY T.D.	316521
STAINLESS STEEL	UNLESS OTHERWISE SPECIFIED:	NIPPLE - INTERMEDIATE HOUSING	CHECKED BY JW	
TYPE #416	±.020 TOL. ON TWO PLACE DECIMALS	MARINE ENGINEERING OUTBOARD MARINE CORP.	APPROVED BY	SCALE FULL / DATE 9·17
.875 HEX. STOCK	±.010 TOL. ON THREE PLACE DECIMALS	USED BY EE / BT	EXP. NO.	RELEASED 29460 WYRICK 9-23
	ZERO TOL. ON FOUR PLACE DECIMALS	SIMILAR 310904		
	ANGLE TOL. ±1°			
	DESIGNED FOR 120-155-210 H.P. STERN DRIVE			

DIV.	QTY.
CORP. IBM	1
A-AUST.	
B-BELG.	✓
C-CAN. P-PION.	✓
G-GALE	✓
J-JOHN.	✓
E-EVIN.	✓
L-LN-BOY	
CU-CUSH.	
OE-OMC ENG	
ME-ENGINES	
BT-OMC BTS.	
BT-CAN.	
CHIEF INSP.	
PROD.	
SUPT.	
TIME STUDY	
TOOL DESIGN	
TOTAL	1

Fig. 12-BPR-2. Industry blueprint. (Outboard Marine Corp.)

Blueprint Reading Activity 12-BPR-2
SECTIONS

Refer to the blueprint in Fig. 12-BPR-2 and answer the study questions below.

1. What is the name of the part? 1. ____________________

2. What is the print number? 2. ____________________

3. What are the general tolerance limits? 3. ____________________

4. What material is to be used? 4. ____________________

5. Give the stock size of material. 5. ____________________

6. What is the scale of the original plan? 6. ____________________

7. The general shape of the part is: 7. ____________________

8. What type of a section is shown? 8. ____________________

9. Surface D in the front view is represented by what line in the right side view? Surface E? Line G? 9. D __________ E __________ G __________

10. The draftsman has used the section lining generally used for all materials. Sketch the section lining for steel. 10. ____________________

11. Give the diameter of the hole thru the part. How is this to be machined? 11. ____________________

12. The draftsman has chosen to eliminate all hidden lines. (a) Why? (b) There are four surfaces or edges which could have been represented. Identify these. 12. a. ____________________ b. ____________________

13. Give the overall length of the part. 13. ____________________

14. What is the measurement across the (a) hex. flats? (b) the width of the flat? 14. a. ____________________ b. ____________________

15. Give the thread specification. 15. ____________________

16. What is the diameter of H? 16. ____________________

For additional Blueprint Activities for Unit 12 see pages 238 and 241.

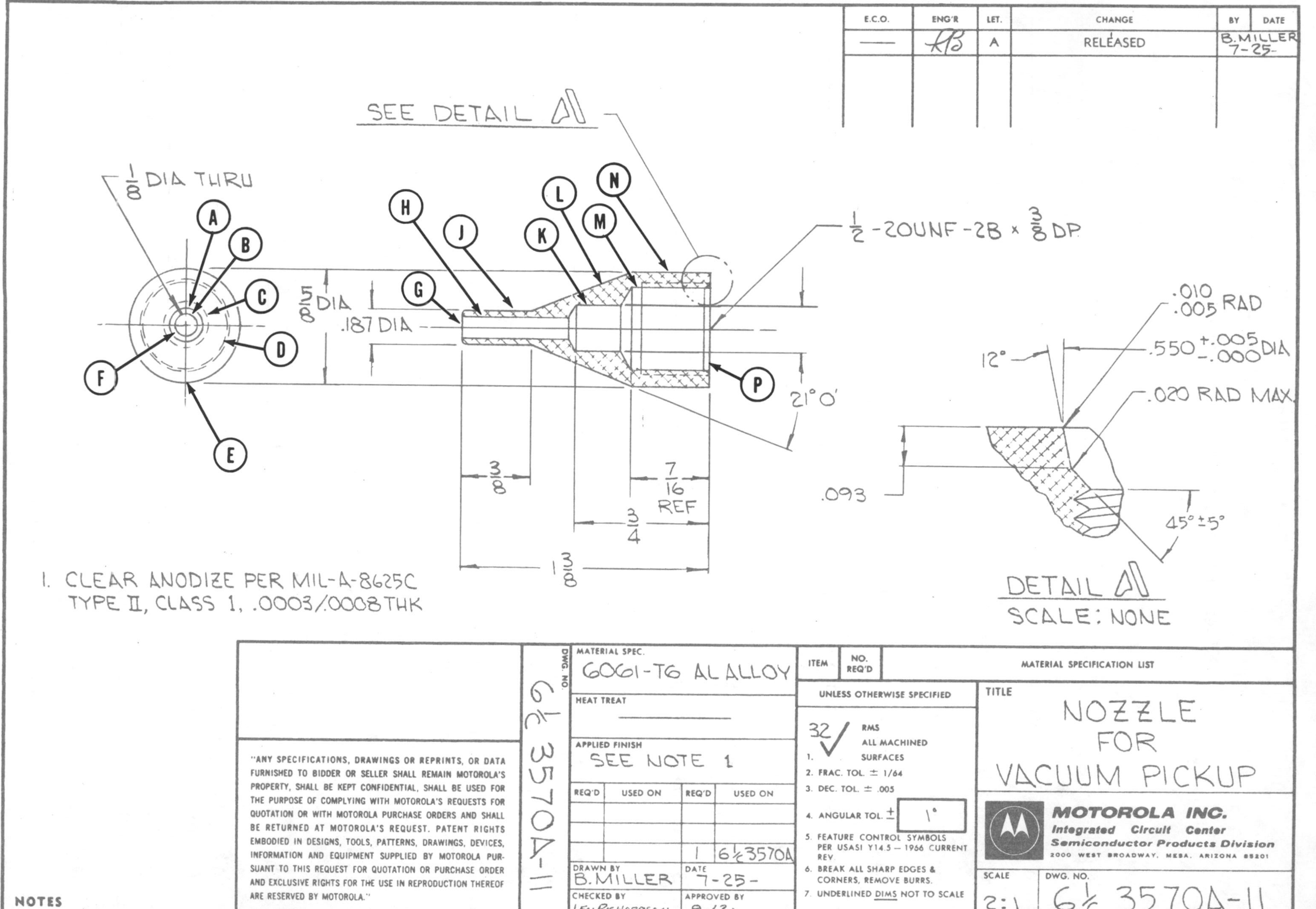

Fig. 12-BPR-3. Industry blueprint. (Motorola Inc.)

Blueprint Reading Activity 12-BPR-3
SECTIONS

Refer to Fig. 12-BPR-3 and answer the following questions.

1. Give the name of the part. 1. ______

2. What is the number of the print? 2. ______

3. What general tolerance limits are given? 3. ______

4. List the material to be used. 4. ______

5. What is the scale of the original plan? 5. ______

6. What is the general shape of the part? 6. ______

7. What type of section is the front view? 7. ______

8. Give the line or surface in the front view that represent the following in the left side view: 8. A ______ B ______ C ______ D ______ E ______ F ______

9. Give the overall dimensions of the nozzle: (a) length, (b) diameter. 9. a. ______ b. ______

10. Give the dimension of feature A. 10. ______

11. What is the diameter of H? J? 11. H ______ J ______

12. Give the dimension of J expressed as tolerance limits. 12. ______

13. What angle does L make with the center line? 13. ______

Unit 13
Pictorial Drawings

Pictorial drawings show an object much as it would appear in a photograph - - as if you were viewing the actual object, Fig. 13-1. Several sides of the object are visible in the one pictorial view. Pictorial drawings are quite easy to understand. They can be used in making or servicing simple objects but are usually not adequate for complicated parts.

You will find the technique of pictorial sketching useful in helping you to visualize two and three view orthographic drawings as well as in communicating your ideas on technical problems to others.

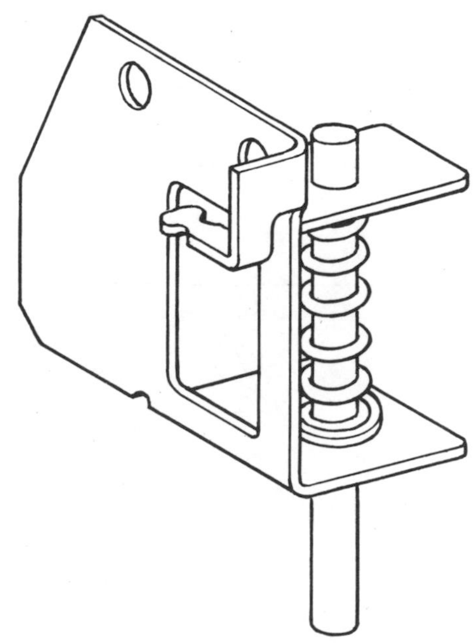

Fig. 13-1. A pictorial assembly drawing. (National Lock)

There are three common types of pictorial drawings in general use: (1) Isometric (i • so • met'rik), (2) Oblique (ob • lek') (Cavalier and Cabinet), and (3) Perspective (per • spek • tiv). A fourth type is the exploded pictorial drawing which is a special use of one of the three more common types.

Isometric Drawing

Isometric drawings are used more often than the other types of pictorial drawings and sketches because direct measurements can be used in their construction. Although they have more of a distorted appearance than perspective drawings, they are more easily constructed. An isometric drawing is constructed with its two faces projected at angles of 30 deg. with the horizontal, Fig. 13-2.

The dark lines forming a Y in the center of the drawing are equally spread at 120 deg. between each set and are known as the isometric axes, Fig. 13-2. Because of this equal spread, the same scale of measure is used along each axis. The word "isometric" means equal measure - - in this type drawing, equal measure on all axes.

Lines that are horizontal in an orthographic drawing are drawn at an angle of 30 deg. in an isometric drawing and vertical lines remain vertical. Slant lines (non-isometric lines) are drawn by locating

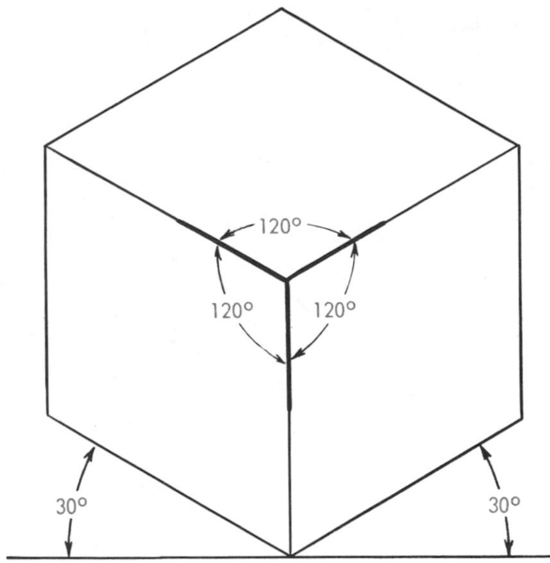

Fig. 13-2. Isometric axes.

their end points on the isometric axes and connecting the two points.

Constructing an Isometric Sketch

Let's look at the steps in constructing an isometric sketch in Fig. 13-3.

PROCEDURE FOR CONSTRUCTING AN ISOMETRIC SKETCH:

1. Given 3-views of a V-block, Fig. 13-3(a).
2. Select the position of the object to best describe its shape.
3. Start the sketch by laying out the axes for the lower corner, Fig. 13-3(b).
4. Make overall measurements in their true length on the isometric axes or on lines parallel to the axes, Fig. 13-3(c).
5. Construct a "box" to enclose the object, Fig. 13-3(d).
6. Sketch the isometric lines of the object, Fig. 13-3(e).
7. Sketch the nonisometric lines (slant lines) by first locating the end points of these lines and then sketching the line between, Fig. 13-3(f).
8. Darken all visible lines and erase the construction lines to complete the isometric sketch, Fig. 13-3(g).

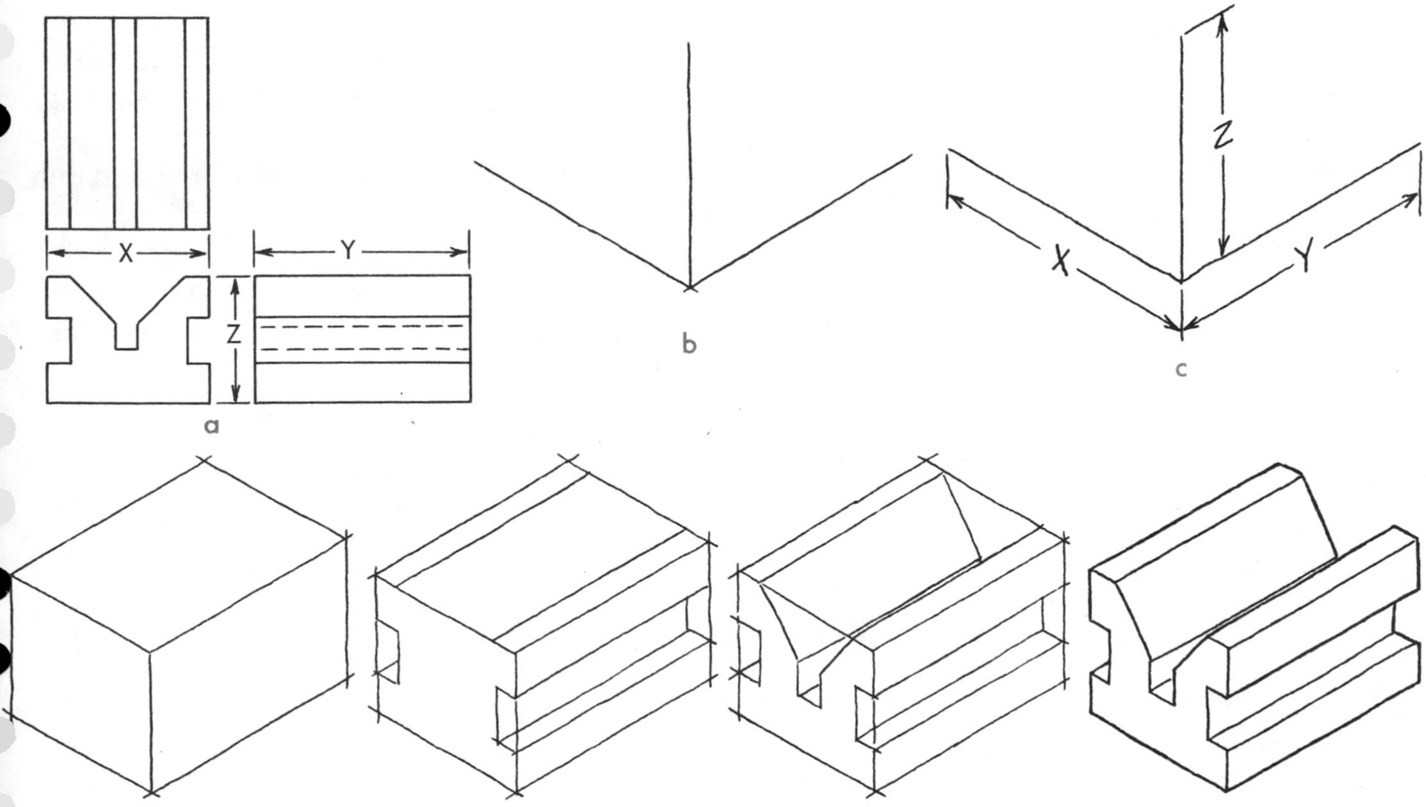

a b c d e f g

Fig. 13-3. Steps in constructing an isometric sketch.

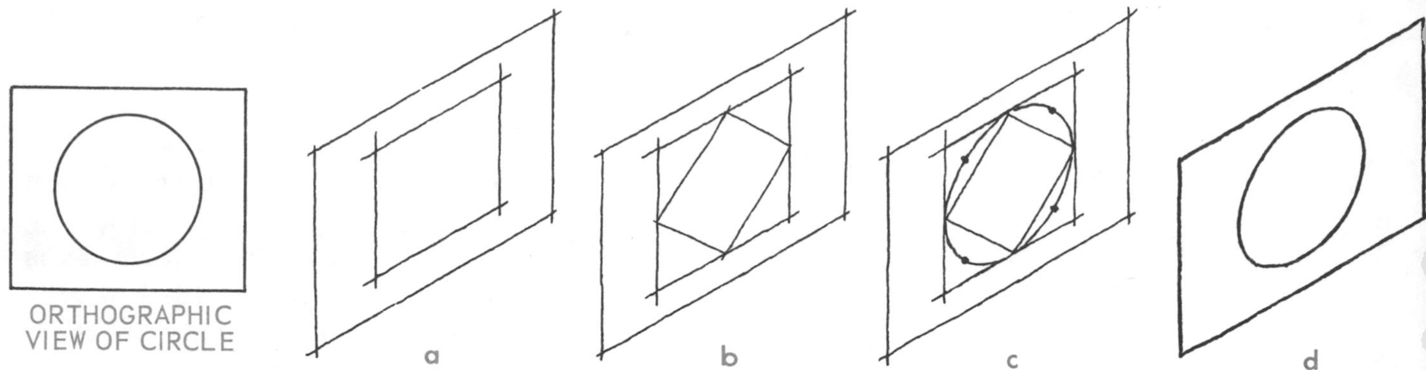

Fig. 13-4. Steps in sketching an isometric circle.

Invisible lines are not shown in isometric views unless required to clarify drawing details.

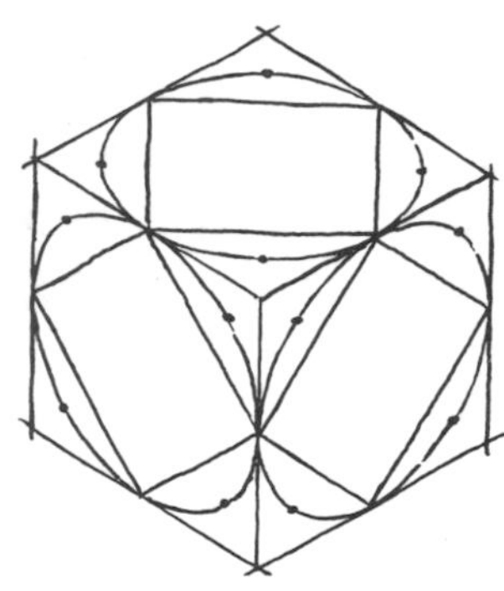

Fig. 13-5. Sketched isometric circles.

Circles and Arcs in Isometric

Circles and arcs in isometric are sketched in the same manner you learned to sketch circles and arcs in Unit 4, except you start with an isometric square.

PROCEDURE FOR SKETCHING ISOMETRIC CIRCLES:

1. Sketch an isometric square to enclose location of circle, Fig. 13-4(a).
2. Locate the midpoints of the sides of the isometric square and connect these midpoints, Fig. 13-4(b).
3. Locate the midpoints of the triangles formed and then sketch isometric arcs through each to form a circle, Fig. 13-4(c).
4. Erase construction lines and darken circle, Fig. 13-4(d).

Circles may be sketched in all three planes of the isometric drawing in the same manner, Fig. 13-5.

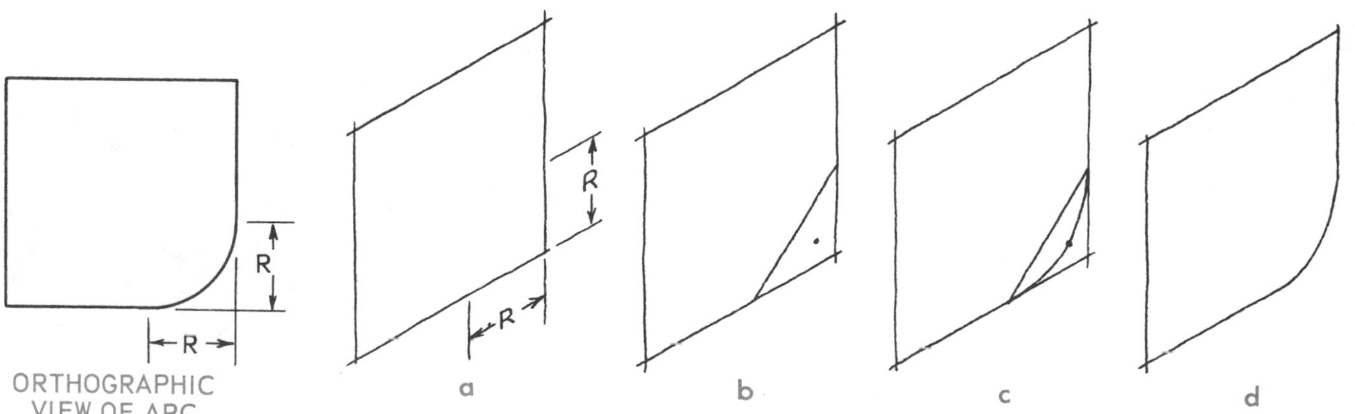

Fig. 13-6. Sketching an isometric arc.

PROCEDURE FOR SKETCHING AN ISOMETRIC ARC:

1. Lay off the radius of the arc from the corner, Fig. 13-6(a).
2. Draw a slant line connecting the two points, forming a triangle, Fig. 13-6(b).
3. Locate midpoint of triangle and sketch an arc through this point to join smoothly with sides, Fig. 13-6(c).
4. Erase construction lines and darken arc, Fig. 13-6(d).

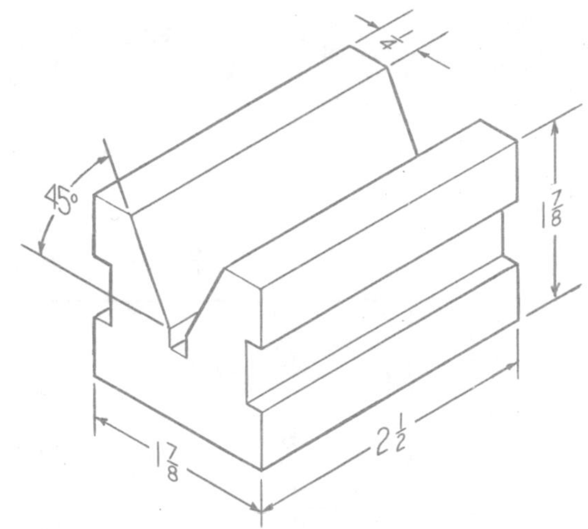

Fig. 13-7. Isometric dimensioning.

Isometric Dimensioning

The dimension lines on an isometric drawing or sketch are parallel to the isometric axes, and extension lines are extended in line with these axes, Fig. 13-7. Note that the dimension figures are also aligned with the isometric axes.

Sketching Assignment 13-BPR-1
TIMING POINTER

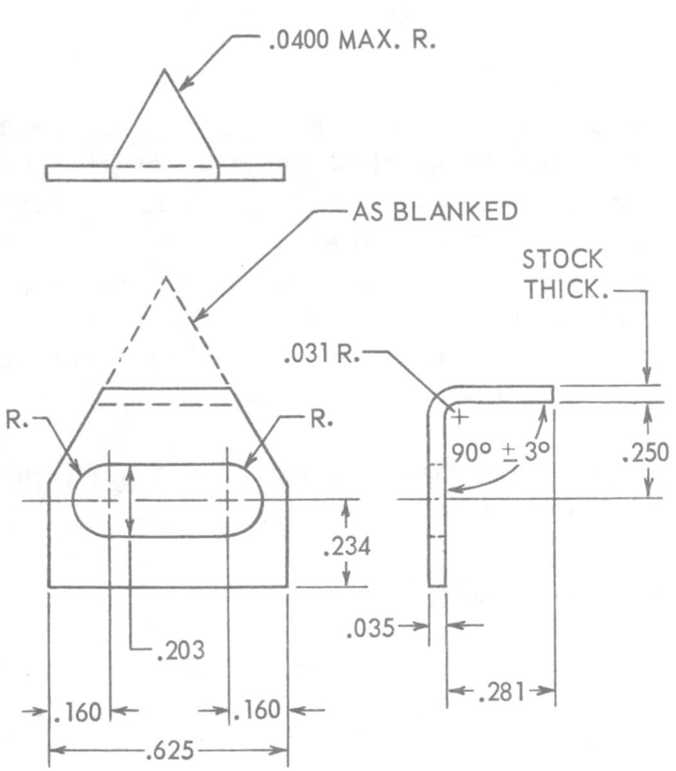

Fig. 13-BPR-1. Make an isometric sketch of the Timing Pointer. Omit the dimensions.

Sketching Assignment 13-BPR-2
MOTOR BRACKET

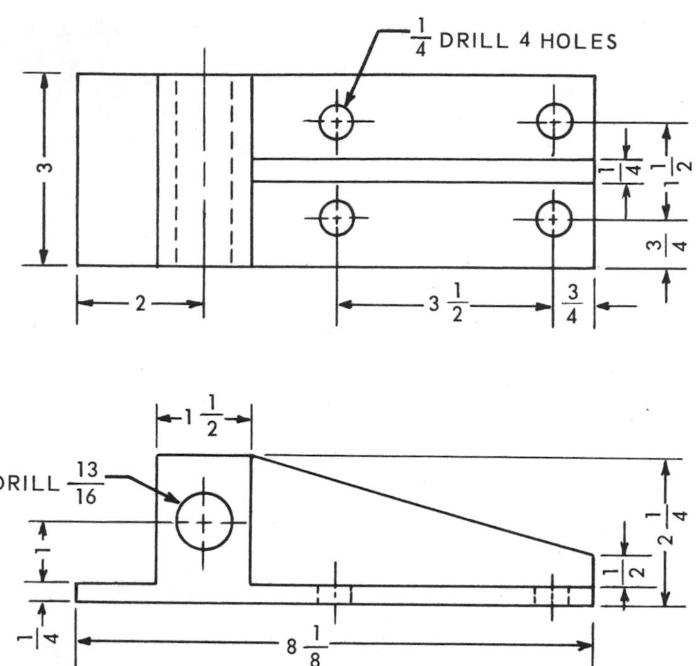

Fig. 13-BPR-2. Make an isometric sketch of the Motor Bracket. Dimension your sketch.

Oblique Drawing

The oblique drawing is drawn with the front face in its true shape and size (unless reduced or enlarged) just as in orthographic projection. This is an advantage when the front view contains circles or arcs which can be represented as true circles and arcs. However, the top and side view are projected back from the front view at an angle, usually 45 deg., and this tends to give a distorted appearance to the drawing, Fig. 13-8.

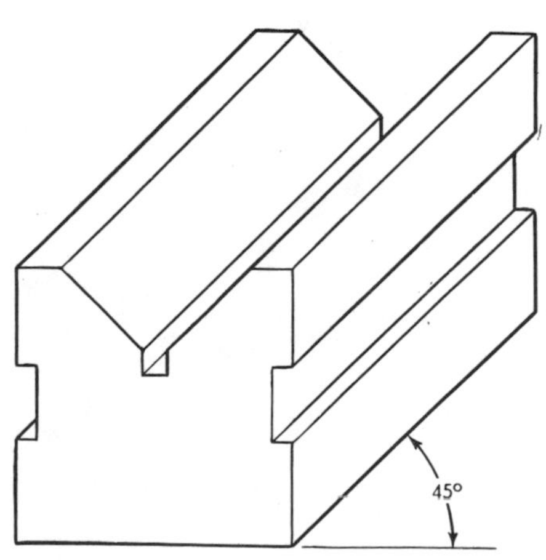

Fig. 13-8. An oblique drawing.

There are two principal types of oblique drawings - - cavalier and cabinet.

Cavalier Oblique

Cavalier oblique drawings are drawn with their receding sides to the same scale as the front view, Fig. 13-9(a). This creates a severe distorted appearance but does have the advantage of using one scale throughout.

Cabinet Oblique

Cabinet oblique drawings differ from the Cavalier only by the fact that measurements made on the receding axes are reduced by half. This gives a much more pleasing appearance, Fig. 13-9(b), and is frequently used in the drawing of cabinet work.

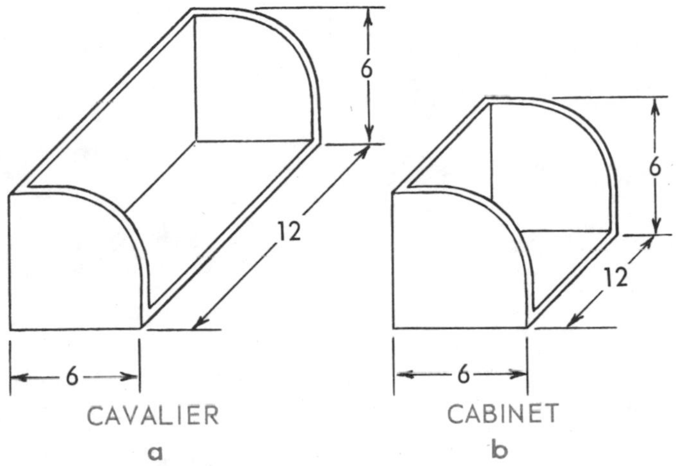

Fig. 13-9. Types of oblique drawings.

Circles and Arcs in Oblique

Circles and arcs in oblique are sketched as true circles or arcs in the front plane as mentioned earlier, Fig. 13-10. When these occur in the receding planes, they are sketched in the same manner as isometric circles and arcs. Fig. 13-10 illustrates the procedure for sketching oblique circles and arcs.

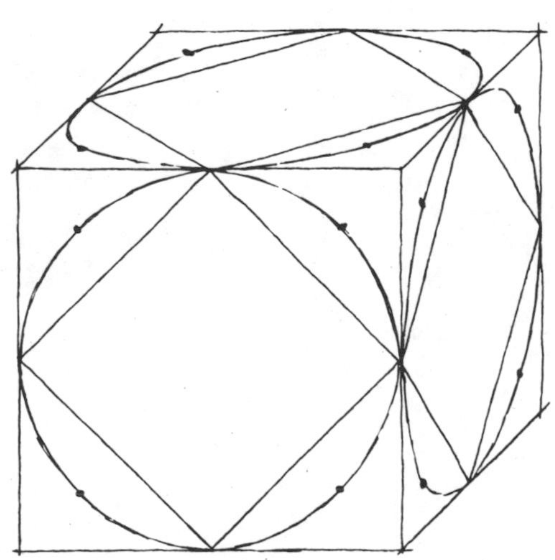

Fig. 13-10. Sketching oblique circles and arcs.

PROCEDURES FOR SKETCHING OBLIQUE CIRCLES AND ARCS:

1. Sketch an oblique square to contain circle, Fig. 13-10.
2. Locate midpoints of sides of the oblique square and connect these midpoints.
3. Locate midpoints of triangles and sketch oblique circles or arcs through these midpoints to join smoothly with sides of oblique square.
4. Erase construction lines and darken circles or arcs.

The procedure is the same for circles and arcs in cabinet oblique as it is for cavalier oblique.

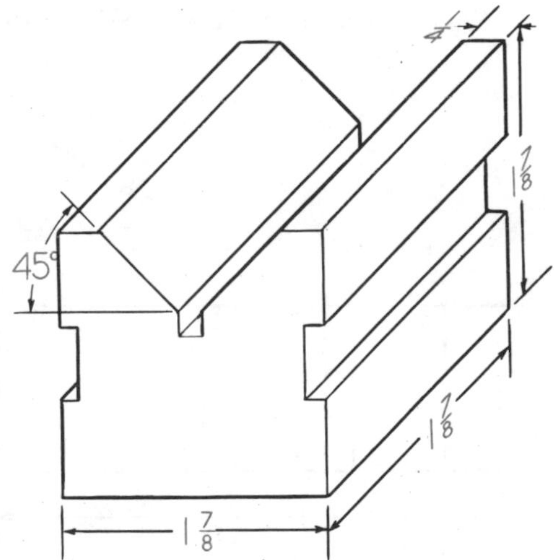

Fig. 13-11. Dimensions on a cavalier oblique drawing.

Oblique Dimensioning

Oblique dimensioning must be done in the same plane as the surface or feature appears just as in isometric dimensioning. Fig. 13-11 illustrates the placement of dimensions on a cavalier oblique drawing.

Sketching Assignment 13-BPR-3
MOTOR BRACKET REST

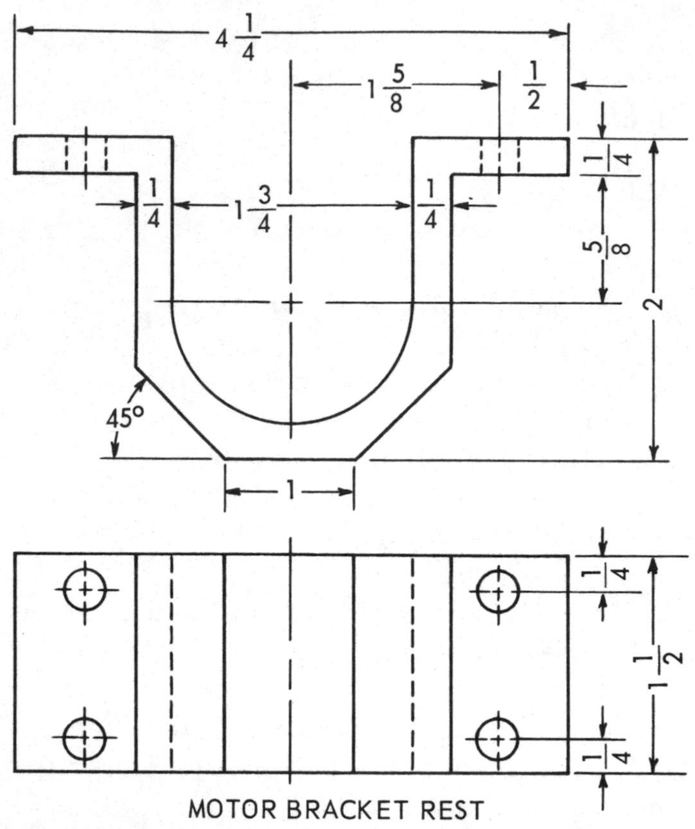

Fig. 13-BPR-3. Make a cavalier oblique sketch of the Motor Bracket Rest. Do not dimension.

Sketching Assignment 13-BPR-4
COUPLING

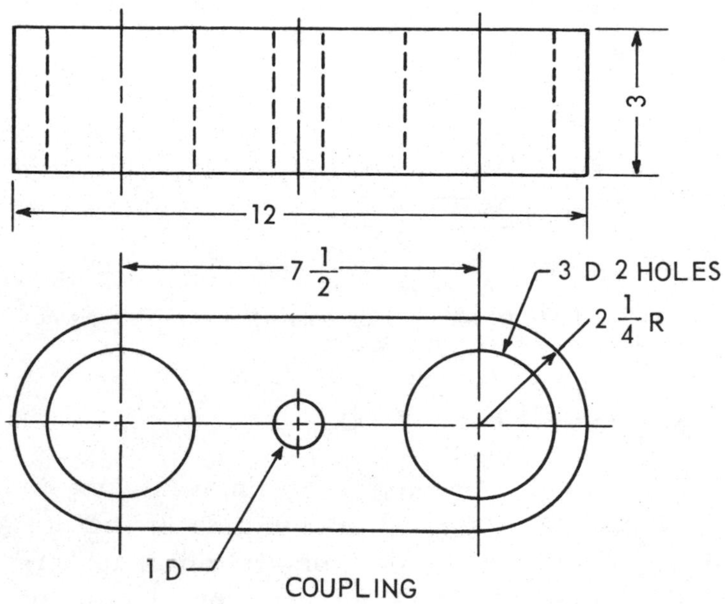

Fig. 13-BPR-4. Make a cabinet oblique sketch of the Coupling. Dimension the sketch.

Perspective Drawing

The perspective drawing is the most realistic of all the pictorial drawings. Instead of the receding lines remaining parallel as they do in the isometric or oblique drawings, receding lines in the perspective drawing or sketch tend to converge (meet at one point), Fig. 13-12. This eliminates entirely the distorted appearance which occurs at the back part of most other pictorial drawings.

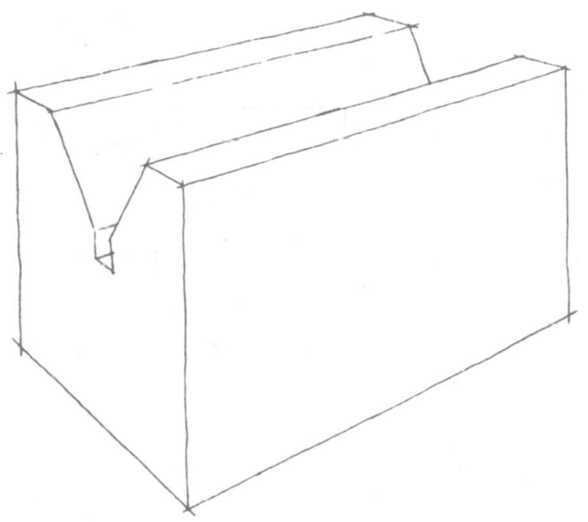

Fig. 13-12. Perspective drawing.

Two types of perspective sketches are discussed and illustrated here to assist you in developing the technique of perspective sketching: parallel and angular perspective.

Parallel or Single-Point Perspective

When one face of an object appears in its true shape and size (unless reduced or enlarged) and is parallel to the picture plane as in orthographic projection, the perspective is known as parallel or single-point perspective. That is, lines parallel to the front picture plane remain parallel and the receding lines of the other two faces converge in the direction of a single vanishing point, Fig. 13-13.

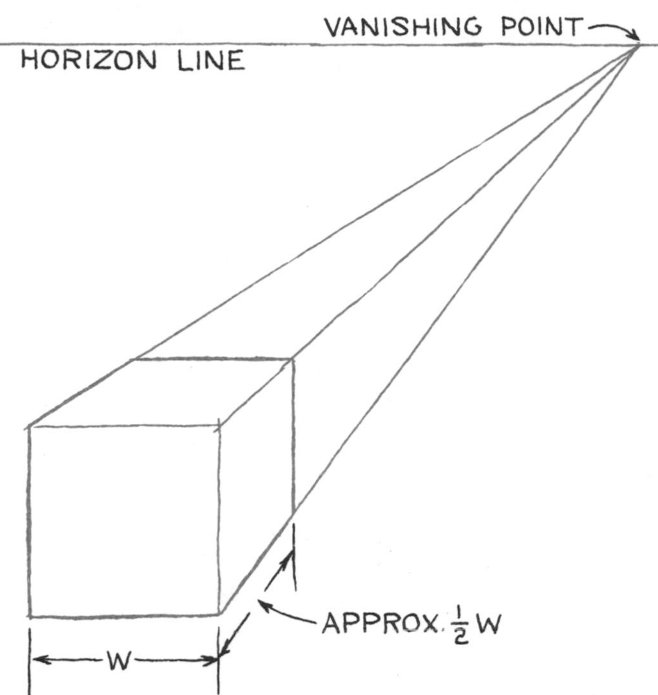

Fig. 13-13. Parallel or single-point perspective.

PROCEDURE FOR SKETCHING A PARALLEL OR SINGLE-POINT PERSPECTIVE:

1. Sketch the front view in its true size and shape as in an orthographic sketch, Fig. 13-14(a).
2. Sketch horizontal line (called Horizon) at the assumed eye level of the viewer. This line may be above, behind or below the object, depending on how you want to view the object, Fig. 13-14(b).
3. Select vanishing point (VP) on the horizon as far to the right or left as desired, Fig. 13-14(c).
4. Sketch lines from front view to VP, Fig. 13-14(d).
5. Enclose object in "box" by sketching rear vertical and horizontal lines, Fig. 13-14(d). To estimate depth of side and top view, reduce these distances by about one-half and adjust until it pleases the eye.
6. To sketch slant lines, locate their end points on the perspective axes and sketch lines between, Fig. 13-14(e).

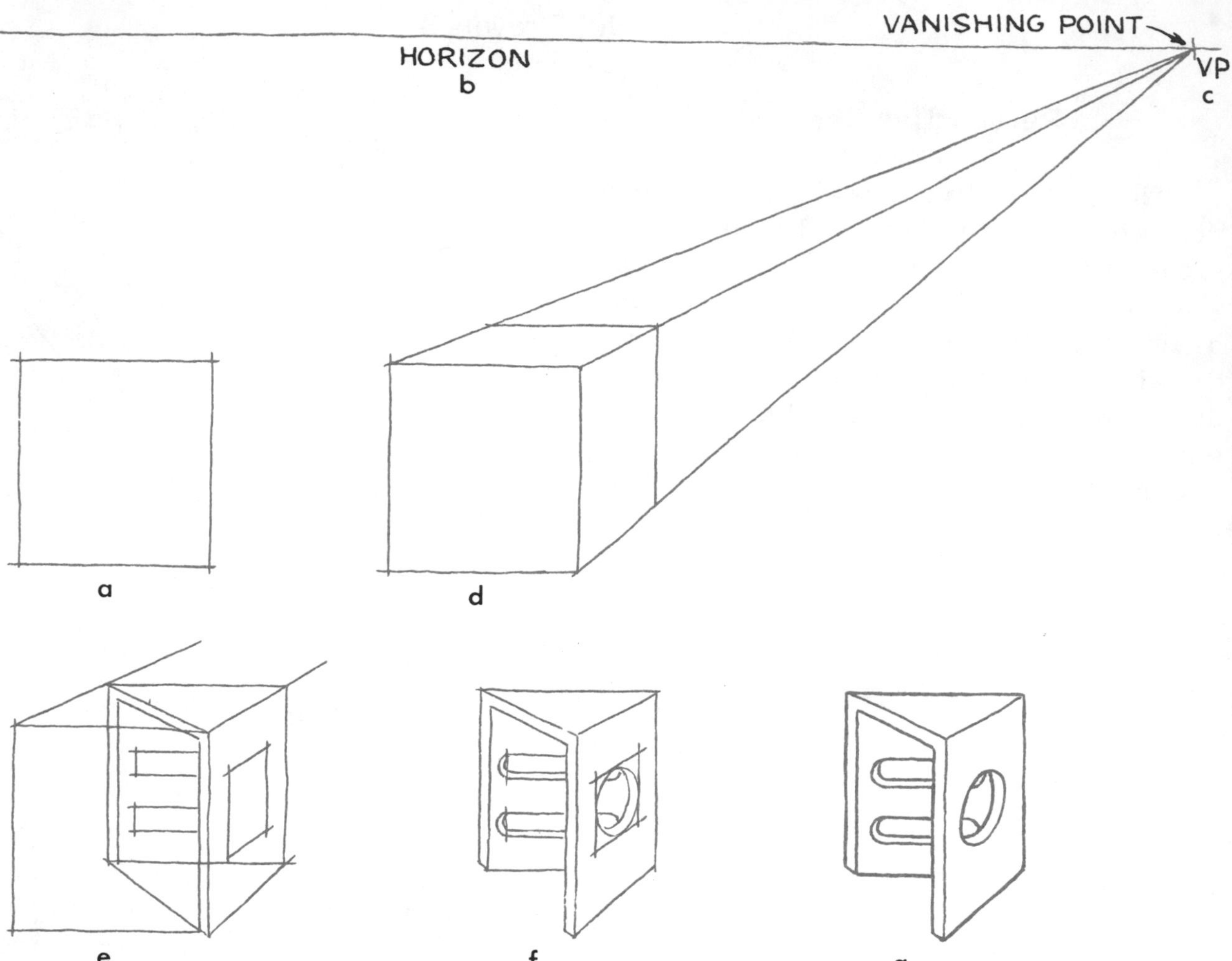

Fig. 13-14. Steps in sketching a parallel or single-point perspective.

7. Block in features such as slots and holes, Fig. 13-14(e).
8. Sketch circles and arcs, Fig. 13-14(f).
9. Darken visible lines and erase construction lines, Fig. 13-14(g).

Angular or Two-Point Perspective

The angular or two-point perspective gets its name from the fact that the two side faces of the object meet the front picture plane at an angle and recede toward two vanishing points on the horizon, Fig. 13-15.

PROCEDURE FOR SKETCHING AN ANGULAR OR TWO-POINT PERSPECTIVE:

1. Sketch horizontal line (horizon) at the assumed eye level of the viewer. This line may be above, behind or below the object, depending on the level from which you want to view the object, Fig. 13-15(a). (See Fig. 13-16 for perspective view with horizon below the object.)
2. Select position in which object is to be viewed and sketch vertical line for front corner of "box" to enclose object, Fig. 13-15(b).
3. Establish right and left vanishing points on horizon, Fig. 13-15(c). If the object is to be positioned so as to view equally the two sides, then the vanishing points will be equidistant on each side of the object. If one side is to be favored, the vanishing point for that side will be extended out and the vanishing point on the other side will be shortened. Both VP's must remain on the line of horizon, however.
4. Sketch receding lines from front verti-

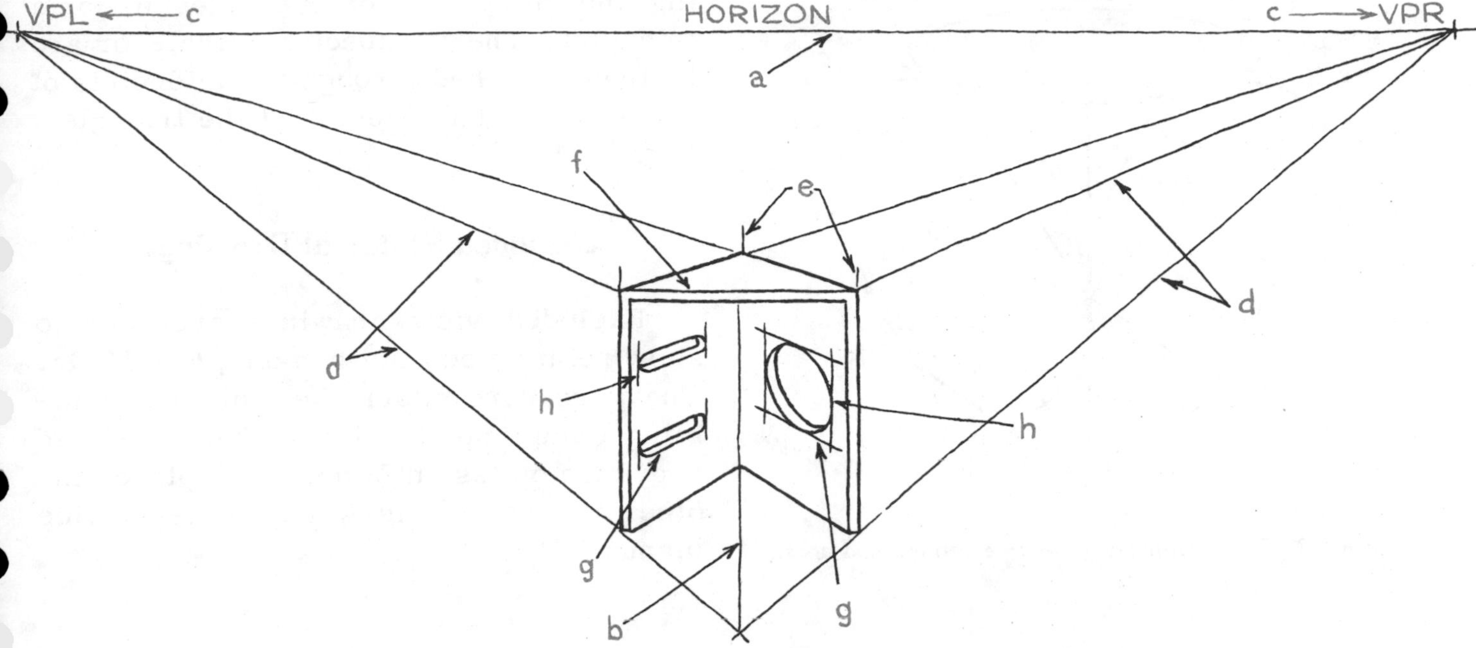

Fig. 13-15. Steps in sketching an angular or two-point perspective.

cal line to the two vanishing points, Fig. 13-15(d).

5. Enclose object in "box" by sketching rear vertical lines, Fig. 13-15(e). To estimate depth of side and top views, reduce these distances by about one-half and adjust until it pleases the eye.
6. Sketch slant lines by locating two end points on the perspective axes and sketch lines between, Fig. 13-15(f).
7. Block in features such as slots and holes, Fig. 13-15(g).
8. Sketch circles and arcs, Fig. 13-15(h).
9. Darken visible lines and erase construction lines.

Circles and Arcs in Perspective

Circles and arcs in perspective are sketched in the same way as isometric circles and arcs by sketching first the perspective square or block and then join-

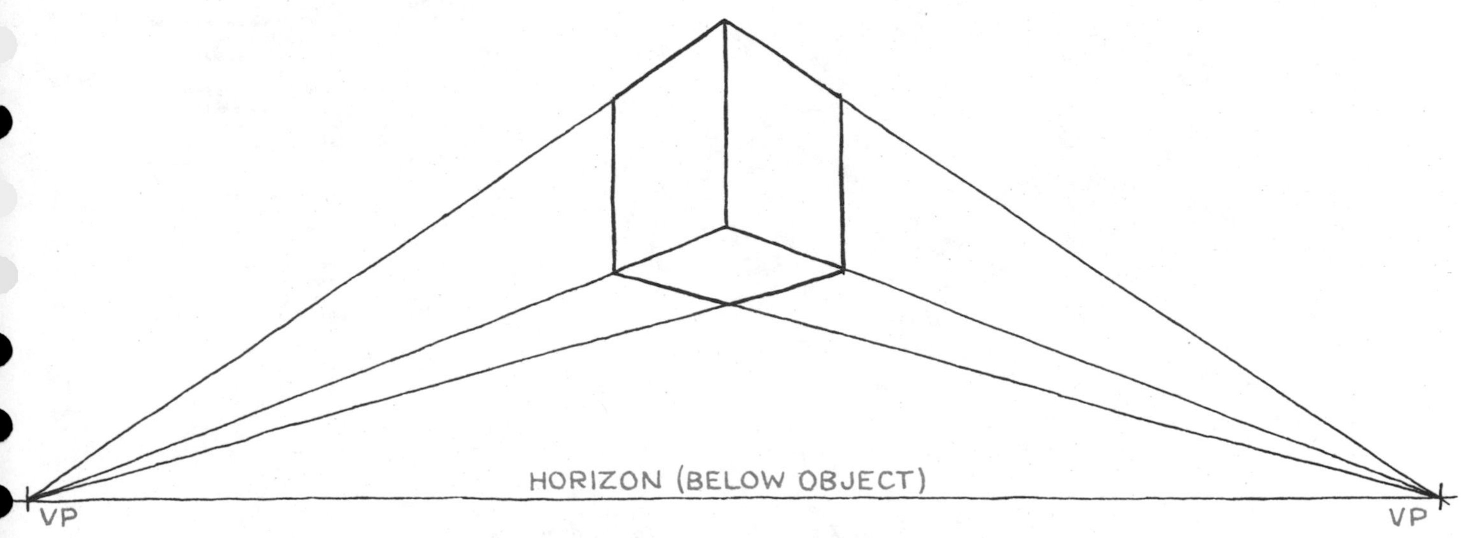

Fig. 13-16. Perspective view with horizon below object.

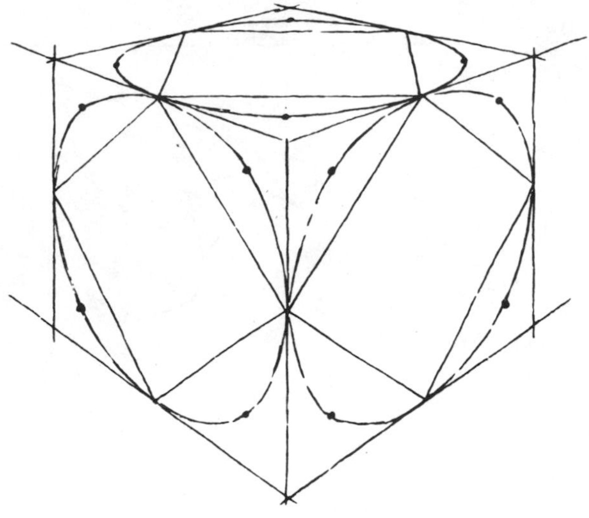

Fig. 13-17. Sketching perspective circles and arcs.

ing the midpoints of the sides to form triangles. The perspective circle or arc is then sketched through the midpoints of the sides and the center of the triangles, Fig. 13-17.

Exploded Pictorial Drawings

Exploded view drawings are used to show relative position of parts, Fig. 13-18. They are particularly helpful in assembling complicated objects when a definite sequence of assembly must take place. Another typical use is in appliance service manuals.

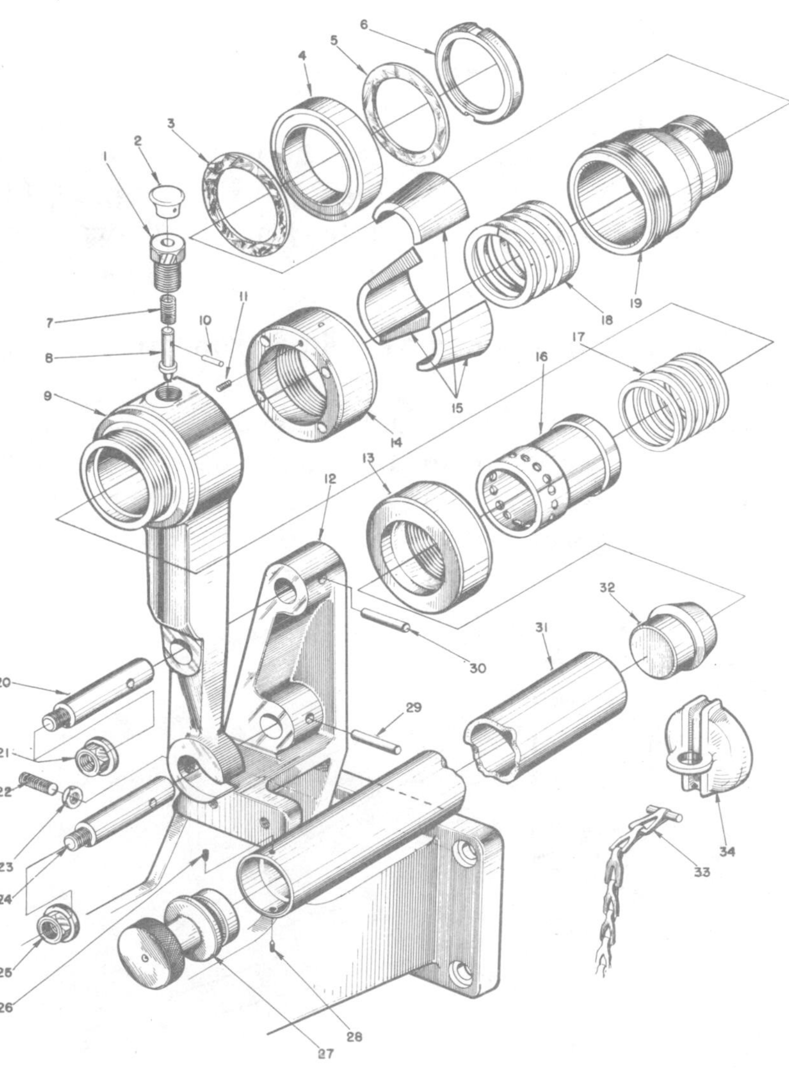

OUTSIDE FEEDING ARRANGEMENT FOR 2⅜" CAPACITY MACHINES

PARTS LIST

Index No.	Part No.	Part Name
1	42-20325	Feed Bracket Plunger (Includes 2, 7, 8 and 10)
2	91-70-66	Feed Latch Knob
3	42-20304	Bearing Oil Shield
4	432-12	Feed Roll Ball Bearing
5	42-20304	Bearing Oil Shield
6	42-20305	Feed Tube Nut
7	92-1000-1228-22	Feed Latch Spring
8	42-7415	Retaining Plunger
9	42-20324	Feed Bracket
10	92-900-312	Feed Latch Plunger Pin
11	472-238	Adjusting Nut Set Screw
*12	42-20309	Feed Bracket Support
13	42-20328	Ball Thrust Nut
14	42-20330	Adjusting Nut
15	710-36224-226	Outside Master Feed Finger Pad
16	42-2501	Feed Ball Retainer
17	91-100-116	Feed Ball Retainer Spring
18	91-100-284	Pad Spring
19	42-20329	Outside Master Feed Finger Body (Includes 3, 4, 5, 6, 14, 15 and 18)
*20	42-20327	Feed Bracket Fulcrum
*21	92-6071-2020	Feed Bracket Fulcrum Nut
22	92-5391-1232	Feed Bracket Fulcrum Set Screw
23	92-6015-1214	Feed Bracket Fulcrum Set Screw Nut
*24	42-20327	Feed Bracket Fulcrum
*25	92-6071-2020	Feed Bracket Fulcrum Nut
26	92-5100-808	Feed Rod Set Screw
27	42-15645	Feed Rod Knob
28	92-5100-808	Feed Rod Set Screw
*29	92-900-844	Feed Bracket Fulcrum Pin
*30	92-900-844	Feed Bracket Fulcrum Pin
31	42-20326	Feed Rod Complete (Includes 27 and 32)
32	42-20455	Feed Rod End
33	42-20443	Pusher Cap Chain
34	42-20442	Bar Pusher Cap

* Furnished with Stock Loader

Fig. 13-18. An exploded pictorial drawing. (Brown and Sharpe)

Sketching Activity 13-BPR-5
OIL STONES

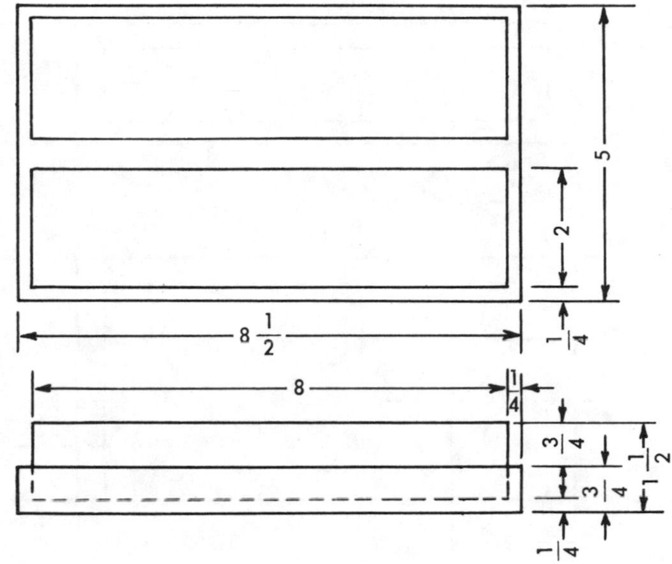

Fig. 13-BPR-5. Make a parallel perspective sketch of the Oil Stones. Do not dimension.

Sketching Activity 13-BPR-6
LOCK HOUSING

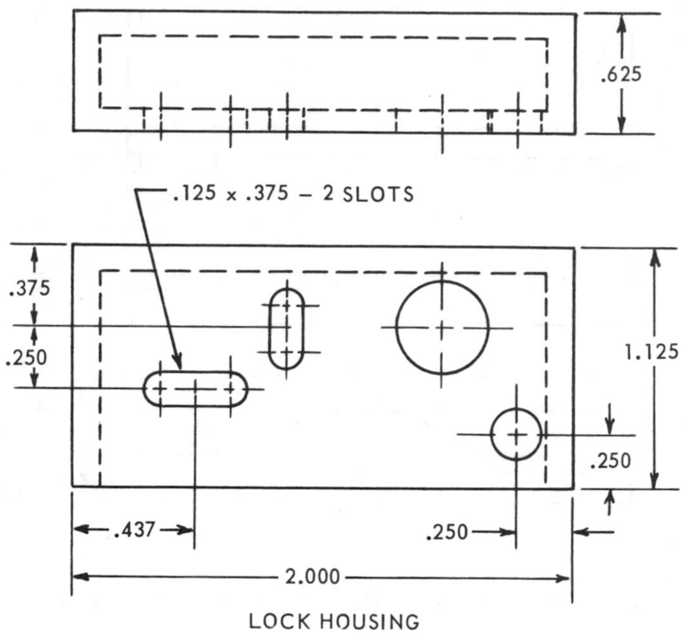

Fig. 13-BPR-6. Make an angular perspective sketch of the Lock Housing. Dimension the sketch.

For Evaluation Blueprint Activity for Part 2 see page 311.

Industry photo. Thousands of blueprints are needed to make and assemble modern aircraft. (McDonnell Douglas)

PART 3
TITLE BLOCK, MATERIALS, NOTES AND DRAWING CHANGES

Unit 14
The Title Block

The purpose of the title block is to provide supplementary information on the part or assembly to be made and to include in one section of the print, information which will aid in identification and filing of the print. The title block is usually located in the lower right-hand corner of the blueprint so that, when the print is correctly folded, the title block may be seen for easy reference and for filing in a storage file.

Title Block Elements

Title block information which is included on plans by most industries is explained in this unit in order that you may properly interpret this section of a blueprint.

THE DRAWING NUMBER is used to identify and control the blueprint. It is also used to designate the part or assembly shown on the blueprint, Fig. 14-1(A). The number is usually coded to indicate department, model, group, serial number and dash numbers.

DASH NUMBERS (a number preceded by a dash after the drawing number) indicate right- or left-hand parts as well as neutral parts and/or detail and assembly drawings. When coded, these are usually special to the particular industry and are not universally applicable to all industries. The drawing number may be repeated at the fold portion of the sheet or in the upper

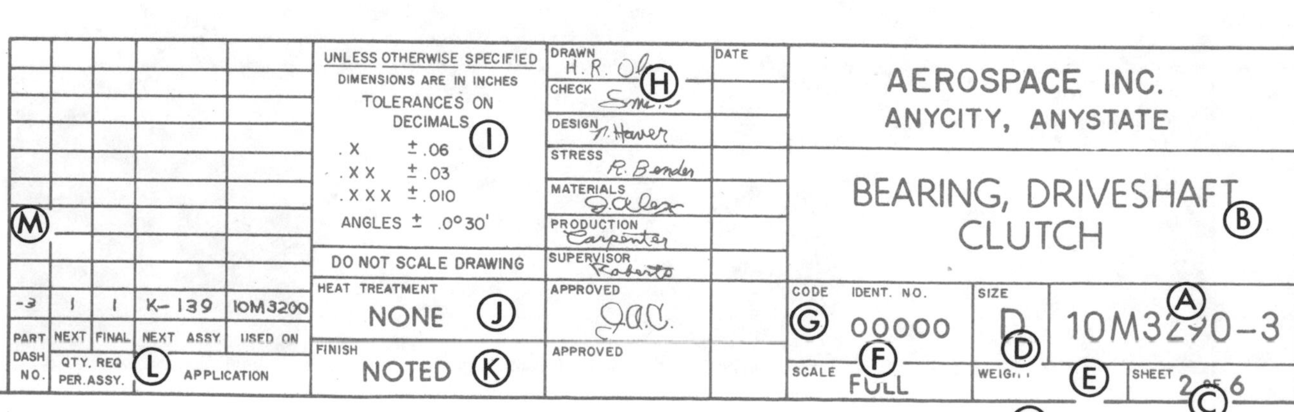

Fig. 14-1. Title block elements.

left-hand corner (in an inverted position) so the print may be easily located in the file drawer should it be filed upside down.

THE DRAWING TITLE is descriptive, brief and clearly stated in its identification of the part or assembly, Fig. 14-1(B). The title starts with the name of the part or assembly, followed by descriptive modifiers, Fig. 14-2.

VALVE ASSY, PRESSURE REGULATING

BASIC NAME — DESCRIPTIVE MODIFIERS

COMPLETE TITLE

Fig. 14-2. Components of a drawing title.

The title is read with the descriptive modifiers first - - PRESSURE REGULATING VALVE ASSEMBLY.

SHEET: The sheet numbering is used on multisheet blueprints to indicate the consecutive order and the total number of prints, Fig. 14-1(C)

SIZE: Drawings are prepared on standard size sheets, Fig. 14-3, in multiples of 8.5 x 11 and 9 x 12 inches and are designated by letter to indicate size as shown in Fig. 14-1(D).

LETTER SIZE	AMERICAN STANDARD	MILITARY STANDARD
A	8.5 x 11 or 9 x 12	8.5 x 11
B	11 x 17 or 12 x 18	11 x 17
C	17 x 22 or 18 x 24	17 x 22
D	22 x 34 or 24 x 36	22 x 34
E	34 x 44 or 36 x 48	
J		36 x ANY LENGTH
R		48 x ANY LENGTH

Fig. 14-3. Standard size drafting sheets and prints.

WEIGHT: This block provides either the actual or calculated weight of the part or assembly as indicated, Fig. 14-1(E). Calculated weight is used during design stages to control weight of finished part or assembly. Actual weight is obtained after manufacture of part or assembly.

SCALE: The drawing scale indicates the comparative size between the part as drawn and the actual part. Typical scale notations are: 1/2" = 1" (one-half actual size), FULL (actual size), 1:1 (actual size) and 2:1 (twice size), Fig. 14-1(F). When the scale is shown as NOTED, it means that several scales have been used in making the drawing and each is indicated below the particular view to which it pertains.

Measurements on a blueprint should NEVER be used because the print may have been reduced in size or stretched. Work from the dimensions given on the print. If you believe these to be in error, report it to your supervisor.

CODE IDENTIFICATION NUMBER is a number assigned to the industry by the Department of Defense, Fig. 14-1(G). Not all industries have such a number.

SIGNATURES: This block is for the signatures and date of release by those who have responsibility for making or approving all or certain facets of the drawing or the manufacture of the part, Fig. 14-1(H).

DRAWN carries the signature of the draftsman who made the drawing and the date completed.

CHECK is for the engineer who checked the drawing for completeness, accuracy and clarity.

DESIGN: This line is for the engineer who has the responsibility for the design of the part.

STRESS is the place for the engineer

who ran the stress calculations for the part.

MATERIALS is where the person signs whose responsibility it is to see that the materials needed to make the part are available. He also indicates the day on which these calculations were made.

PRODUCTION: This is where the engineer who approved the producibility of the part approves the drawing.

SUPERVISOR: The person in charge of drafting indicates his approval and date in this block.

APPROVED: These lines are to record any other required approvals.

TOLERANCES: This block indicates the general tolerance limits for one, two and three place decimal and angular dimensions, Fig. 14-1(I). These limits are applicable unless the tolerance is given along with the dimension callout.

HEAT TREATMENT: The heat treatment and hardness requirements are specified in this block, Fig. 14-1(J). The entry could be AS REQUIRED or NOTED which means the part must conform to the specification block notation or to the callout on the drawing detail. If heat treatment is not required, the word NONE or a diagonal line is entered in the block.

FINISH: General finish requirements (paint, chemical or other) are indicated in this block, Fig. 14-1(K). Specific finish requirements would be a callout on the drawing with the word NOTED in the finish block.

APPLICATION BLOCK: The purpose of the application block (sometimes called the USAGE BLOCK) is to assist in determining the equipment in which the part or assembly shown on the drawing is used. The block reveals the diversity of uses a part or an assembly may have, and this aids in determining the effects of a change in the part or assembly.

NEXT ASSEMBLY indicates the number of the next drawing or assembly on which the part or subassembly is to be used, Fig. 14-1(L).

USED ON indicates the model number on which the part or assembly is used.

QUANTITY REQUIRED PER ASSEMBLY: The column NEXT tells the quantity of each part required to make the subassembly. The column FINAL lists the total number of units required per final assembly or completed product, such as an aerospace vehicle.

DASH NUMBER: The dash numbers and other detail parts numbers required are included in this block on assembly drawings, Fig. 14-1(M). They occur in numerical order. No entries are required on detail drawings.

SECURITY CLASSIFICATIONS: In some industries, prints are classified for security reasons, Fig. 14-1(N). When this is the case, security classification markings are placed outside the border below the title block and at the top of the sheet. These markings will be indicated to you by the industry where you work.

ZONING is used on larger size blueprints to aid in locating details or parts. The system is similar to that used on highway maps as an aid in locating cities and points of interest. Zones are indicated outside the border, numerically at intervals from right to left and alphabetically from bottom to top, Fig. 14-4.

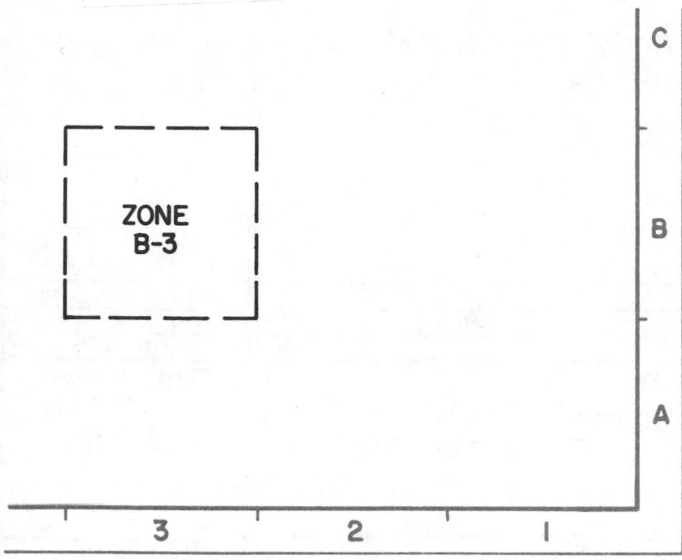

Fig. 14-4. Zoning on blueprints.

OTHER LISTINGS: A few industries will include information indicating the drawing has been SUPERSEDED BY or drawing SUPERSEDES. These blocks show the drawing number and issue of the drawings affected. CUSTOMER for whom product is being made is sometimes listed as is the CONTRACT NUMBER. STANDARDS USED FOR INSPECTION may be used to refer to specific sets of inspection standards used by the company.

When a block contains a diagonal line or is X'd out, this means the item is not required.

Blueprint Reading Activity 14-BPR-1
TITLE BLOCK

QTY PER END ITEM	USED ON	NEXT ASSY	
1	NA265-40	265-230002	(A2)
1	T-39D	265-230002	(A1)
1	T-39B	265-230002	
1	T-39A	265-230002	
APPLICATION			

		265-230428	SUPPORT	7075-T6 AL ALY BAR	$3/4 \times 1\,5/8 \times 4\,13/16$		QQ-A-277 TEMPT6	001
REQD RH	REQD LH	PART NUMBER	DESC	MATERIAL	SIZE	ZONE	MATL SPEC	LINE
LIST OF MATERIAL								

DRILLED HOLE TOLERANCES
.040 TO .1285: +.002, −.001
.136 TO .228: +.003, −.001
.234 TO 1/2: +.004, −.001
33/64 TO 3/4: +.005, −.001
49/64 TO 1: +.007, −.001
1-1/64 TO 2: +.010, −.001

TOLERANCES EXCEPT AS NOTED
ANGLES ±1/2° FRACTIONS ±1/32 DECIMALS ±.010

✓ SURFACE ROUGHNESS PER MIL-STD-10 (FA6-219)

HEAT TREAT

FINISH 246-00024

DATE	5 2
DR BY	T. LONG
CHK BY	Van Bella 5/4/
APPD BY	
APPD BY	

INSPECT PER MIL – I–6870 (LQ0501-007 CLASS I)

SUPPORT–HINGE, INTERMEDIATE, UPPER VERT STAB

SCALE FULL SIZE WT

NORTH AMERICAN AVIATION, INC.
ENGINEERING
INTERNATIONAL AIRPORT
LOS ANGELES 45, CALIF.

DWG SIZE D 265-230428

CODE IDENT 43999 SHEET 1 OF 1

Fig. 14-BPR-1. Refer to the title block above and answer the following questions.

1. What is the name of the part or assembly as it would be read? 1. ____________________
2. Give the drawing number. 2. ____________________
3. How many sheets are in the set? 3. ____________________
4. To what scale was the drawing made? 4. ____________________
5. What general tolerance is required for a drilled hole 1/4 in. in diameter? 5. ____________________
6. The general tolerance for: 6.
   a. Three place decimals. a. ____________________
   b. Fractions. b. ____________________
   c. Angles. c. ____________________
7. How would tolerances which are different from those in the general tolerance block be indicated? 7. ____________________
8. What heat treatment is specified? 8. ____________________
9. What would a diagonal line in a block indicate? 9. ____________________
10. How is the finish specified? 10. ____________________
11. What is the meaning of NOTED in a block? 11. ____________________
12. How many of the Support-Hinge, No. 265-230428 are used on Model T-39B? 12. ____________________
13. What blueprint would contain the assembly application of this part to Model T-39B? 13. ____________________
14. Inspection is to be done per what standard? 14. ____________________

For additional Blueprint Activities for Unit 14 see pages 242 and 245.

Blueprint Reading Activity 14-BPR-2
TITLE BLOCK

PROCESS SPEC GOV'T AND/OR GAC	FINISH OR PROCESS	NO. OF COATS	PARTS AFFECTED
MIL-M-10578	CONDITIONER	X	-101 & -103
MIL-E-7729 FED STD 595-17875	ENAMAL GLOSS WHITE	2	-101 & -103
TT-P-636	PRIMER	1	-101 & -103

DASH NO.	NEXT ASSY (APPLICATION)	USED ON (APPLICATION)	NEXT ASSY (QTY REQD)	FINAL ASSY (QTY REQD)
-103	3015100-005	GTV-1	2	2
-101	3015100-005	GTV-1	2	2

UNLESS OTHERWISE SPECIFIED
DIMENSIONS ARE IN INCHES
TOLERANCES ON:
DECIMALS TENTHS ±.1
HUNDREDTHS ±.04
THOUSANDTHS ±.010
ANGLES ±1°

MATERIAL
FINISH PER GERA 1523

SEE GR60-9-60 FOR DRAWING INTERPRETATION STANDARDS

PROJ ENGR	R.L. STEVENS	10-17-
WEIGHTS	E. Twitchell	10/14
STRESS	N. MAUL	9/26
MATL ENGR		
GR ENGR		
CHECKER	Renfroe	9/26
DFTSMAN	Spitler	9/9

GAC APPROVAL ______ DATE

CUSTOMER APPROVAL ______ DATE

GOODYEAR AEROSPACE
CORPORATION
LITCHFIELD PARK ARIZONA

STRUT, ASSEMBLY OF — AFT JACK

SIZE	CODE IDENT NO.	
E	99696	3015100-007

SCALE 1/1 | SHEET 1 OF 1

Fig. 14-BPR-2. Refer to the title block above and answer the following questions.

1. Give the name of the assembly. 1. ____
2. What is the drawing number? 2. ____
3. How many sheets are in the set? 3. ____
4. Give the scale of the drawing. 4. ____
5. What size drawing was made? 5. ____
6. List the general tolerances. 6. Tenths ____ Thousandths ____ Hundredths ____ Angles ____
7. How many individuals have signed indicating their approval of the drawing besides the draftsman? 7. ____
8. What finish is specified? 8. ____
9. The drawing is to be interpreted according to what standard? 9. ____
10. What is the next assembly on which dash number — 101 is used? 10. ____
11. What is the model number on which — 101 is used? How many are required on the final assembly? 11. ____

Unit 15
List of Materials

The List of Materials is a tabular form usually appearing immediately above the title block on the print. It is used primarily for assembly or installation drawings and provides specific information on quantity and types of materials used in the manufacturing and assembling of parts of a machine or structure. The block is sometimes called PARTS LIST, MATERIALS LIST, BILL OF MATERIALS, or SCHEDULE OF PARTS.

The List of Materials enables a purchasing department to requisition the quantity of materials needed to produce a given number of the assemblies.

The following items are usually included in the List of Materials.

NUMBER REQUIRED PER ASSEMBLY: In Fig. 15-1, the columns marked A are used to indicate the number of parts re-

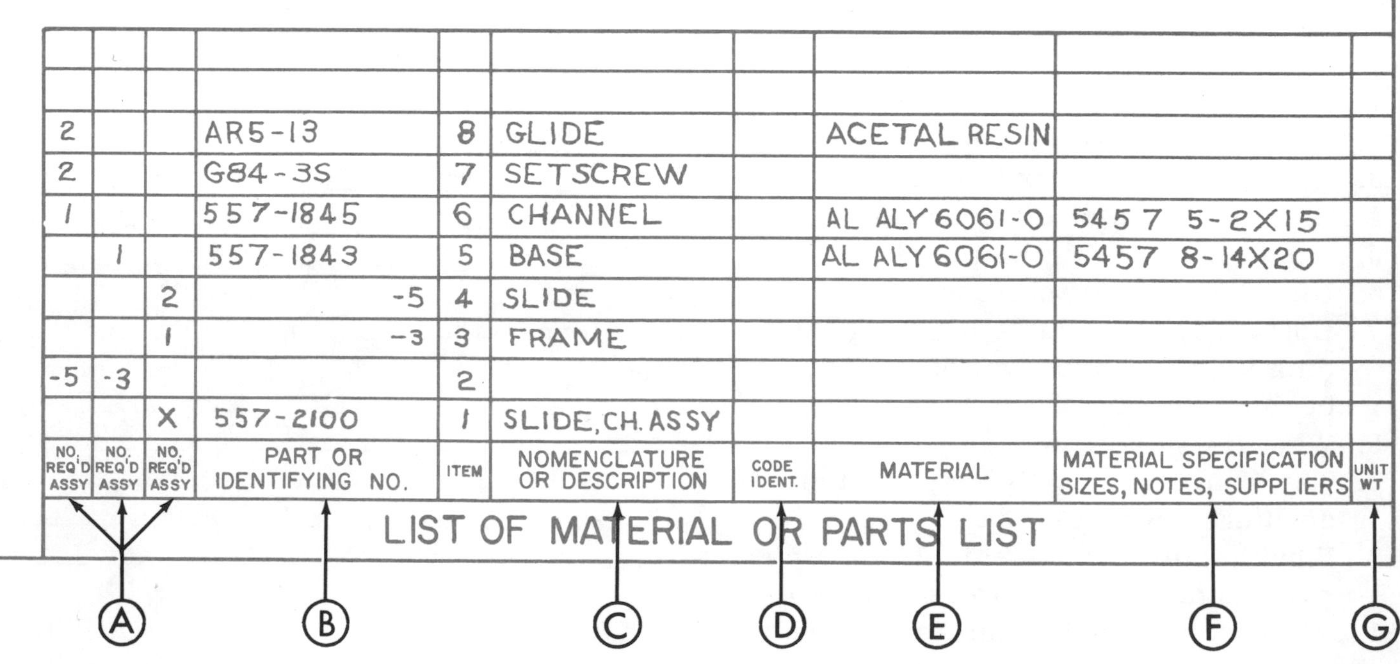

NO. REQ'D ASSY	NO. REQ'D ASSY	NO. REQ'D ASSY	PART OR IDENTIFYING NO.	ITEM	NOMENCLATURE OR DESCRIPTION	CODE IDENT.	MATERIAL	MATERIAL SPECIFICATION SIZES, NOTES, SUPPLIERS	UNIT WT
2			AR5-13	8	GLIDE		ACETAL RESIN		
2			G84-3S	7	SETSCREW				
1			557-1845	6	CHANNEL		AL ALY 6061-O	5457 5-2X15	
	1		557-1843	5	BASE		AL ALY 6061-O	5457 8-14X20	
		2	-5	4	SLIDE				
		1	-3	3	FRAME				
-5	-3			2					
		X	557-2100	1	SLIDE, CH. ASSY				

LIST OF MATERIAL OR PARTS LIST

Fig. 15-1. List of Materials block on a drawing.

quired for an assembly. When the letters AR are found in the column instead of a number, it means AS REQUIRED. The column on the right is used for the basic assembly which is listed on the first (bottom) line. When dash-numbered assemblies are required in the basic assembly, columns are added for these assemblies on the left and the quantity of each part required is recorded in the appropriate column on lines above those used for the basic assembly.

PART OR IDENTIFICATION NUMBER: Dash numbers and other detail parts numbers required for the assembly are shown in Fig. 15-1(B). All company and subcontractor manufactured parts' numbers are listed in the List of Materials block. Standard parts, such as bearings, bolts and screws are listed by number. Rivets usually are not listed in the block.

NOMENCLATURE OR DESCRIPTION: When the main assembly is listed on the first line of the List of Materials, frequently, the only entry is the part number and sometimes an abbreviated name, Fig. 15-1 (C). The names of all other subassemblies and parts are usually limited to the basic noun (lines 3 through 8).

CODE IDENTIFICATION NUMBERS: Column D gives the industry code of the supplier if the part is purchased from the outside.

MATERIAL: This column contains the commercial name of the material used in making the part, Fig. 15-1(E).

MATERIALS SPECIFICATION, SIZES, NOTES, SUPPLIERS: Column F furnishes the commercial specification and stock size of the material and is frequently referred to as the procurement specification. The names and addresses of manufacturers of purchased parts are sometimes included here.

UNIT WEIGHT: Column G gives the actual weight of the part when required by contract.

DETAIL CALLOUT is the name and number of each part on an assembly drawing that appears adjacent to the part on the face of the drawing, Fig. 15-2.

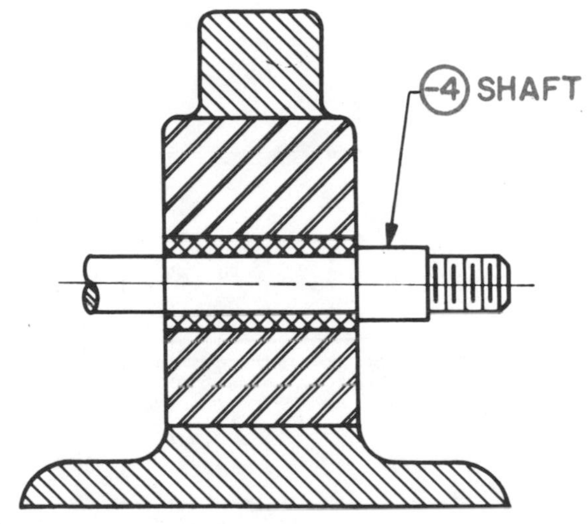

Fig. 15-2. A detail callout on an assembly drawing.

This is the only information that is duplicated in the "detail callout" and the List of Materials block. Occasionally, part dimensions are given in the detail callout and when this occurs, they are true dimensions which are not found elsewhere on the drawing. Dimensions in the List of Materials are "stock sizes" as noted earlier.

ADDITIONAL LISTINGS: Some industries further detail the material specifications by adding columns for size: DIAMETER, THICKNESS, WIDTH and LENGTH of material. TENSILE STRENGTH (TS/1000/PSI) is required for some manufactured parts and an entry of 42 in this column would mean the material must have a tensile strength of 42,000 pounds per square inch.

Notes which concern INTERCHANGEABILITY and REPLACEABILITY and are sometimes placed in a column to indicate parts that are interchangeable with parts in other models, and parts that are likely to wear rapidly and need replacement. REPLACEMENT PARTS is another designation for replaceability and is an indication that additional parts must be stocked.

ZONE: Detail parts on size D and larger prints may be located by a zone designation code for ease in locating the parts on prints.

		AR	TYPE 57C	36	SEALANT	△9	04552 EMERSON & CUMING	
		12	MS20470B3-12	35	RIVET		MIL-R-5674	
		96	MS20470B3-8	34	RIVET		MIL-R-5674	
		52	MS20470B3-6	33	RIVET		MIL-R-5674	
		40	MS20426B3-9	32	RIVET, C'SUNK HD		MIL-R-5674	
		64	MS20426B3-8	31	RIVET, C'SUNK HD		MIL-R-5674	
		32	MS20426B3-6	30	RIVET, C'SUNK HD		MIL-R-5674	
		15	MS20426B3-5	29	RIVET, C'SUNK HD		MIL-R-5674	
		4	MS20426B3-10	28	RIVET, C'SUNK HD		MIL-R-5674	
		12	MS20600B415	27	RIVET, SELF PLUGGING		MIL-R-7885	
		39	MS20600B413	26	RIVET, SELF PLUGGING		MIL-R-7885	
		30	MS20601B413	25	RIVET, SELF PLUGGING		MIL-R-7885	
		2	SS10-116	24	RIVNUT		03481 B. F. GOODRICH	
		4	MS21209-C4-15	23	INSERT		MIL-S-7742	
		21	LKS-032-2	22	PEM NUT		46384 PENN	
		38	LKS-832-2	21	PEM NUT		46384 PENN	
		1		20	PERFORATED SHEET	△7		
		1		19	PERFORATED SHEET	△7		
		20	MF1001-3	18	NUT ANCHOR		15653 LAYLOCK	
		42	MF1001-08	17	NUT ANCHOR		15653 LAYLOCK	
		1		16	TEE	1 1/4 x 7/8 x 1/8 THK	6063-T5 QQ-A-200/9	
		1	MS20257P5-2L18	15	HINGE			
		1		14	RECEIVER DECK	.125 THK AL ALY	5052-H32 QQ-A-250/8	
		1		13	FILTER BOX	.125 THK AL ALY	5052-H32 QQ-A-250/8	
		1		12	PANEL. RIGHT SIDE	.063 THK AL ALY	5052-H32 QQ-A-250/8	
		1		11	PANEL. LEFT SIDE	.063 THK AL ALY	5052-H32 QQ-A-250/8	
		1		10	PANEL. BOTTOM	.063 THK AL ALY	5052-H32 QQ-A-250/8	
		1		9	PANEL. TOP	.063 THK AL ALY	5052-H32 QQ-A-250/8	
		1	557-21000-1105	8	BRACKET, PRESSURE SWITCH			
		1	557-21000-1104	7	BRACKET, POST AMP			
		6	557-21000-1103	6	GUSSET			
		14	557-21000-1102	5	GUSSET EXTRUDED			
		4		4	CHANNEL EXTRUSION	M/F ALS x 22.5LG	98587 AMCO	
		4		3	CORNER EXTRUSION	M/F ALE-2 x 64.5LG	98587 AMCO	
		4		2	CORNER EXTRUSION	M/F ALE-2 x 19.06LG	98587 AMCO	
		4		1	CORNER EXTRUSION	M/F ALE-2 x 25.5LG	98587 AMCO	
QTY. REQUIRED			PART NUMBER OR IDENTIFYING NUMBER	ITEM NO.	DESCRIPTION	MATERIAL	MATL SPECIFICATION OR MFR'S. CODE IDENT. NO.	CHG.
DASH NUMBER			PARTS LIST					

UNLESS OTHERWISE SPECIFIED
DIMENSIONS ARE IN INCHES AND INCLUDE PLATED AND CHEMICALLY APPLIED FINISHES.
GEOMETRIC TOLERANCES – ASA Y 14.5
ITEM TO BE FREE OF BURRS AND SHARP EDGES
MACHINE SURFACES 125 ASA B-16.1

DECIMAL TOLERANCES			ANGLES
1 PLACE	2 PLACE	3 PLACE	± 0° 30'
.X ± .1	.XX ± .03	.XXX ± .010	

COMMERCIAL TOL. APPLY TO STK SIZES
APPLY STD MFG REQUIREMENTS 100S0001
MATERIAL SEE L/M
FINISH △4

SIGNATURES		DATE
DFT	RA Aчan	1/21
SUPVSN		1/22
CHK		1/22
EE/ME		1/22
STRUCT/WT		
PROJ		1/23
QA/REL		
CONTRACT NO.		

FAIRCHILD HILLER
SPACE AND ELECTRONICS SYSTEMS DIVISION

T/R RACK ASSEMBLY-MECHANICAL
CABINET-ELECT EQUIPMENT
CY-6001/SPO-10

SIZE	CODE IDENT NO.	
E	86360	557-21000-1030
SCALE:		SHEET 1 OF 4

DETAIL WEIGHT	
CALC	ACT.
ASSY WEIGHT	
CALC	ACT.

DASH NO.	NEXT ASSY APPLICATION	QTY REQD
△1	557-21000-1000	1

Fig. 15-BPR-1. Industry blueprint. (Fairchild Hiller)

Blueprint Reading Activity 15-BPR-1
LIST OF MATERIALS

Refer to Fig. 15-BPR-1 and answer the questions below.

1. What is the drawing title? 1.__________

2. What is the drawing number? 2.__________

3. How many sheets are in the set? 3.__________

4. What is the block called that lists the materials? 4.__________

5. How many different parts are listed? 5.__________

6. Give the name of item No. 4. 6.__________

7. For the Channel Extrusion, give the (a) material and (b) material specification. 7. (a)__________ (b)__________

8. What is the name of Part No. 557-21000-1105? 8.__________

9. Give the (a) material and (b) specification for the Receiver Deck. 9. (a)__________ (b)__________

10. Give the (a) quantity required, (b) part name, and (c) specification for Part No. LKS-832-2. 10. (a)__________ (b)__________ (c)__________

11. What is the (a) part name, and (b) material specification for item No. 24. 11. (a)__________ (b)__________

12. Give the (a) part No. and (b) the specification for the Sealant.* 12. (a)__________ (b)__________

* AR in the QTY REQUIRED column means AS REQUIRED.

For additional Blueprint Activities for Unit 15 see pages 246 and 249.

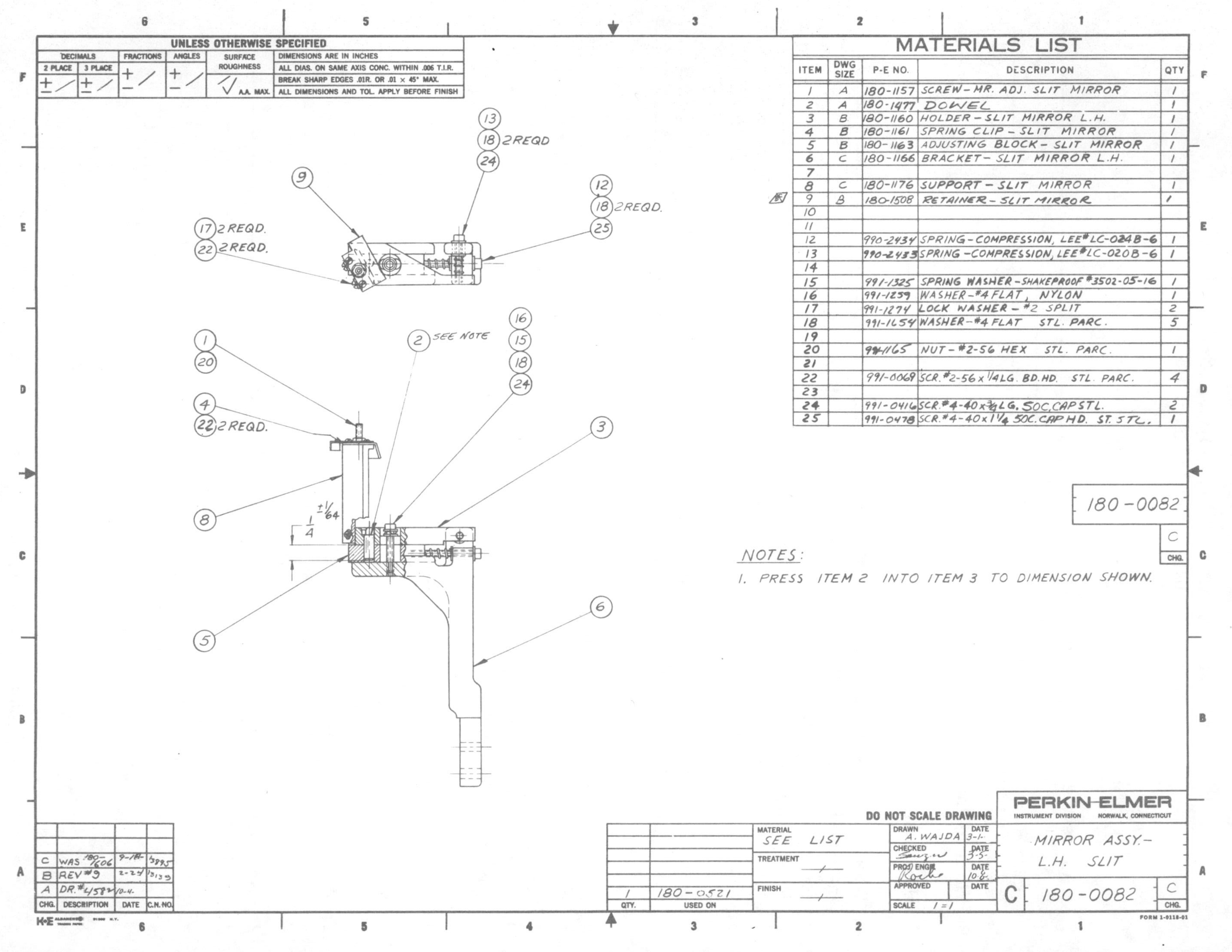

Fi 15-BPR-2. Indus blueprint. Perkin-Elm

Blueprint Reading Activity 15-BPR-2
LIST OF MATERIALS

Refer to Fig. 15-BPR-2 and answer the questions below.

1. What is the name of the part? 1. ______________________
____
2. What is the print number? 2. ______________________
3. What is the scale of the original plan? 3. ______________________
4. What is the name of the block called that lists the materials? 4. ______________________
5. How many different parts are listed? 5. ______________________
6. What is item 2 and how many are required for this assembly? What is to be done with it? 6. ______________________
____
____
7. Give the name of the part and the specification for item 12. 7. ______________________
____
8. Name the part and give the specification for item 15. 8. ______________________
____
9. How many of item 16 are required and what material is called for? 9. ______________________
____
10. What is the print number in the application block on which this assembly is used? How many are required? 10. ______________________
____

Unit 16
Drawing Notes

Notes on drawings provide information and instructions which supplement the graphic presentation as well as the information in the title block and list of materials. Notes eliminate numerous repetition on the face of the drawing, such as the size of holes to be drilled, type fasteners to be used, removal of machining burrs, etc.

When notes become too extensive, as often is the case in architectural and structural drawings, they are typed or printed on separate sheets and included along with the set of drawings - - hence, the term DRAWINGS and SPECIFICATIONS.

Types of Notes

Notes are classified as GENERAL or LOCAL depending on the application.

GENERAL NOTES are notes which apply to the entire drawing and are usually placed either above or to the left of the title block in a horizontal position. General notes are not referenced in the list of materials nor from specific areas of the drawing. Some examples of general notes are given in Fig. 16-1.

When there are exceptions to general notes on the field of the drawing, the general note will usually be followed by the phrase EXCEPT AS SHOWN or UNLESS OTHERWISE SPECIFIED. These exceptions will be shown by local notes or data on the field of the drawing.

LOCAL NOTES or specific notes apply only to certain features or areas and are located near, and directed to, the feature or area by a leader, Fig. 16-2.

1. THIS PART SHALL BE PURCHASED ONLY FROM SOURCES APPROVED BY THE ENGINEERING DEPARTMENT.
2. BREAK SHARP EDGES .030 R UNLESS OTHERWISE SPECIFIED.
3. REMOVE BURRS.
4. FINISH ALL OVER.
5. METALLURGICAL INSPECTION REQUIRED BEFORE MACHINING.

Fig. 16-1. Examples of general notes on drawings.

Fig. 16-2. A local note directed to feature.

Local notes may also be referenced from the field of the drawing or the list of materials by the note number enclosed in an equilateral triangle (sometimes called a FLAG), Fig. 16-3.

SPOTWELD IN ACCORDANCE WITH MIL-W-6858.

Fig. 16-3. Local note reference.

Some examples of local notes are given in Fig. 16-4.

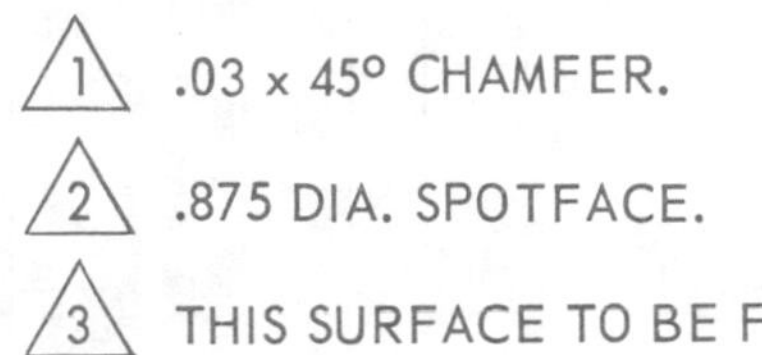

1 .03 x 45° CHAMFER.

2 .875 DIA. SPOTFACE.

3 THIS SURFACE TO BE FLAT WITHIN .003 TOTAL.

4 3 HOLES EQUALLY SPACED AND LOCATED WITHIN .001 R OF TRUE POSITION.

5 RUBBER STAMP PART NUMBER HERE.

Fig. 16-4. Examples of local notes on drawings.

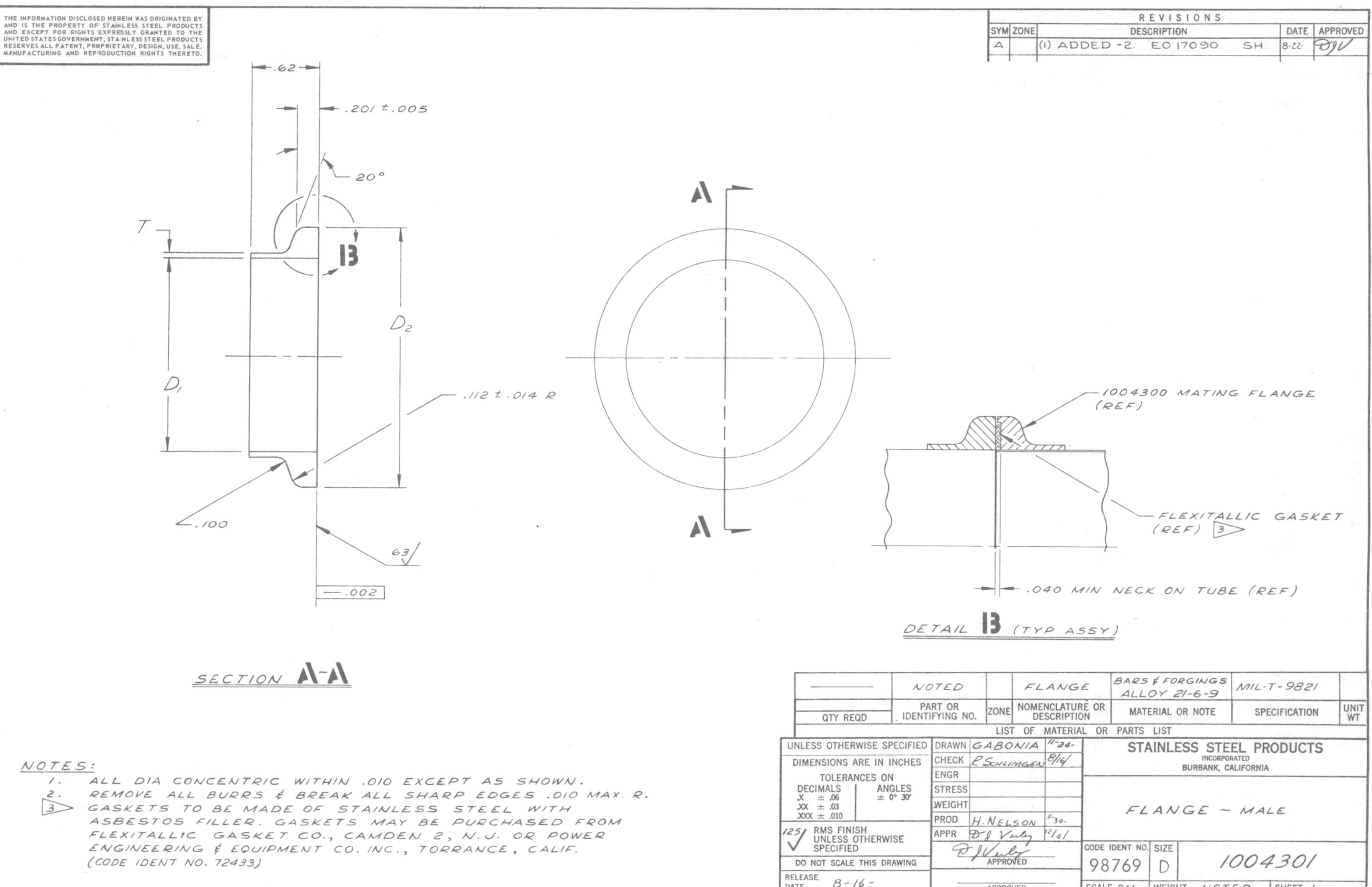

Fig. 16-BPR-1. Industry blueprint. (Stainless Steel Products)

Blueprint Reading Activity 16-BPR-1
DRAWING NOTES

Refer to the blueprint in Fig. 16-BPR-1, and answer the following questions:

1. What is the name of the part? 1. ______
2. Give the part number. 2. ______
3. What is the scale of the original plan? 3. ______
4. What material is required for the part? 4. ______
5. What kind of a drawing is used? 5. ______
6. How many (a) general notes and (b) local notes are shown on the drawing? 6. (a) ______ (b) ______
7. What does "note 1" tell you about diameters D_1 and D_2? 7. ______
8. How are sharp edges to be finished? 8. ______
9. Give the name of the gasket in note 3. 9. ______
10. What is the part number that mates with the MALE FLANGE? 10. ______

For additional Blueprint Activities for Unit 16 see pages 250 and 253.

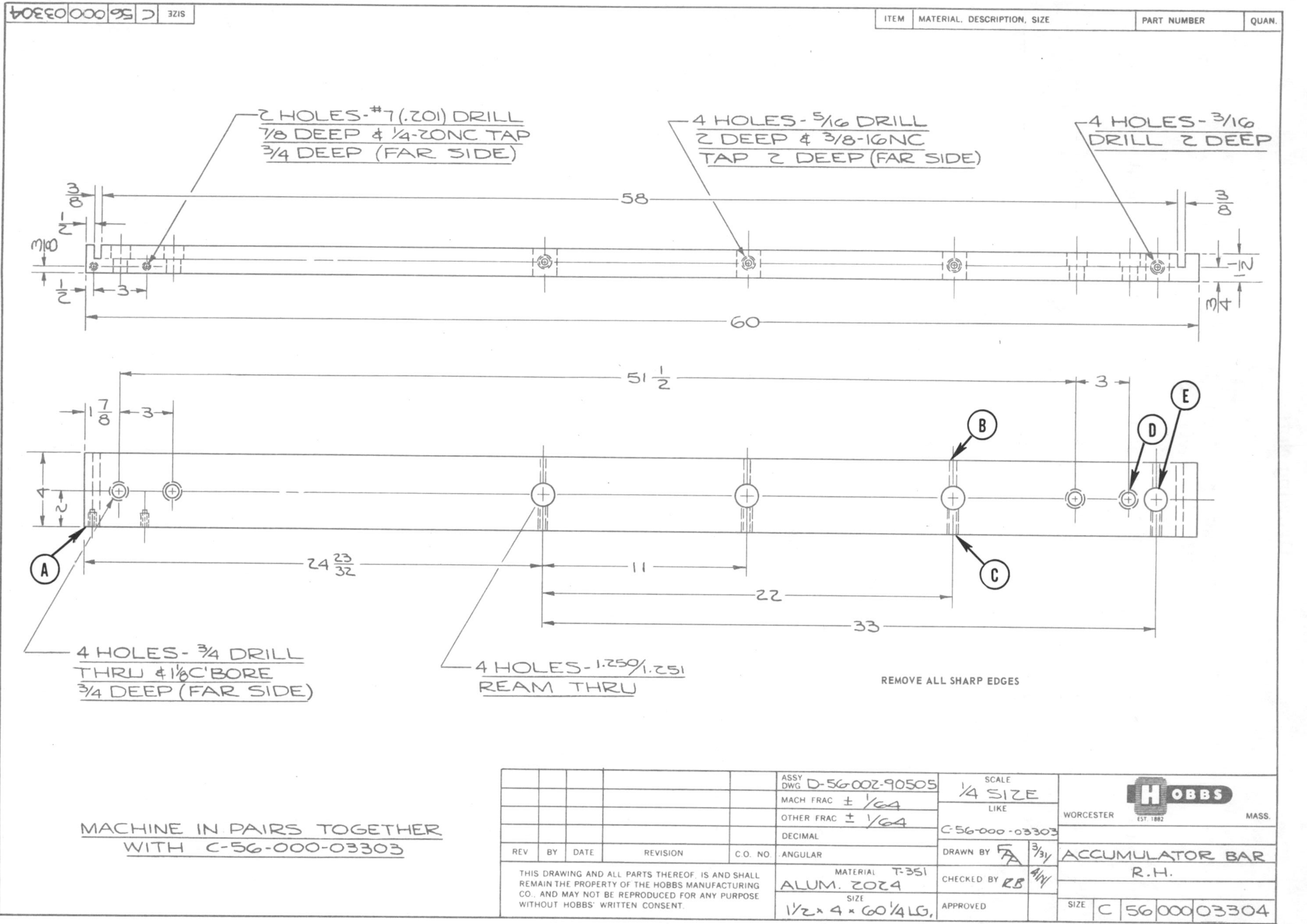

Fig. 16-BPR-2. Industry blueprint. (Hobbs)

Blueprint Reading Activity 16-BPR-2
DRAWING NOTES

Refer to the blueprint in Fig. 16-BPR-2, and answer the following questions:

1. Give the name of the part.	1. ______
2. What is the part number?	2. ______
3. What is the scale of the original plan?	3. ______
4. Give the material (a) specification and (b) stock size.	4. (a) ______ (b) ______
5. What general tolerances are stated?	5. ______
6. How many of each type of note are given?	6. General ______ Local ______
7. The part shown is to be machined in pairs. What is the part number of the other part?	7. ______
8. What are the overall dimensions of the part?	8. ______
9. Give the size and depth of the following drilled holes:	9. A ______ B ______ C ______ D ______
10. Give the size and depth of the counterbore on the far side at D.	10. ______
11. How is the hole at E finally machined and to what size?	11. ______
12. What are the thread specifications at C?	12. ______

FAIRCHILD HILLER
SPACE AND ELECTRONICS SYSTEMS DIVISION

STANDARD ENGINEERING FORM

PROGRAM 557 FUNCTION EAI

EFFECTIVITY
PROTO 1 & 2
PRODUCTION 1 & UP

PREFIX | NUMBER 557-11410-1000 | REV | AMEND

TITLE OR ACTION: ADVANCED DWG CHANGE

SECURITY CLASSIFICATION U (C/S/TS) CHARGE

PREPARED BY	SUPVSN	CHK	EE/ME/WT/STR	SPEC AND STD
K H Simon DATE 9-5	C Harris DATE 9/5/	A Sarn DATE 9/5/	L. Rankin DATE 9/8/	DATE
PROJECT C J Secaro DATE 9-9	QA/REL DATE	CONFIG MGT DATE	MFG DATE	DATA CONTROL (RELEASE) YB Appleby DATE 9-9-

ITEM | DESCRIPTION

IN ZONE F-5 REVISE MARKING
WAS A3
NOW A20

NAME | MAIL POST

SPECIAL DISTRIBUTION

DISPOSITION OF MATERIAL
- REWORK PER: THIS EAI AT TSD
- NOT AFFECTED
- USE TO DEPLETION
- SCRAP
- OTHER, SEE ABOVE

REASON FOR CHANGE
- AUTHORIZING DOCUMENT:
- DESIGN IMPROVEMENT
- [X] RECORD CHG
- DFT ERROR
- OTHER, SEE ABOVE

SHEET 1 OF 1

Fig. 17-1. An Engineering Order (EO).

Unit 17
The Drawing Change System

The term "drawing change" refers to any revision made to a drawing after prints of the drawing have been released to the shop. Changes may occur for a variety of reasons. The change may relate to the improvement of the product's function or reliability, method of manufacture, cost reduction, quality control or the correction of errors in the original drawing and prints. It may be the customer has requested a change to incorporate new or improved design into the product. Whatever the reason, the reader must understand the drawing change systems in order to correctly interpret blueprints involving changes.

Kinds of Changes

Changes made in drawings vary in degree as to whether they represent permanent changes in the design or manufacturing processes, or whether they are temporary changes to accommodate a special situation.

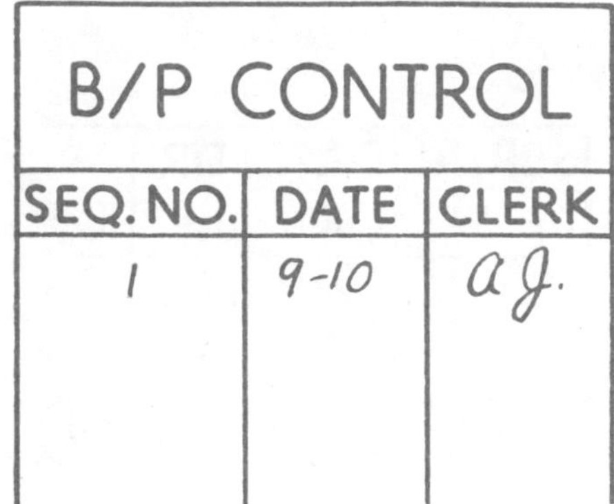

B/P CONTROL		
SEQ. NO.	DATE	CLERK
1	9-10	a.J.

Fig. 17-2. Blueprint Control Block.

ENGINEERING ORDER (EO) is a "permanent" change used to make changes in drawings when:

1. Time does not permit changing and reprinting the drawing.

2. The cost in reprinting the drawing is not justified for this change alone.

An EO is a separate sheet that includes the necessary record data, notes and/or sketches which indicate changes in the drawing such as dimensions, materials, notes and features (chamfers, holes, slots, etc.) that detail the effected change in the drawing. In some industries an Engineering Order is referred to as an Engineering Advance Information (EAI), Fig. 17-1. The EO shows the old or WAS condition of the drawing as well as the new or NOW condition. It is attached to the print and recorded in the BLUEPRINT CONTROL block, Fig. 17-2, which is usually stamped on the back of the print by the recording clerk. A sequence number is assigned and recorded in the B/P Control Block. If an EO is recorded in the B/P Control Block but is not with the drawing, CHECK WITH YOUR SUPERVISOR. EO's remain in effect until removed by a Drawing Change Notice (DCN) which is discussed later in this unit.

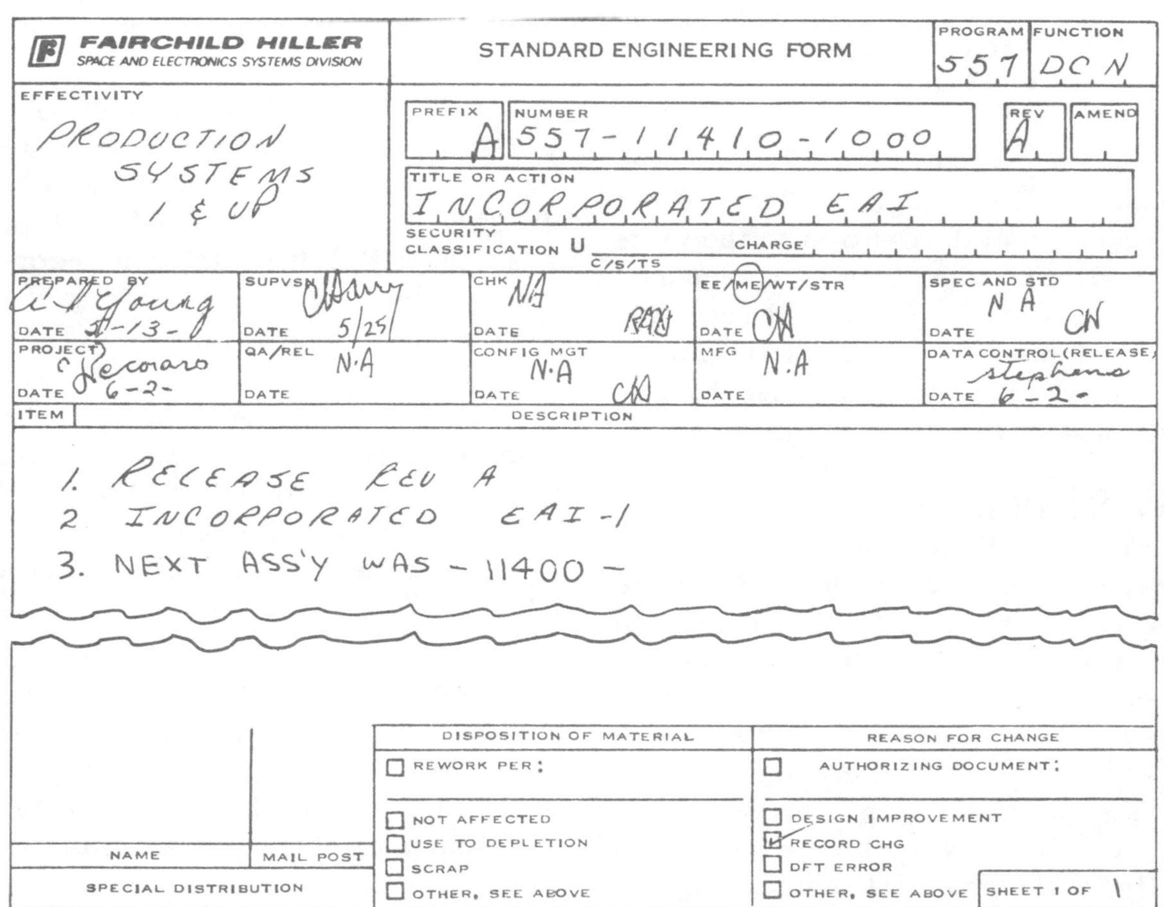

FAIRCHILD HILLER
SPACE AND ELECTRONICS SYSTEMS DIVISION

STANDARD ENGINEERING FORM

PROGRAM: 557 | FUNCTION: DCN

EFFECTIVITY: PRODUCTION SYSTEMS 1 & UP

PREFIX: A | NUMBER: 557-11410-1000 | REV: A | AMEND

TITLE OR ACTION: INCORPORATED EAI

SECURITY CLASSIFICATION: U | C/S/TS | CHARGE

PREPARED BY	SUPVSN	CHK	EE/ME/WT/STR	SPEC AND STD
A. Young	C. Harry	NA		N A
DATE 5-13-	DATE 5/25	DATE RAX	DATE CH	DATE CN
PROJECT: C. Recoraro	QA/REL: N.A	CONFIG MGT: N.A	MFG: N.A	DATA CONTROL (RELEASE): Stephens
DATE 6-2-	DATE	DATE CM	DATE	DATE 6-2-

ITEM | DESCRIPTION

1. RELEASE REV A
2. INCORPORATED EAI-1
3. NEXT ASS'Y WAS -11400-

NAME	MAIL POST

SPECIAL DISTRIBUTION

DISPOSITION OF MATERIAL	REASON FOR CHANGE
☐ REWORK PER:	☐ AUTHORIZING DOCUMENT:
☐ NOT AFFECTED	☐ DESIGN IMPROVEMENT
☐ USE TO DEPLETION	☑ RECORD CHG
☐ SCRAP	☐ DFT ERROR
☐ OTHER, SEE ABOVE	☐ OTHER, SEE ABOVE

SHEET 1 OF 1

Fig. 17-3. Drawing Change Notice (DCN) for EO in Fig. 17-1.

NOTICE OF CHANGE						
LTR	ZONE	DESCRIPTION	SERIAL	DATE	DR	APP
A	G2	ADDED NOTE 1.	06945	9/16/xx	JWT	RS
B	D-4	.06 × 45° CHAMFER REMOVED	07621	3/22/xx	HR	RS
(1)	(2)	(3)	(4)	(5)	(6)	(7)

Fig. 17-4. The Change Block is a record of changes made to the original drawing.

DRAWING DEVIATION (DD) is a TEMPORARY change which authorizes a deviation from the blueprint only as described on the drawing deviation sheet. DD's are issued for changes such as:

1. Temporary change in the standard parts.
2. Temporary substitution in method of manufacture.
3. A design change that requires rework of existing units.

The DD is prepared on the same type sheet as the EO although a colored sheet may be used to call attention to the temporary nature of the change. A sequence number is assigned the DD and recorded in the B/P Control Block. The DD remains in effect until changed or voided by another DD or a revision in the drawing.

DRAWING CHANGE NOTICE (DCN) is a PERMANENT change that results in an actual revision of the drawing and issuance of a revised print. The DCN is issued on a sheet similar to the other changes but usually of a different color to indicate that a change has been made in the drawing. A sample form for a DCN, see Fig. 17-3, indicates the change letter, details of the change, zone location on blueprint, model effectivity, authorization (Engineering Order), disposition of parts already made, and reasons for change.

Outstanding DCN's are attached to the back of sheet 1 of the drawing until the affected sheet of the drawing has been changed. When the change has been included on the drawing, the action will be recorded in the CHANGE BLOCK on sheet 1 and the DCN removed.

Change Block

The CHANGE BLOCK usually appears in the upper right hand corner of the drawing and may be titled ALTERATIONS, NOTICE OF CHANGE, or REVISIONS. Whatever the title, it contains a record of the changes that have been made to the original drawing, Fig. 17-4.

When a Drawing Change Notice has been prepared, the drawing is changed and the pertinent information recorded in the Change Block on the drawing. The following items are usually included in the Change Block.

SEQUENCE LETTER is assigned to the change and recorded in the Change Block, Fig. 17-4 (1). This index letter is also referenced to the field of the drawing next to the change effected, for example:

(A) 1. BREAK ALL SHARP EDGES.

ZONE: This column is used on larger sized prints to aid in locating changes, column (2), Fig. 17-4.

DESCRIPTION column (3), Fig. 17-4, provides a concise description of change; for example, when a note is removed from the drawing, the type of note is referred to in the description block - - PAINTING NOTE REMOVED, or when a dimension is changed WAS .999-1.003.

SERIAL NUMBER of the assembly or machine on which this change becomes effective, Fig. 17-4(4).

DATE column is for the date the change was written, Fig. 17-4(5).

DRAFTSMAN making the change enters his initials here, Fig. 17-4(6).

APPROVED column carries the initials or name of the engineer approving the change, Fig. 17-4(7).

OTHER LISTINGS: A few industries will include other items in their Change Block to further document the changes made in the original drawing. Some of these are:

CHECKED column is for the initials or signature of the person who checks and approves the revision.

AUTHORITY column usually consists of recording approved engineering change request number.

CHANGE NUMBER is a listing of the Drawing Change Notice (DCN) number.

DISPOSITION column carries the coded number indicating the disposition of the change request.

MICROFILM column is used to indicate the date the revised drawing was placed on microfilm.

EFFECTIVE ON column (sometimes a separate block) gives the serial number or ship number of the aircraft, machine, assembly or part on which the change becomes effective. The change may also be indicated as effective on a certain date and would take effect on that date forward.

A drawing which has been extensively revised may be redrawn and carry an entry to that effect in the Change Block or the drawing number may carry a dash letter (-A) indicating a revised drawing.

It should be noted that considerable variation exists among industries in the form of processing and recording changes in prints, Fig. 17-5. The information presented in this unit will enable you to develop an understanding of the change system in general. Each skilled mechanic and technician will want to become thoroughly familiar with the details of the change system in the industry where he works.

REVISIONS				
ZONE	LTR	DESCRIPTION	DATE	APPROVED
F-5	A	INCORP. EAI 1 REVISED MARKING A20 WAS A3 N/A WAS 11400	6-2-	

Fig. 17-5. Change Block with DCN in Fig. 17-3 recorded.

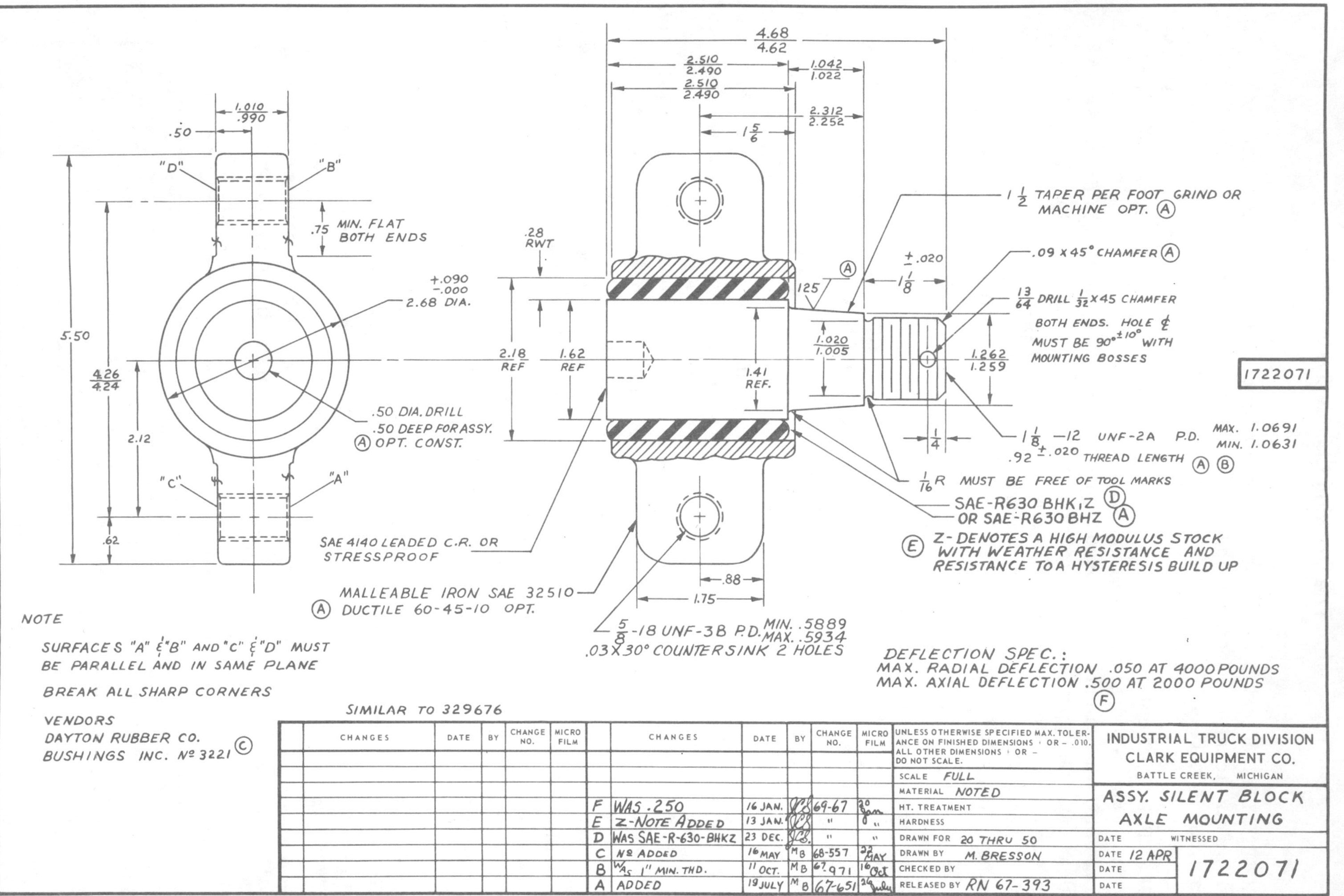

Fig. 17-BPR-1. Industry blueprint. (Clark Equipment Co.)

Blueprint Reading Activity 17-BPR-1
THE CHANGE SYSTEM

Refer to the blueprint in Fig. 17-BPR-1, and answer the following questions.

1. What is the name of the assembly? 1. ____________________
2. What is the drawing number? 2. ____________________
3. What is the scale of the original plan? 3. ____________________
4. How is the material specification handled? 4. ____________________
5. How many changes are listed? 5. ____________________
6. How many separate notes are listed by change A? 6. ____________________
7. What modification of material specification was affected by D? 7. ____________________
8. What revision was made to the Vendors note by change C? 8. ____________________
9. Is the note added by change E a general or a local note? 9. ____________________
10. How many general notes are listed on the print? 10. ____________________
11. Write the words for the listed abbreviations. 11. EO ____________________
    DD ____________________
    DCN ____________________
12. Where would you expect the Blueprint Control Block to appear on a print? 12. ____________________

For additional Blueprint Activities for Unit 17 see pages 254 and 257.

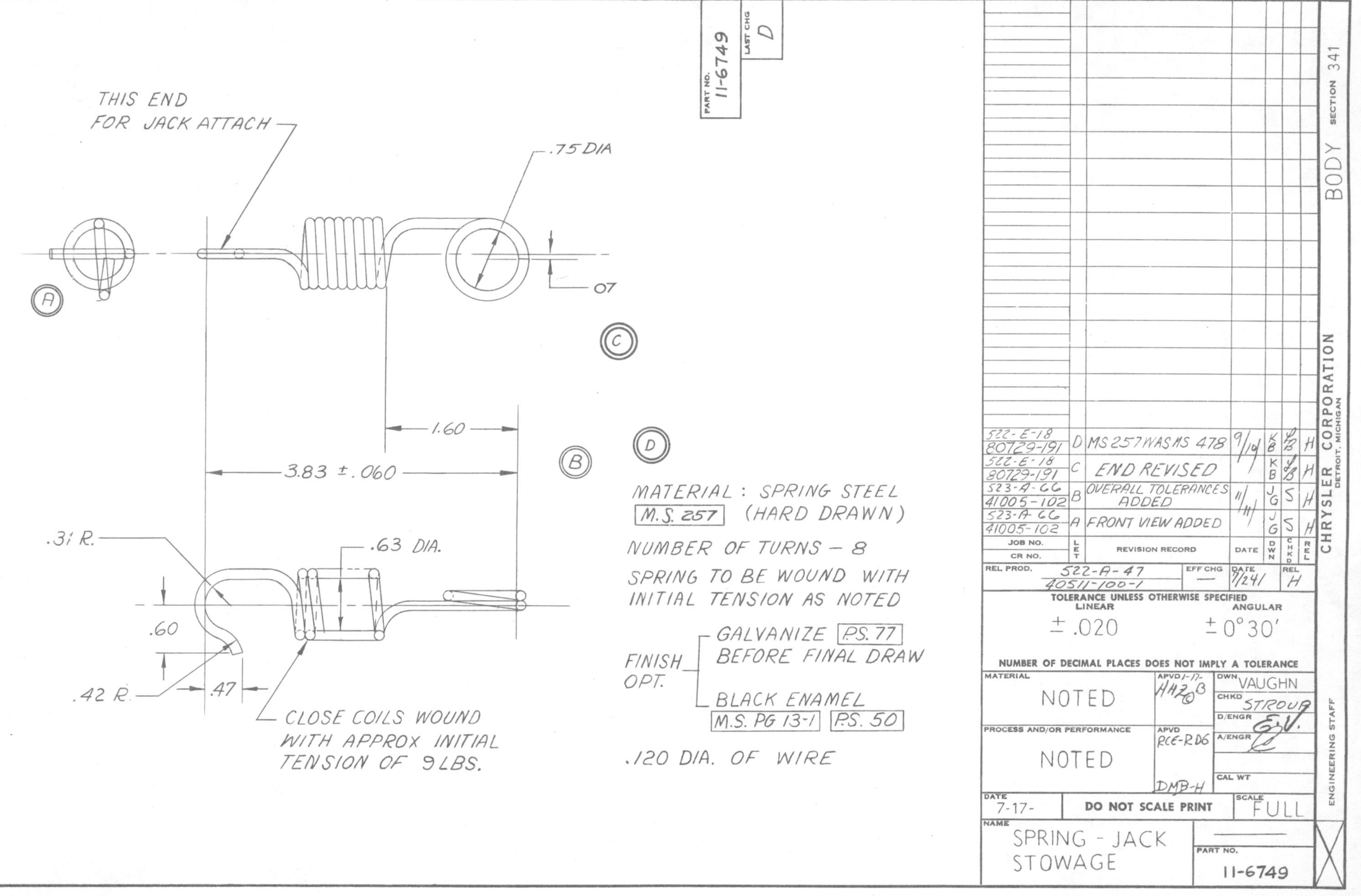

Fig. 17-BPR-2. Industry blueprint. (Chrysler Corp.)

Blueprint Reading Activity 17-BPR-2
THE CHANGE SYSTEM

Refer to the blueprint in Fig. 17-BPR-2, and answer the following questions:

1. Give the name of the part. 1. ____________

2. What is the drawing number? 2. ____________

3. Is this a detail or assembly drawing? 3. ____________

4. What is the scale of the original plan? 4. ____________

5. What material is specified? 5. ____________

6. How many changes have been made on the original drawing? 6. ____________

7. What was the change in A? 7. ____________

8. In change B, what is the tolerance that was added? 8. ____________

9. What sequence letter identifies the change in the material used for the spring? 9. ____________

10. What is the initial of the person who approved the release of the prints after each change? 10. ____________

11. How many general notes are listed? 11. ____________

12. What is the inside diameter of the spring coils? 12. ____________

13. Give the diameter of the spring steel stock. 13. ____________

14. Between the hook and loop, what is the maximum allowable inside distance? 14. ____________

15. What initial tension is to be used in winding the spring coils? 15. ____________

For Evaluation Blueprint Activity for Part 3 see page 315.

PART 4
MACHINING SPECIFICATIONS

Unit 18
Thread Representation and Specification

Screw threads and pipe threads are features of very important fasteners and machine parts used in industry. Threads such as the American National or the Unified Standard (V-type) are used for such fasteners as bolts, screws and nuts, and for fastening machine parts together. Other thread forms are designed for special purposes such as transmitting power along their axis. An example is the lead screw on a machine lathe.

Pipe threads are also standardized and designed for particular functions.

Threads are formed by hand, ground or rolled, on a lathe and on special machines.

Screw Thread Forms

Most thread forms used in industry today are based on the approved American Standards. Unified Screw Thread Standard, ASA B1.1, is the recognized straight thread standard in the United States for screws, bolts, nuts and other threaded fasteners. These threads which are called Unified Standard (or just Unified) are also in common use in Great Britain and Canada. The Unified threads have substantially the same thread form as the former American National thread but fewer difficulties are encountered in production with the Unified Standard due to the flat or rounded top and bottom of the thread form.

Thread forms in use for purposes other than fasteners include the square, the Acme, the buttress, and the Whitworth.

Thread Representation

There are three conventional methods in which threads may be represented on drawings: The DETAILED, SCHEMATIC, and SIMPLIFIED, Fig. 18-1. The schematic and the simplified conventions are commonly used to save drafting time. The detailed convention is a closer representation of the thread as it actually appears and is sometimes used to show the geometry of a thread form as a portion of a greatly enlarged detail on a drawing.

Thread Series

There are four series of Unified screw threads: COARSE, FINE, EXTRA FINE and CONSTANT-PITCH SERIES. The Unified Coarse thread series is designated UNC and is used for nuts, bolts, screws and general use where a finer thread is not required. The Unified Fine thread series (UNF) is used where the length of the threaded engagement is short and where a small lead angle is desired. The Unified

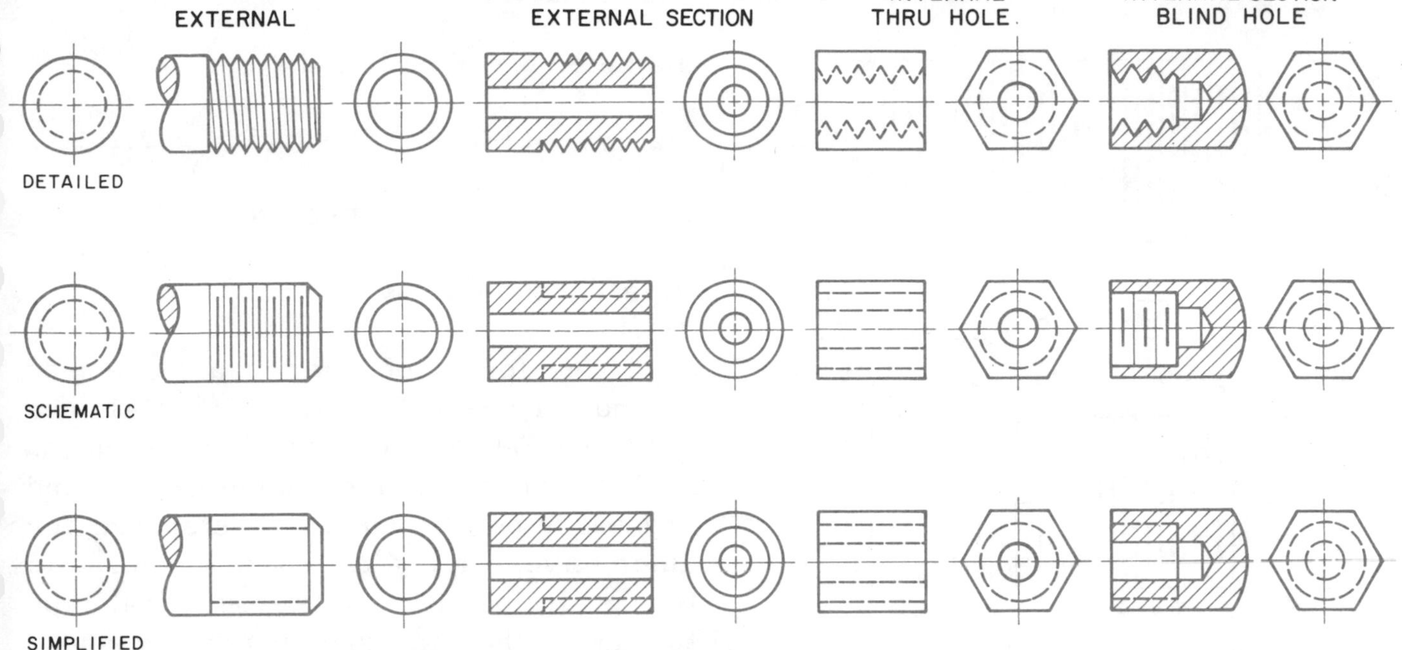

Fig. 18-1. Conventions for screw thread representation.

Extra Fine series (UNEF) is used for even shorter lengths of thread engagements and for thin-wall tubes, nuts, ferrules and couplings. The Unified Constant-Pitch series is designated UN with the number of threads per inch preceding the designation as, 8UN. This series of threads is for special purposes such as high pressure application and large diameters where the other thread series do not meet the requirements.

There are also three special thread series designated UNS, NS and UN that cover special combinations of diameter, pitch and length of engagement.

Thread Classes

Threads are further classified by the amount of tolerance and allowance permitted in manufacturing. These classes are 1A, 2A and 3A for external threads and 1B, 2B and 3B for internal threads. On some older drawings, classes 2 and 3 without a letter designation apply to both external and internal threads.

Classes 1A and 1B replace the older American Standards' class 1 and are used in applications requiring minimum binding to permit frequent and quick assembly-disassembly of parts.

Classes 2A and 2B because of the realistic tolerances, are for general purpose use such as for nuts, bolts, screws and normal applications by mass production industries.

Class 3A and 3B are for applications in industries requiring closer tolerances than the preceding classes of 1A and 1B, or 2A and 2B.

Specification of Screw Threads

A screw thread is specified on a drawing by a standard note with a leader and an arrow pointing to the thread. The following information is specified in sequence for a standard thread. (a) nominal size (major diameter or screw number), (b) number of threads per inch, (c) thread series symbol, and (d) the thread class number or symbol, Fig. 18-2. Threads are right-hand and single lead unless otherwise specified by the letters LH after the class symbol and the word DOUBLE to indicate a left-hand thread and a double lead, Fig. 18-2 (e).

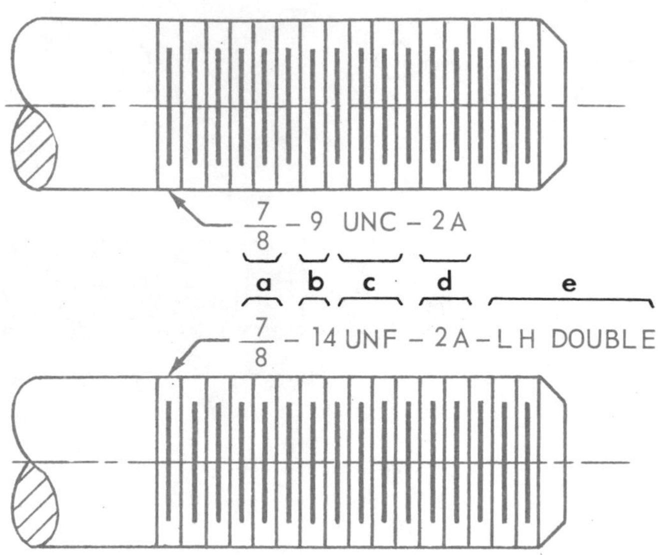

Fig. 18-2. Notes specifying screw threads.

Other specifications for threads may be given such as thread length, hole size (for internal threads), chamfer, and countersink, Fig. 18-3.

DRILL 49/64 - 1 1/4 DEEP
C'SINK 45° TO 7/8 DIA.
7/8 - 9 UNC - 2B 1" DEEP

Fig. 18-3. Specification for internal thread.

If the tolerance of the thread pitch diameter is given, it is placed at the end of the specification for the thread:

3/4 - 10UNC - 2B PD 0.6850 to 0.6927

The specification for a constant-pitch series thread with a tolerance for the pitch diameter would be listed:

2 1/4 - 8UN - 3A PD 2.1688 to 2.1611

Pipe Threads

Pipe Thread Forms

Three forms of pipe threads are used in industry - - REGULAR, DRYSEAL and AERONAUTICAL. The regular pipe thread is the standard for the plumbing trade and dryseal pipe thread is the standard for automotive, refrigeration and hydraulic tube and pipe fittings. Aeronautical pipe thread is the standard in the aerospace industry.

Regular and aeronautical pipe thread forms must be filled with a lute or sealer to prevent leakage in the joint. Dryseal pipe thread form does not allow leakage due to metal to metal contact at the crest and root of the threads.

Representation and Specification

Straight and taper pipe threads are represented (graphically) in the same manner as screw threads except the taper pipe threads are shown tapered at an angle of approximately 3 degrees to the axis.

The specifications for American Standard Pipe Threads are listed in sequence - - the nominal size, number of threads per inch, and the symbols for the thread series and form: 1/2 - 14NPT.

The following symbols are used to designate American Standard Pipe Threads:

NPT - - American Standard Taper Pipe Thread.
NPTR - - American Standard Taper Pipe Thread for Railing Joints.
NPSC - - American Standard Pipe Thread for Couplings.
NPSM - - American Standard Pipe Thread for Free-Fitting Mechanical Joints.
NPSL - - American Standard Pipe Thread for Loose-Fitting Mechanical Joints with Locknuts.
NPSH - - American Standard Pipe Thread for Hose Couplings.

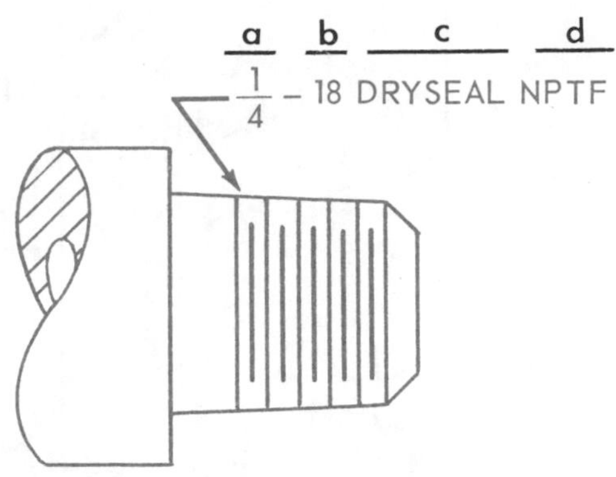

Fig. 18-4. Note specifying pipe threads.

Symbols for Dryseal Pipe Threads are designated as follows:

NPTF - - Dryseal American Standard Pipe Thread.
PTF – SAE SHORT - - Dryseal SAE Short Taper Pipe Thread.
NPSF - - Dryseal American Standard Fuel Internal Straight Pipe Thread.
NPSI - - Dryseal American Standard Intermediate Internal Straight Pipe Thread.

A typical specification for the Dryseal pipe thread would include: (a) nominal size, (b) number of threads per inch, (c) form (Dryseal), and (d) symbol, Fig. 18-4.

Metric Threads

Metric threads are designated in a manner similar to that used for Unified and American National Standard, but with some slight variations. A diameter and pitch are used to designate the metric series, as in the inch system, with the following modifications:

1. Metric coarse threads are designated by simply giving prefix M and diameter. For example, in Fig. 18-5, M8 is a coarse thread designation representing a nominal thread diameter of 8 mm with a pitch of 1.25 mm understood. Thread designation is for a coarse thread unless otherwise noted.
2. Metric fine threads are designated by listing pitch as a suffix. A fine thread for a part would be M8 x 1.0, or 8 mm diameter with a pitch of 1.0 mm. Most common metric thread is coarse, which generally falls between coarse and fine series of inch system for a comparable diameter.

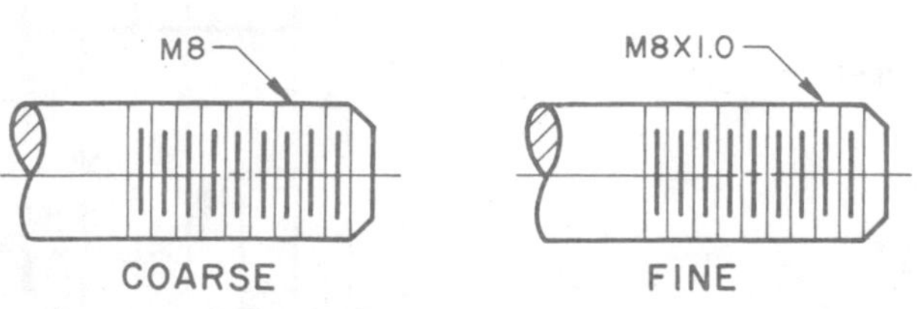

Fig. 18-5. The metric coarse thread designation gives only the diameter, the pitch is understood. The fine thread designation gives the pitch, following the diameter.

In the metric series, the pitch is really the pitch: M x 1.0. That is, the pitch is actually 1 mm.

The tolerance and class of fit in metric threads are designated by adding numbers and letters in a certain sequence to the callout. The thread designation in Fig. 18-6 calls for a fine thread of 6 mm diameter, 0.75 mm pitch (no pitch is given in the designation for a coarse thread) with a pitch diameter tolerance grade 6 and an allowance "h", crest diameter tolerance grade 6 and allowance "g".

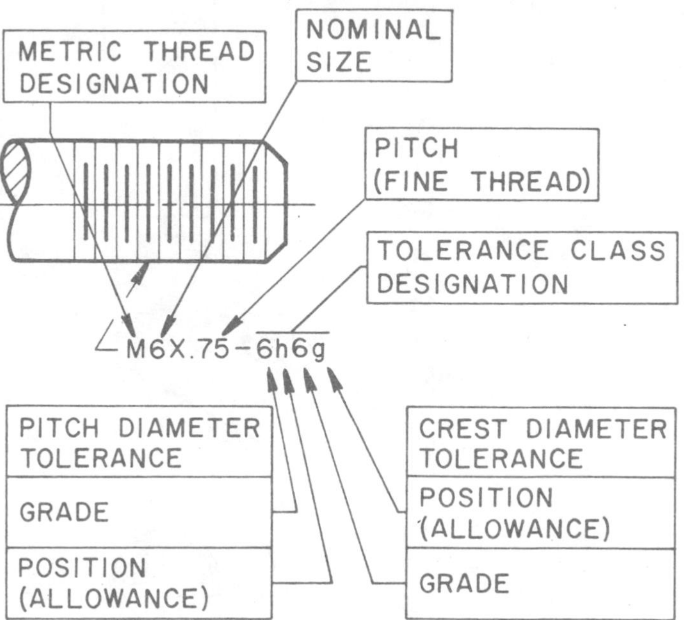

Fig. 18-6. A complete designation for an ISO metric thread.

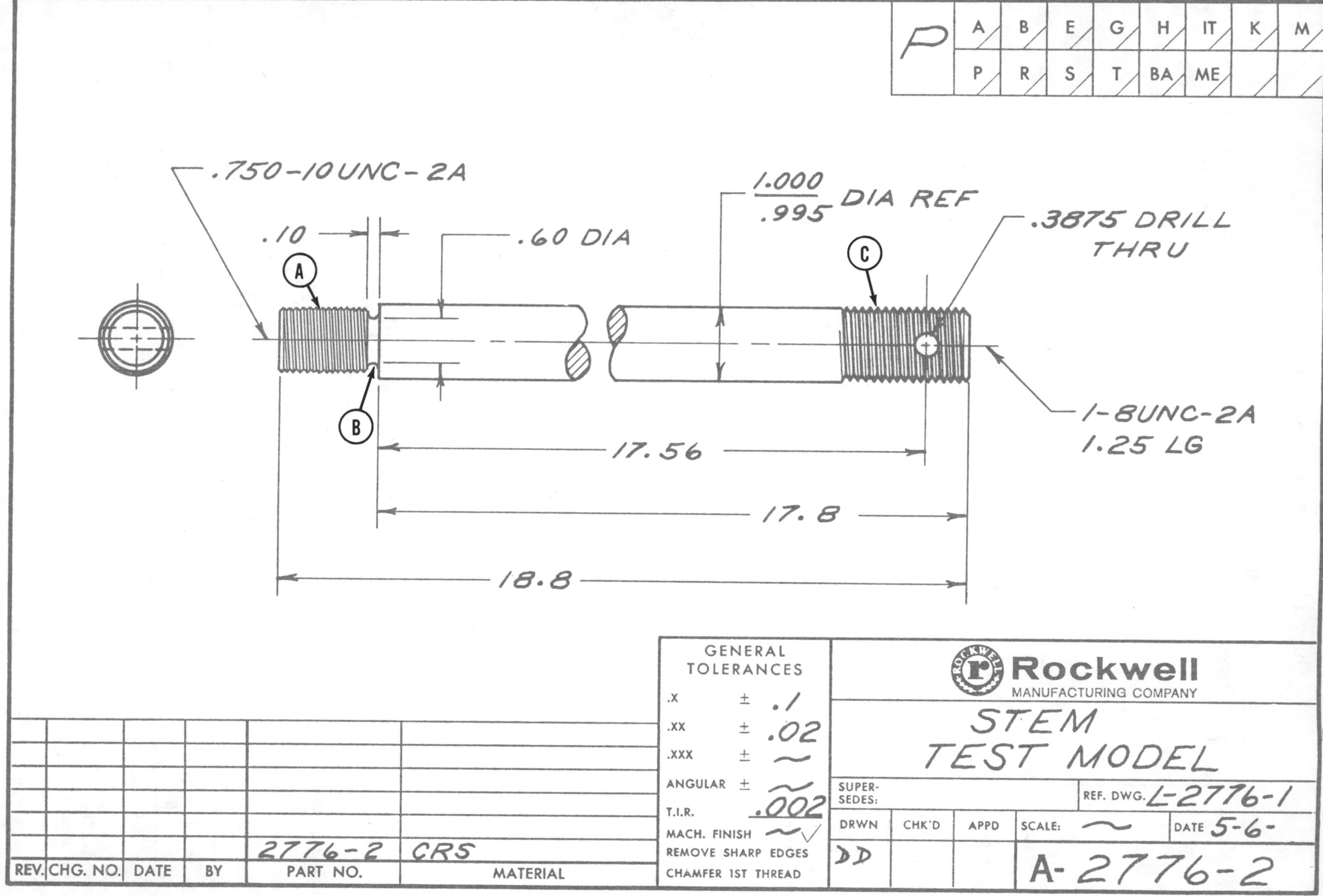

Fig. 18-BPR-1. Industry blueprint. (Rockwell Mfg. Co.)

Blueprint Reading Activity 18-BPR-1
EXTERNAL THREADS

Refer to the blueprint in Fig. 18-BPR-1, and answer the following questions:

1. What is the name of the object? 1. ____________

2. Give the drawing number. 2. ____________

3. Is this a detail or assembly drawing? 3. ____________

4. What material is required? 4. ____________

5. Give the finished length including the tolerance permitted. 5. ____________

6. What size hole is drilled through the piece? 6. ____________

7. Interpret the thread specification for A. 7. .750 ____________
   10 ____________
   UNC ____________
   2A ____________

8. Give the dimensions for the groove at B. 8. ____________

9. Interpret the thread specification for C. 9. 1 ____________
   8 ____________
   UNC ____________
   2A ____________
   1.25 LG ____________

10. How are the ends of the piece to be finished? 10. ____________
   ____

For additional Blueprint Activities for Unit 18 see pages 258 and 261.

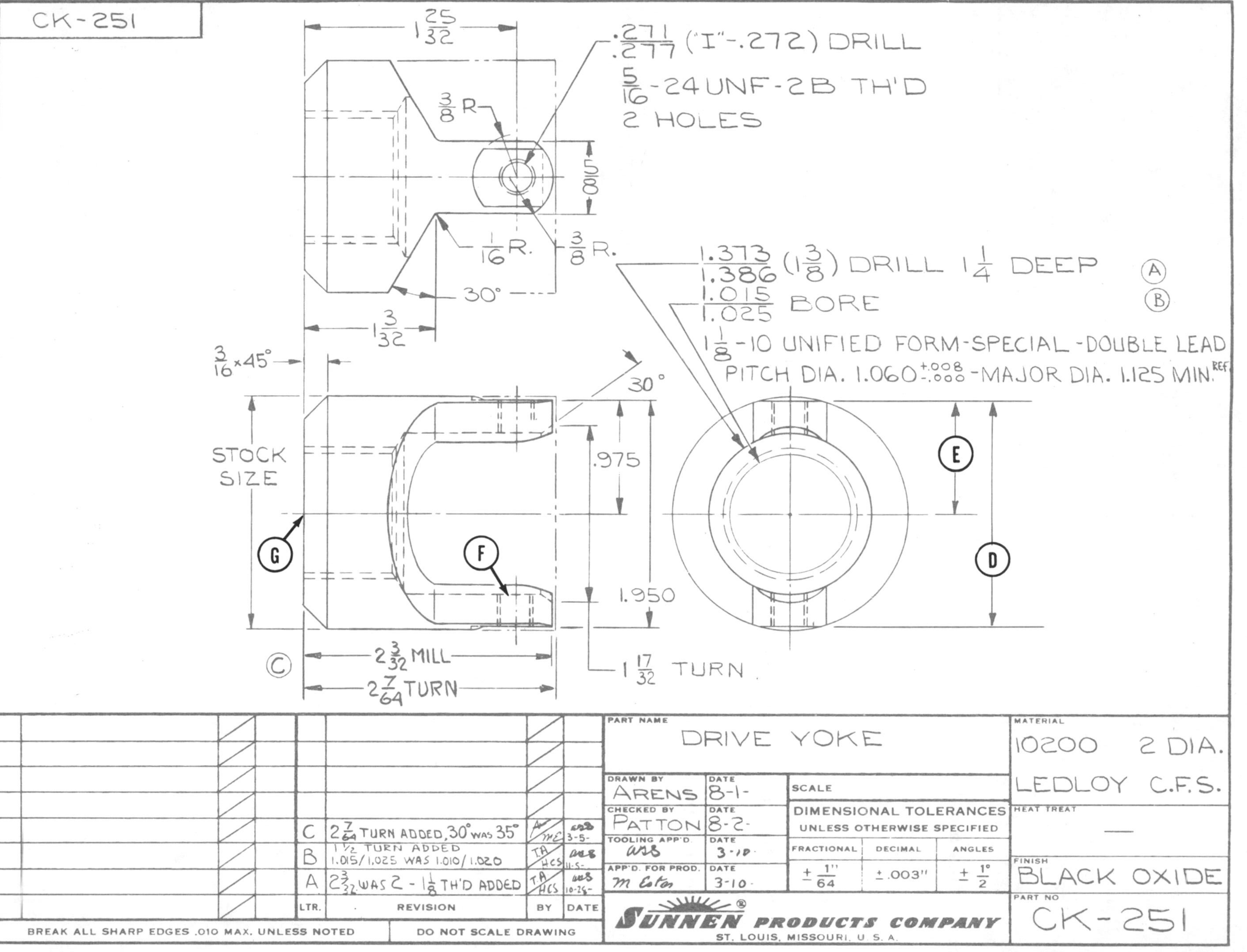

Fig. 18-BPR-2. Industry blueprint. (Sunnen Products Co.)

Blueprint Reading Activity 18-BPR-2
INTERNAL THREADS

Refer to the blueprint in Fig. 18-BPR-2, and answer the following questions:

1. Give the part name. — 1. ______

2. List the number of the part. — 2. ______

3. What material is specified? — 3. ______

4. What is the nominal stock size? — 4. ______

5. What is the finished length dimension and how is it machined? What tolerance is permitted? — 5. ______

6. Give the dimension and tolerance for D and E. — 6. D ______ E ______

7. How many holes are to be threaded? — 7. ______

8. What size tap drill is noted for holes F? — 8. ______

9. Give the thread specification for holes F. — 9. ______

10. Give the interpretation of the thread specification in No. 9 above. — 10. 5/16 ______ 24 ______ UNF ______ 2B ______

11. How is hole G machined prior to cutting threads and to what diameter? — 11. ______

12. Interpret the thread specification for hole G. — 12. 1-1/8 ______ 10 ______ UNIFIED FORM SPECIAL ______ DOUBLE LEAD ______ PITCH DIA. 1.060 $^{+.008}_{-.000}$ ______ MAJOR DIA. 1.125 MIN. REF ______

13. What changes were made in the drawing with Revision C? — 13. ______

14. What finish is specified? — 14. ______

Unit 19
Specification and Callouts for Machining Processes

You have already learned that a blueprint provides considerable information for the craftsman in addition to the shape and size of a detailed part or assembly. It is important in the making or servicing of a mechanism that this information is properly understood.

Many of the large industries have process specification manuals which specify how machine processes are to be performed; machine, tools, and cutters to be used; and tolerances to be held. You should become familiar with the manual in your industry and the processes involved in the work you are required to perform.

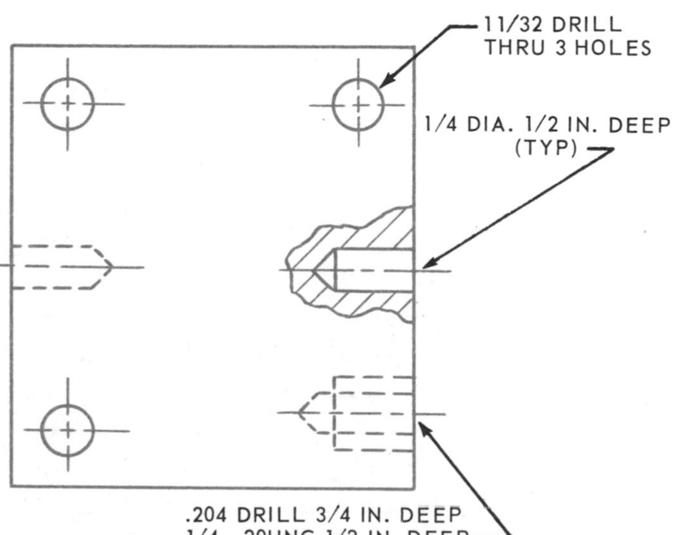

Fig. 19-1. Specifications involving drilled holes.

This unit will familiarize you with the "callouts" for most of the commonly used machine processes, also specifications found on industrial blueprints.

DRILLED HOLES are drilled with a twist drill and may be specified for diameter of drill to use, depth of hole or "thru" and number of holes of this specification to be drilled, Fig. 19-1.

The term typical (TYP) means the specification in the note holds for similar features on print unless otherwise noted.

REAMED HOLES are specified when a hole must be truer, smoother and more accurate in size than a drilled hole. The hole is drilled slightly smaller (.010 to .025 in.) than the finished size and then reamed to size. The specification for a reamed hole is shown in Fig. 19-2. Size of the drilled hole may be omitted.

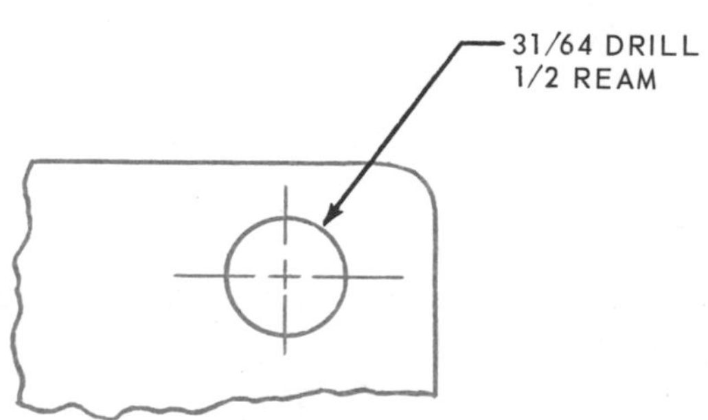

Fig. 19-2. Specification for a reamed hole.

COUNTERBORED HOLES are holes that have been machined to a larger diameter for a certain distance to form a recessed flat shoulder which receives a bolt or screw head below the surface. See Fig. 19-3.

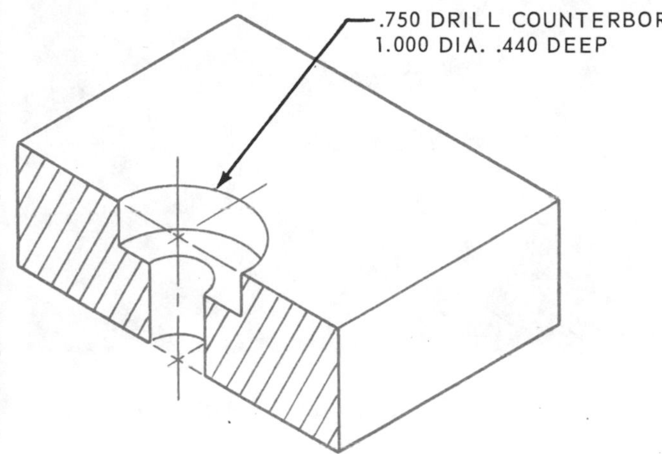

Fig. 19-3. A counterbored hole.

COUNTERSINKING is a process of removing metal around the edge of a hole to provide a seat for conical shaped screw heads and rivets; or to provide a seat for work held between centers in a lathe, mill or similar machine. Countersinking is also used to remove burrs and chamfer holes, Fig. 19-4.

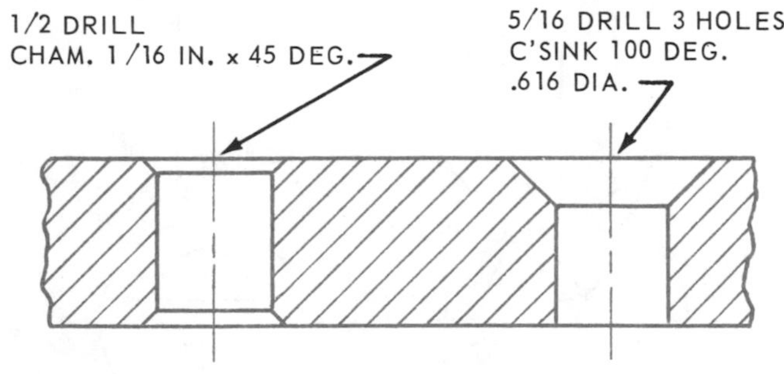

Fig. 19-4. A chamfered hole and a countersunk hole.

SPOTFACING is a process similar to counterboring only not as deep. It provides a square surface in a rough piece of work as a bearing or seat for a bolt head, nut, etc., Fig. 19-5.

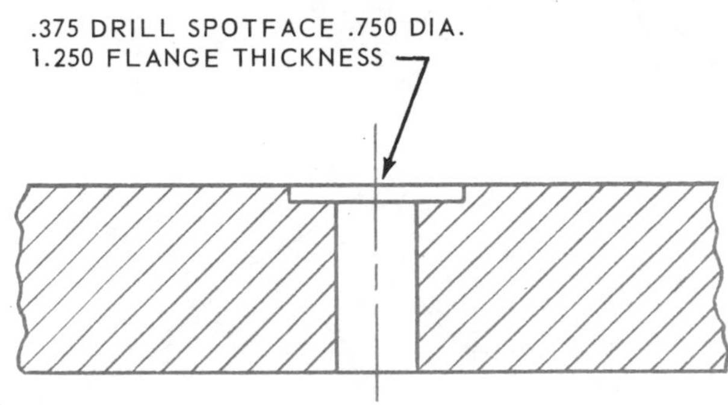

Fig. 19-5. Spotfacing callout.

BROACHING is done with a machine operated cutter called a broach. This passes through or over the stationary part to produce internal or external machined surfaces. The machined surfaces include holes of circular, square or irregular outline, keyways, internal gear teeth, splines, and flat or varied external contours. The broach has a series of cutting teeth, set so each is a few thousandths of an inch higher than the preceding tooth. Sizes increase to the exact finish size required. Broaching is fast, accurate, and produces a finish of good quality. The callout for a typical broaching operation is shown in Fig. 19-6.

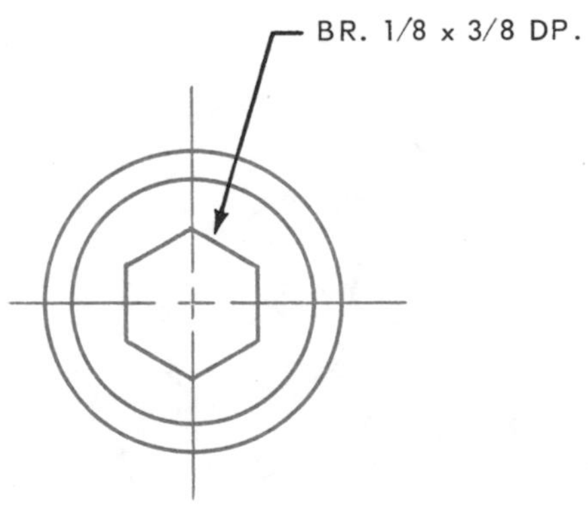

Fig. 19-6. Broaching callout.

KEYS are fasteners used to prevent rotation of gears, pulleys and rocker arms on rotating shafts. The key is a piece of metal which fits into a keyseat in the shaft and keyway in the hub. These are shown with their specification callouts in Fig. 19-7.

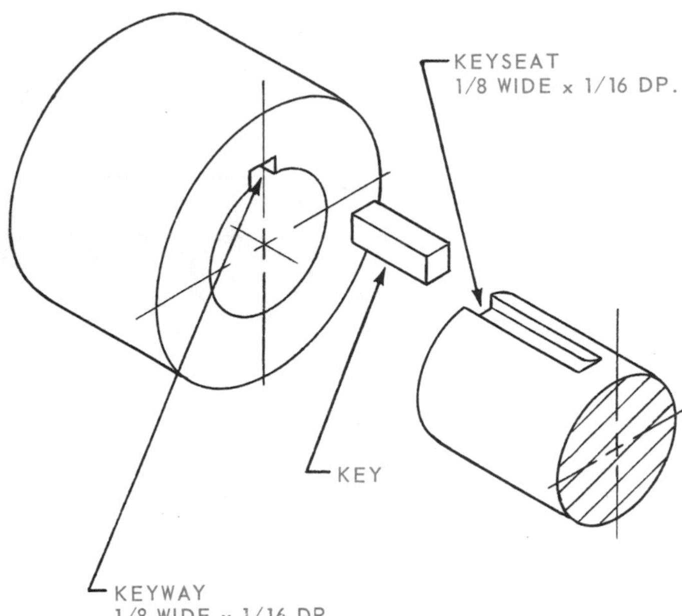

Fig. 19-7. Keyway, key, and keyseat.

Some standard types of keys are shown in Fig. 19-8.

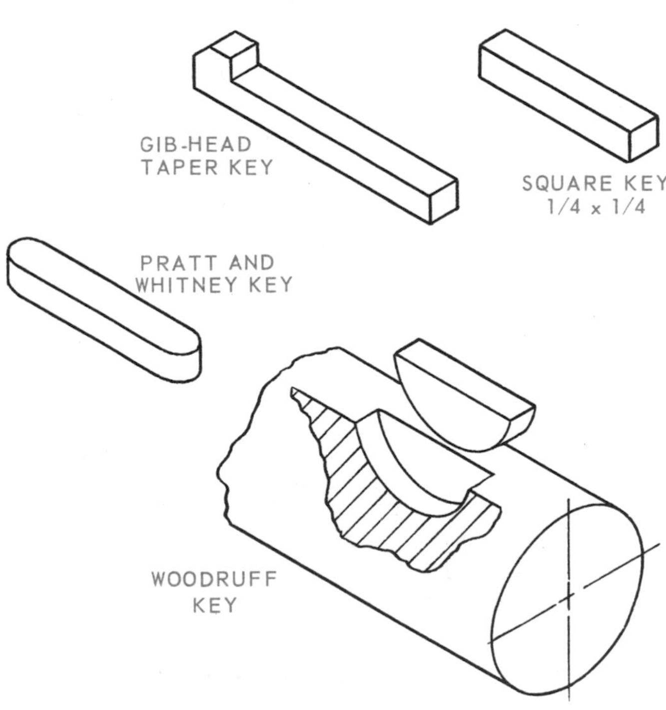

Fig. 19-8. Some standard types of keys.

TOTAL INDICATOR READING (TIR) specifies the relationship between a feature (cylinder, cone, sphere, hexagon, etc.) and its axis or the axes of one or more other features when the indicator readings are taken while the part is rotated 180°, Fig. 19-9. Eccentricity $\left(\frac{\text{TIR}}{2}\right)$ is the con-

Fig. 19-9. Total indicator reading (TIR) of a feature.

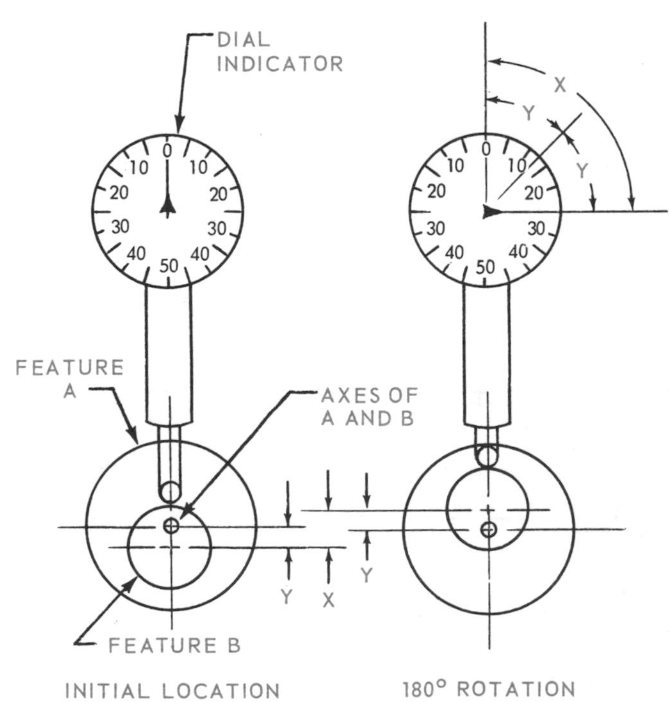

Fig. 19-10. Taking an eccentricity reading.

dition where the axis of one regular feature is offset with respect to the axis of another such feature, as shown in Fig. 19-10.

SURFACE MISMATCH. When two or more machine cutter passes are required to machine a surface, it is difficult to blend the surfaces. If a "mismatch" is permitted, it is called out in a general note on the drawing or may be applied to specific surfaces, Fig. 19-11.

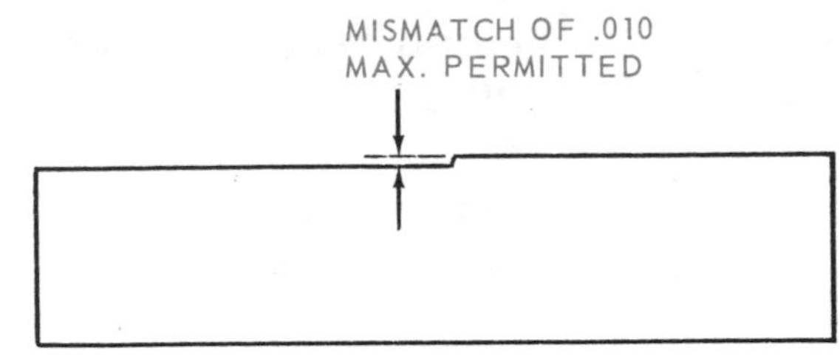

Fig. 19-11. Calling out a mismatch in surface.

SURFACE TEXTURE (FINISH) refers to the roughness, waviness, lay and flaws of a surface. Surface roughness which is sometimes stated as surface finish, is the specified smoothness required on the finished surface of a part obtained by machining, grinding or lapping.

SURFACE ROUGHNESS (fine irregularities in the surface texture) is a result of the production process used. Type 1 roughness height is measured by a profilometer in microinches (millionths of an inch).

SURFACE WAVINESS is the usually widely-spaced component of surface texture due to such factors as machine chatter, vibrations, work deflections, warpage and heat treatment. Waviness is rated in inches.

LAY is the term used to describe the direction of the predominant surface pattern, Fig. 19-12. The lay symbol is used

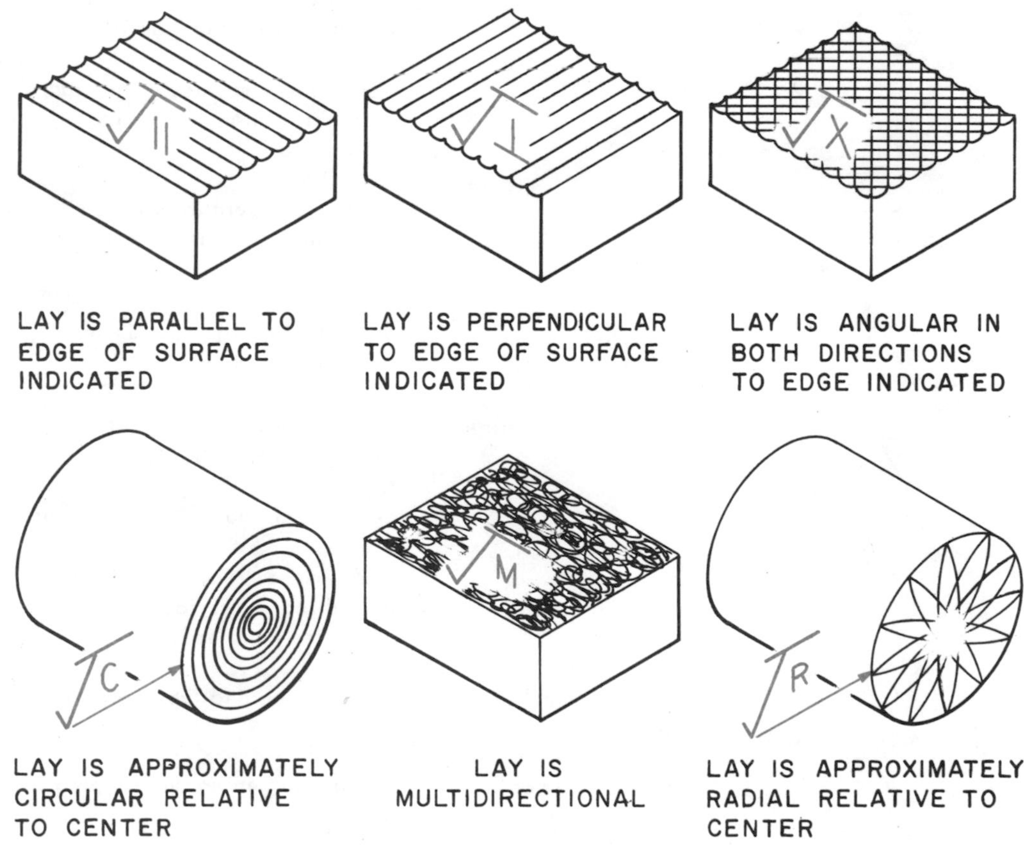

Fig. 19-12. Designation of lay for surface texture.

if considered essential to a particular surface finish.

The symbol used to call out the details of the surface texture resembles a check mark, as shown in Fig. 19-13. Usually, only the surface roughness height figure is used with the symbol but when other characteristics of a surface texture are specified, they are shown in the callout as indicated in Fig. 19-13. The point of the symbol is placed on the line indicating the surface, on an extension line, or is directed to the surface by a leader.

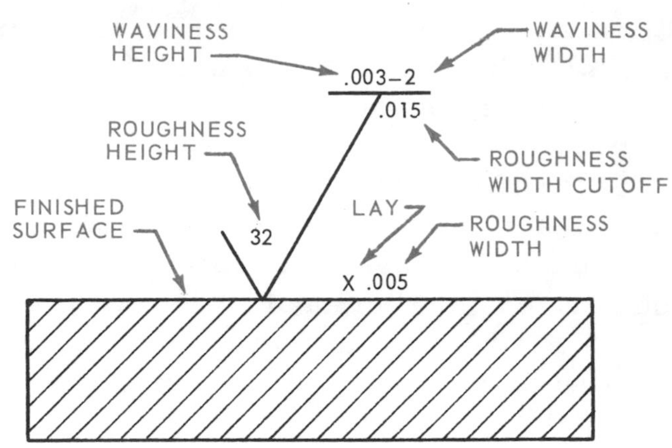

Fig. 19-13. Surface texture symbol.

The table in Fig. 19-14 indicates some recommended roughness height values in microinches, a description of the surface and the process by which the surface may be produced.

ROUGHNESS HEIGHT RATING	SURFACE DESCRIPTION	PROCESS
1000	Very rough	Saw and torch cutting, forging or sand casting.
500	Rough machining	Heavy cuts and coarse feeds in turning, milling and boring.
250	Coarse	Very coarse surface grind, rapid feeds in turning, planning, milling, boring and filing.
125	Medium	Machining operations with sharp tools, high speeds, fine feeds and light cuts.
63	Good machine finish	Sharp tools, high speeds, extra fine feeds and cuts.
32	High grade machine finish	Extremely fine feeds and cuts on lathe, mill and shapers required. Easily produced by centerless, cylindrical and surface grinding.
16	High quality machine finish	Very smooth reaming or fine cylindrical or surface grinding, or coarse hone or lapping.
8	Very fine machine finish	Fine honing and lapping of surface.
2-4	Extremely smooth machine finish	Extra fine honing and lapping of surface.

Fig. 19-14. Description of Roughness Height Values.

FINISH ALL OVER (OR SHORTENED TO F.A.O.)

SURFACE ROUGHNESS 125√ UNLESS OTHERWISE NOTED.

Fig. 19-15. General surface finish notes.

The value specified is the maximum average roughness height permissible. When a maximum and a minimum limit is specified, the average roughness height must lie within the two limits.

When a surface is to be machined and the control of the texture is not necessary or is specified in the title block, the finish symbol may consist of the italic "*f*", the √ check mark without the horizontal bar, or the standard finish symbol √ without the microinch value given. When an entire part is to be machined, usually a general note is given as in Fig. 19-15.

Type 2 Surface Texture Specifications

Although the Type 2 Surface Texture Symbols are not an industry wide standard, they are being used by some large industries to compare surface texture produced to a replica (a copy or reproduction) of a Master Surface by sight and/or feel, Fig. 19-16. There are five grades of surface - - D, F, H, K and N. The replica "D" is the smoothest and roughness increases with each letter to a maximum with letter "N."

Fig. 19-17. Type 2 surface texture symbol.

When the letters HFS-SF appear across the symbol bar, they indicate a stock surface: HFS - Hot Finished Surface; SF - Scale Free, Fig. 19-17. If CFS were used, it would mean a Cold Finished Surface.

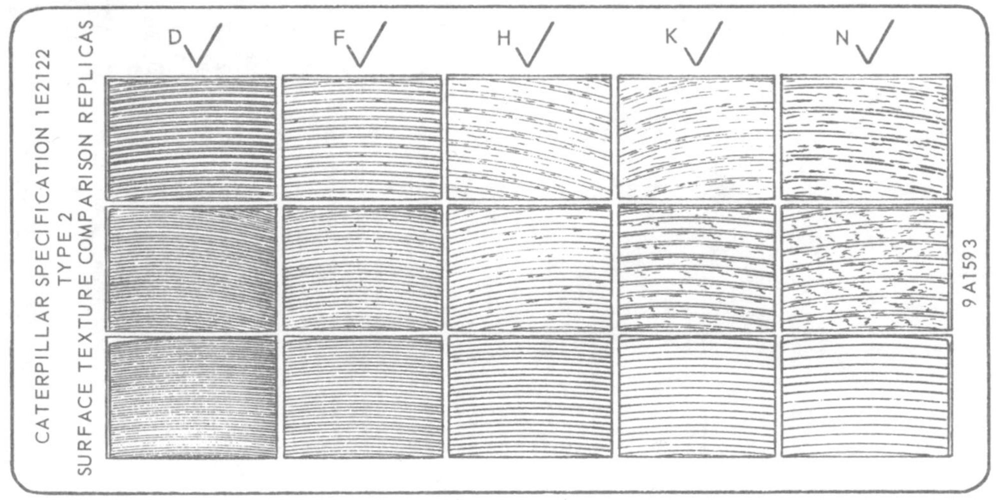

Fig. 19-16. Type 2 comparison replicas.

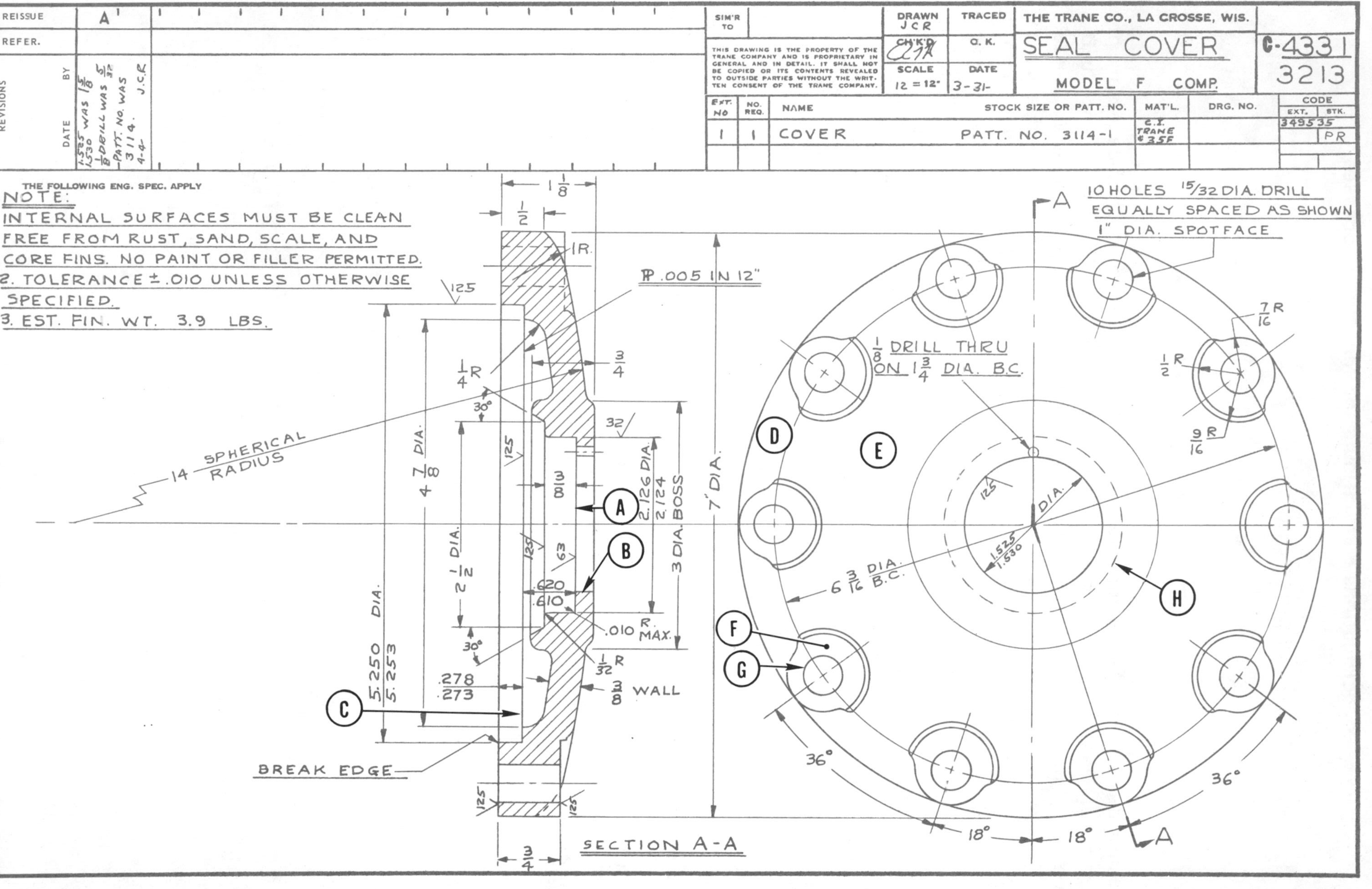

Fig. 19-BPR-1. Industry blueprint. (The Trane Co.)

Blueprint Reading Activity 19-BPR-1
CALLOUTS FOR MACHINE PROCESSES

Refer to the blueprint in Fig. 19-BPR-1, and answer the following questions.

1. What is the name of the part? 1. ______________________

2. What is the number of the print? 2. ______________________

3. What is the scale of the original plan? 3. ______________________

4. What is the material specification? 4. ______________________

5. List the surface texture required at A? B? 5. A ______________________
B ______________________

6. Give the diameter of B. 6. ______________________

7. What surface texture is called out at C? F? 7. C ______________________
F ______________________

8. What is the machine process specification called for at G and at F? 8. G ______________________
____
F ______________________

9. Give the thickness of the casting at G after the spotfacing operation. 9. ______________________

10. What dimensions are given for surfaces D and E? 10. D ______________________
E ______________________

11. Give the diameter of the invisible surface H. 11. ______________________

12. What treatment is to be given internal surfaces? 12. ______________________
____
____

For additional Blueprint Activities for Unit 19 see pages 262 and 265.

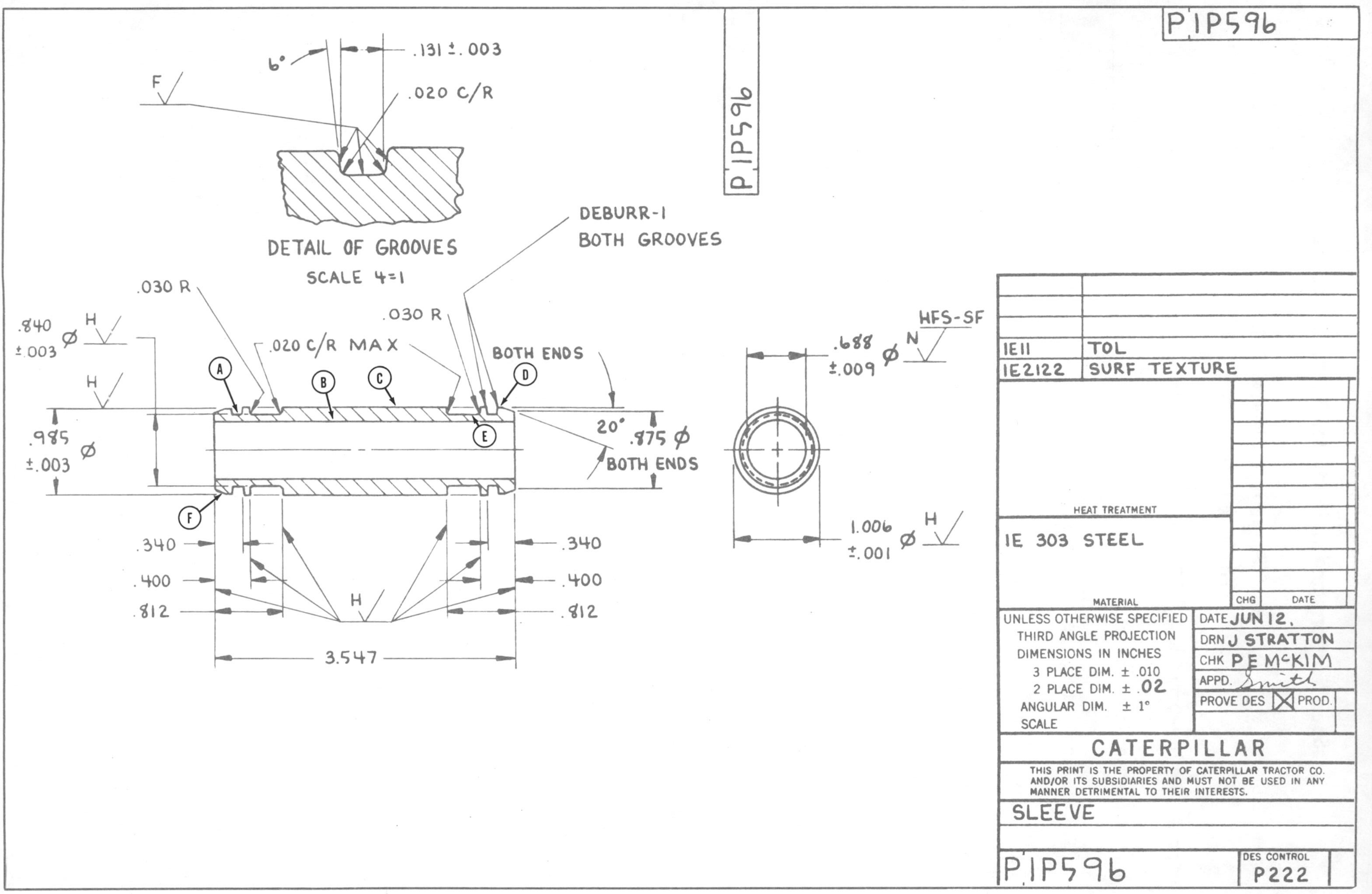

Fig. 19-BPR-2. Industry blueprint. (*Caterpillar Tractor Co.*)

Blueprint Reading Activity 19-BPR-2
CALLOUTS FOR MACHINE PROCESSES

Refer to the blueprint in Fig. 19-BPR-2, and answer the following questions:

1. What is the name of the part? 1. ______

2. Give the print number. 2. ______

3. Are there general notes on the drawing? 3. ______

4. Have any changes been made to this drawing? 4. ______

5. What is the overall size of the part? 5. ______

6. What type of section is the front view? 6. ______

7. The detail view would be called what type section? 7. ______

8. What material is to be used in making the part? 8. ______

9. Give the diameter at E as a tolerance limit dimension. 9. ______

10. What surface texture requirement is specified at E? 10. ______

11. Give the diameter at: A, B, C and D. 11. A ______ B ______ C ______ D ______

12. What surface texture is called out for B? 12. ______

13. Give the chamfer dimensions at F. 13. ______

14. What is the tolerance on the 20° chamfer? 14. ______

Industry photo. Chordal tooth thickness measurement after rough cutting on a Gleason Generator – Speed Blade Method. (Western Gear Corp.)

Unit 20
Tolerances of Position and Form

Advancing technology in industry, along with the subcontracting of parts to other industries, have brought the need for more preciseness (exactness) in the designing and reproducing of machined parts. To accomplish this preciseness, industry is making greater use of close tolerances to control the position (location) and form (shapes) of machined parts.

You should become familiar with the essential elements of the system as discussed in this Unit so you can read and interpret blueprints which use notes and/or symbols used in form and positional tolerancing.

Definition of Terms

GEOMETRIC TOLERANCING controls the function or relationship of part features by indicating the permitted variation from perfect form or true position.

POSITIONAL TOLERANCING is the permitted variation of a feature from the exact or true position indicated on the drawing.

FORM TOLERANCING is the permitted variation of a feature from the perfect form indicated on the drawing.

PROFILE TOLERANCE is the specified variation in the profile of a surface or line from the perfect profile indicated on the drawing.

FEATURE refers to a portion of a part such as a diameter, hole, keyway, flat surface or thread.

DATUM is a point, line, surface or plane used as the location from which form or positional tolerances are checked.

BASIC DIMENSION is an exact dimension with a zero tolerance which describes size, form or location of a feature. Tolerances may be applied to the basic dimension or given in notes.

TRUE POSITION is the basic or theoretically exact position of a feature.

MAXIMUM MATERIAL CONDITION (MMC) exists when a feature contains the maximum amount of material, that is: minimum hole diameter and maximum shaft diameter. When MMC is not specified on a drawing, true position and related datum references apply at MMC.

REGARDLESS OF FEATURE SIZE (RFS) means that tolerances of position or form must be met irrespective of where the feature lies within its size tolerance. When RFS is not specified on a drawing; angularity, parallelism, perpendicularity, concentricity and symmetry tolerances, including related datum references, apply RFS.

Symbols for Dimensioning and Tolerancing

Symbols included on drawings of parts which are to be machined using true position dimensioning and tolerancing are shown in Fig. 20-1. Differences in symbolization for positional and form tolerances in the International, American, British and Canadian standards are shown in the Appendix on page 332.

GEOMETRIC CHARACTERISTIC SYMBOLS

		CHARACTERISTIC	SYMBOL	NOTES
INDIVIDUAL FEATURES	FORM TOLERANCES	STRAIGHTNESS	⏤	1
		FLATNESS	⏥	1
		ROUNDNESS (CIRCULARITY)	○	
		CYLINDRICITY	⌭	
INDIVIDUAL OR RELATED FEATURES		PROFILE OF A LINE	⌒	2
		PROFILE OF A SURFACE	⌓	2
RELATED FEATURES		ANGULARITY	∠	
		PERPENDICULARITY (SQUARENESS)	⊥	
		PARALLELISM	//	3
	LOCATION TOLERANCES	POSITION	⌖	
		CONCENTRICITY	◎	3,7
		SYMMETRY	⌯	5
	RUNOUT TOLERANCES	CIRCULAR	↗	4
		TOTAL	↗	4,6

1) The symbol ∼ formerly denoted flatness.
The symbol ⌒ or — formerly denoted flatness and straightness.
2) Considered "related" features where datums are specified.
3) The symbol || and ◉ formerly denoted parallelism and concentricity, respectively.
4) The symbol ↗ without the qualifier "CIRCULAR" formerly denoted total runout.
5) Where symmetry applies, it is preferred that the position symbol be used.
6) "TOTAL" must be specified under the feature control symbol.
7) Consider the use of position or runout.

Where existing drawings using the above former symbols are continued in use, each former symbol denotes that geometric characteristic which is applicable to the specific type of feature shown.

Fig. 20-1. Geometric symbols for positional and form tolerances. (ANSI–American National Standards Institute)

DATUM features are identified on drawings by a reference letter preceded and followed by a dash and enclosed in a frame, Fig. 20-2.

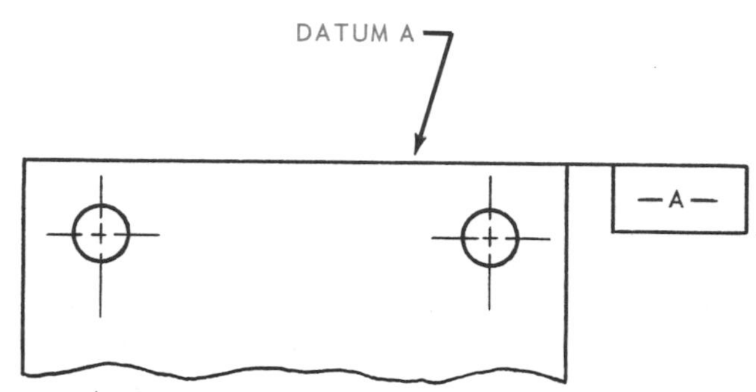

Fig. 20-2. Datum identifying symbol.

BASIC OR TRUE POSITION dimension symbol is indicated by the word BASIC or BSC following or just below the dimension figure, or by enclosing the basic dimension in a rectangular frame, Fig. 20-3.

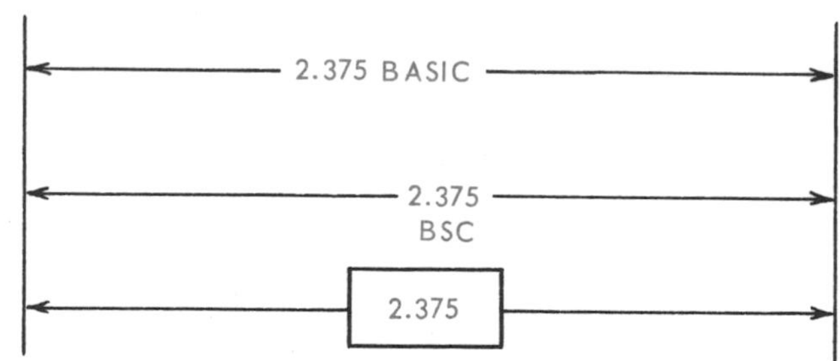

Fig. 20-3. Means of indicating basic or true position dimension.

MAXIMUM MATERIAL CONDITION (MMC) symbol consists of the letter (M) enclosed in a circle. It is used only as a modifier in feature control symbols.

REGARDLESS OF FEATURE SIZE (RFS) symbol consists of the letter S enclosed in a circle. It is also used only as a modifier in feature control symbols.

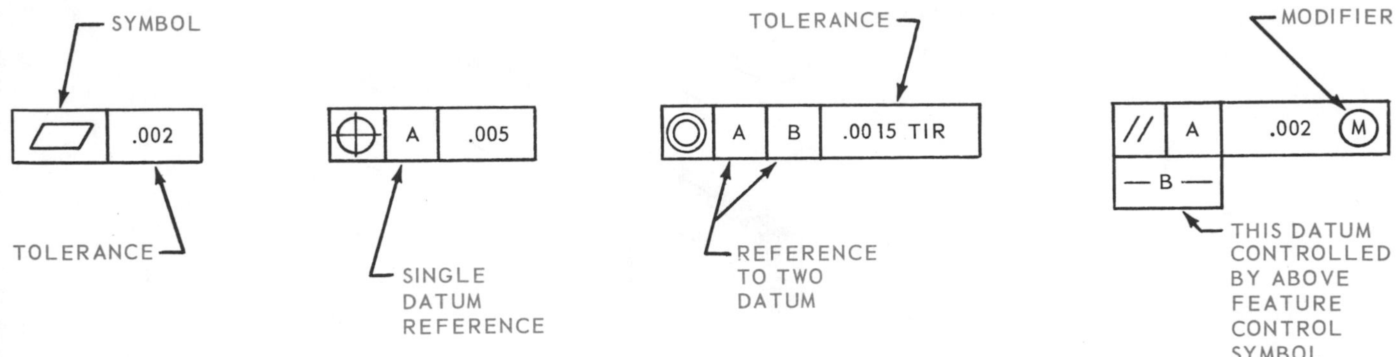

Fig. 20-4. Examples of feature control symbols.[1]

FEATURE CONTROL SYMBOLS is a combined symbol which consists of a frame containing the geometric characteristic symbol followed by the datum reference (if any), by the permissible tolerance, and in some cases by the modifier (M) or (S), Fig. 20-4.

Positional Tolerancing

The tolerances assigned to dimensions which locate the position of one or more features in relation to another feature are known as positional tolerances. These include tolerances assigned to TRUE POSITION (the exact specified location of a feature such as the center distances between holes, bosses or slots) and tolerances assigned to CONCENTRICITY (the location of two or more features having a common axis).

A circular feature located by true position dimensions with a tolerance of plus or minus .010 on the diameter is shown in Fig. 20-5.

The interpretation of the illustration in Fig. 20-5 means that the axis of the feature must lie within a cylindrical tolerance

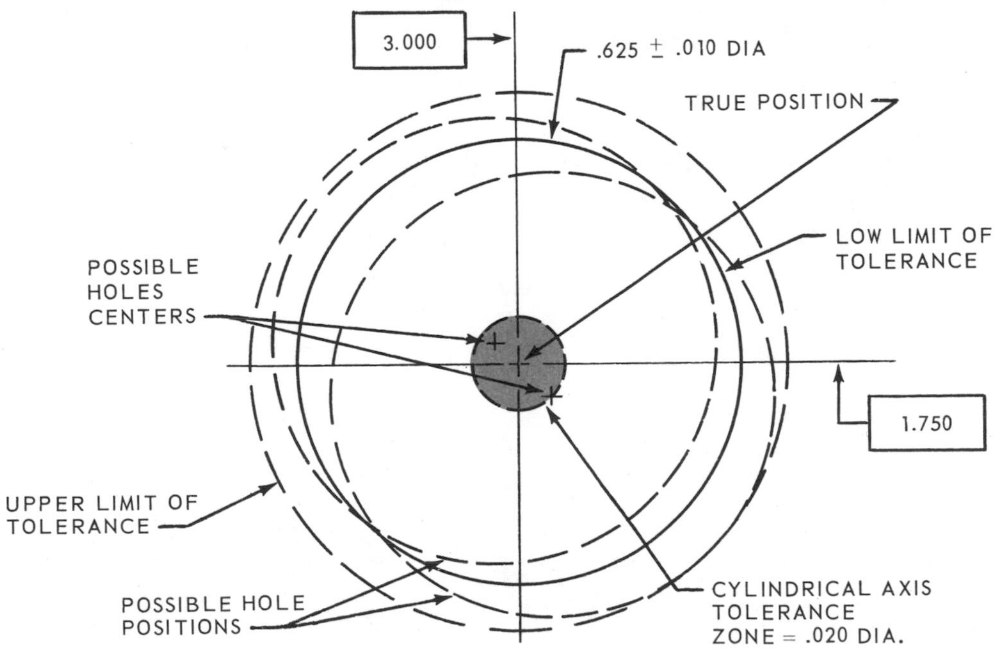

Fig. 20-5. Cylindrical tolerance zone.

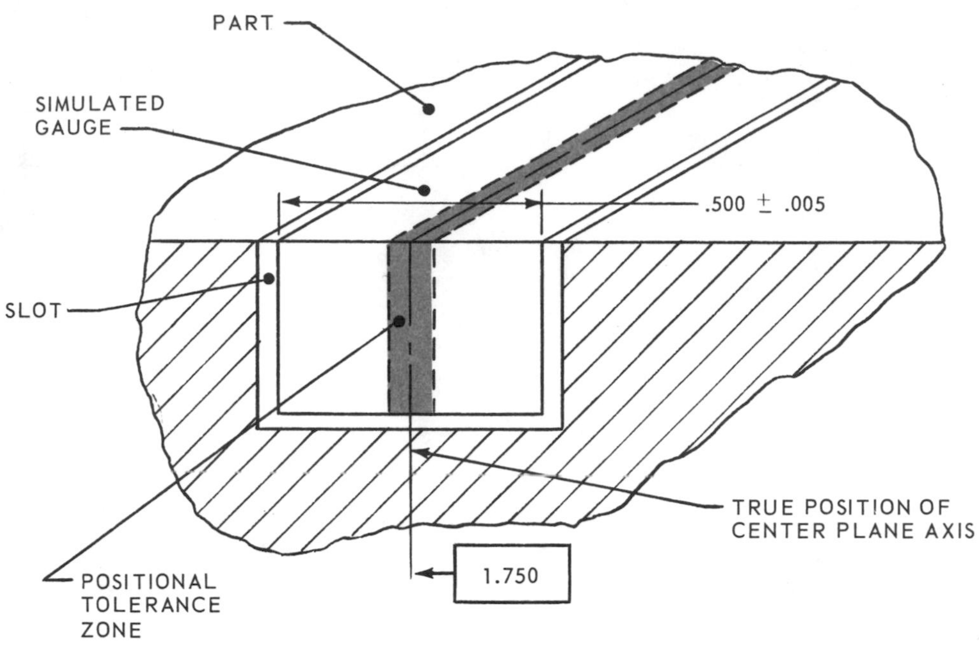

Fig. 20-6. Slot tolerance zone.

zone of .020 diameter whose center is at true position.

A slot located by true position dimensions with a tolerance zone of ±.005 on the center plane is shown in Fig. 20-6.

The interpretation of the illustration in Fig. 20-6 means that the axis of the center plane of the feature must lie within a rectangular tolerance zone of .010 in width whose center plane is at true position.

The notation, symbol and interpretation are given in Fig. 20-7 for true position, concentricity and symmetry.

Form Tolerancing

Tolerances that apply to the control of form for the various geometrical shapes and the control of free-state variations are called form tolerances. Form tolerance specifies a tolerance zone within which the controlled feature, its axis or center plane must lie.

Notes and symbols for form tolerances are shown and interpreted in Fig. 20-8 for each of the standard form tolerances. Study these carefully to understand the meaning of each control symbol.

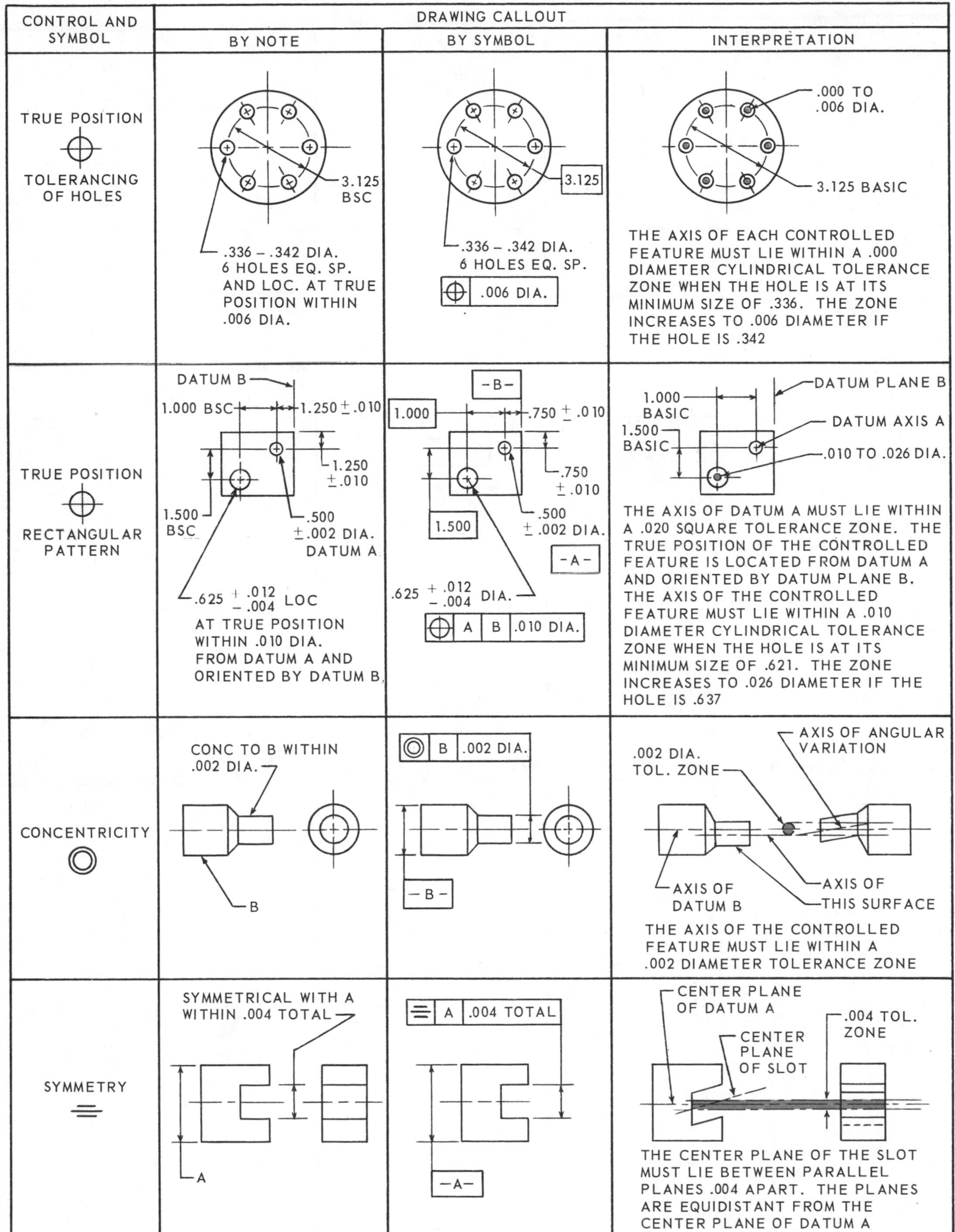

TOLERANCES OF POSITION

CONTROL AND SYMBOL	DRAWING CALLOUT BY NOTE	DRAWING CALLOUT BY SYMBOL	INTERPRETATION
TRUE POSITION TOLERANCING OF HOLES	3.125 BSC .336 – .342 DIA. 6 HOLES EQ. SP. AND LOC. AT TRUE POSITION WITHIN .006 DIA.	3.125 .336 – .342 DIA. 6 HOLES EQ. SP. .006 DIA.	.000 TO .006 DIA. 3.125 BASIC THE AXIS OF EACH CONTROLLED FEATURE MUST LIE WITHIN A .000 DIAMETER CYLINDRICAL TOLERANCE ZONE WHEN THE HOLE IS AT ITS MINIMUM SIZE OF .336. THE ZONE INCREASES TO .006 DIAMETER IF THE HOLE IS .342
TRUE POSITION RECTANGULAR PATTERN	DATUM B 1.000 BSC 1.250 ± .010 1.250 ± .010 1.500 BSC .500 ± .002 DIA. DATUM A .625 $^{+.012}_{-.004}$ LOC AT TRUE POSITION WITHIN .010 DIA. FROM DATUM A AND ORIENTED BY DATUM B	-B- 1.000 .750 ± .010 .750 ± .010 1.500 .500 ± .002 DIA. -A- .625 $^{+.012}_{-.004}$ DIA. A B .010 DIA.	DATUM PLANE B 1.000 BASIC DATUM AXIS A 1.500 BASIC .010 TO .026 DIA. THE AXIS OF DATUM A MUST LIE WITHIN A .020 SQUARE TOLERANCE ZONE. THE TRUE POSITION OF THE CONTROLLED FEATURE IS LOCATED FROM DATUM A AND ORIENTED BY DATUM PLANE B. THE AXIS OF THE CONTROLLED FEATURE MUST LIE WITHIN A .010 DIAMETER CYLINDRICAL TOLERANCE ZONE WHEN THE HOLE IS AT ITS MINIMUM SIZE OF .621. THE ZONE INCREASES TO .026 DIAMETER IF THE HOLE IS .637
CONCENTRICITY	CONC TO B WITHIN .002 DIA. B	B .002 DIA. -B-	.002 DIA. TOL. ZONE AXIS OF ANGULAR VARIATION AXIS OF DATUM B AXIS OF THIS SURFACE THE AXIS OF THE CONTROLLED FEATURE MUST LIE WITHIN A .002 DIAMETER TOLERANCE ZONE
SYMMETRY	SYMMETRICAL WITH A WITHIN .004 TOTAL A	A .004 TOTAL -A-	CENTER PLANE OF DATUM A .004 TOL. ZONE CENTER PLANE OF SLOT THE CENTER PLANE OF THE SLOT MUST LIE BETWEEN PARALLEL PLANES .004 APART. THE PLANES ARE EQUIDISTANT FROM THE CENTER PLANE OF DATUM A

Fig. 20-7. Positional tolerancing.

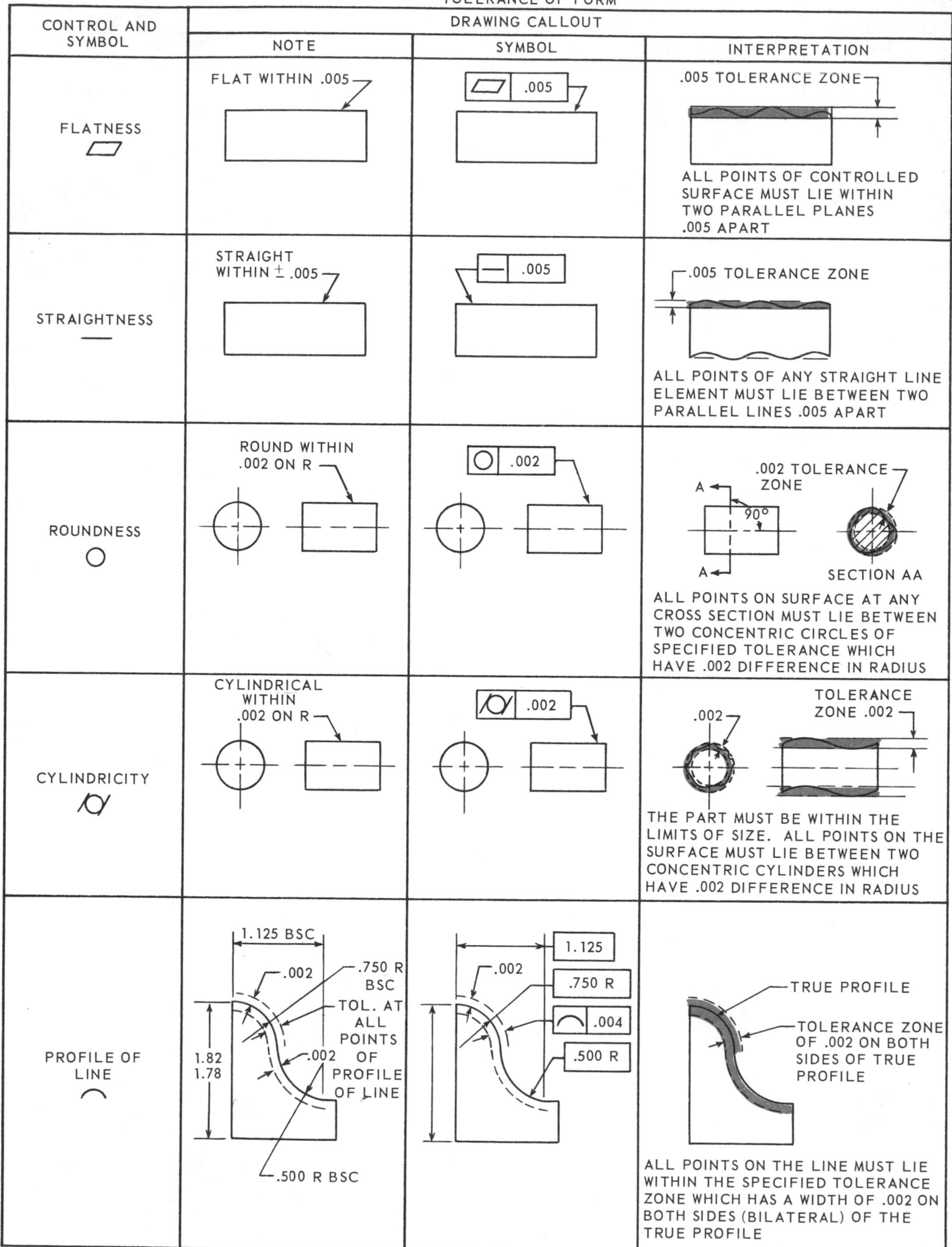

TOLERANCE OF FORM

CONTROL AND SYMBOL	DRAWING CALLOUT		INTERPRETATION
	NOTE	SYMBOL	
FLATNESS ⏥	FLAT WITHIN .005	⏥ .005	.005 TOLERANCE ZONE ALL POINTS OF CONTROLLED SURFACE MUST LIE WITHIN TWO PARALLEL PLANES .005 APART
STRAIGHTNESS —	STRAIGHT WITHIN ± .005	— .005	.005 TOLERANCE ZONE ALL POINTS OF ANY STRAIGHT LINE ELEMENT MUST LIE BETWEEN TWO PARALLEL LINES .005 APART
ROUNDNESS ○	ROUND WITHIN .002 ON R	○ .002	A A 90° .002 TOLERANCE ZONE SECTION AA ALL POINTS ON SURFACE AT ANY CROSS SECTION MUST LIE BETWEEN TWO CONCENTRIC CIRCLES OF SPECIFIED TOLERANCE WHICH HAVE .002 DIFFERENCE IN RADIUS
CYLINDRICITY ⌭	CYLINDRICAL WITHIN .002 ON R	⌭ .002	.002 TOLERANCE ZONE .002 THE PART MUST BE WITHIN THE LIMITS OF SIZE. ALL POINTS ON THE SURFACE MUST LIE BETWEEN TWO CONCENTRIC CYLINDERS WHICH HAVE .002 DIFFERENCE IN RADIUS
PROFILE OF LINE ⌒	1.125 BSC .002 .750 R BSC TOL. AT ALL POINTS OF PROFILE OF LINE 1.82 1.78 .002 .500 R BSC	1.125 .002 .750 R ⌒ .004 .500 R	TRUE PROFILE TOLERANCE ZONE OF .002 ON BOTH SIDES OF TRUE PROFILE ALL POINTS ON THE LINE MUST LIE WITHIN THE SPECIFIED TOLERANCE ZONE WHICH HAS A WIDTH OF .002 ON BOTH SIDES (BILATERAL) OF THE TRUE PROFILE

Fig. 20-8. Form tolerancing.

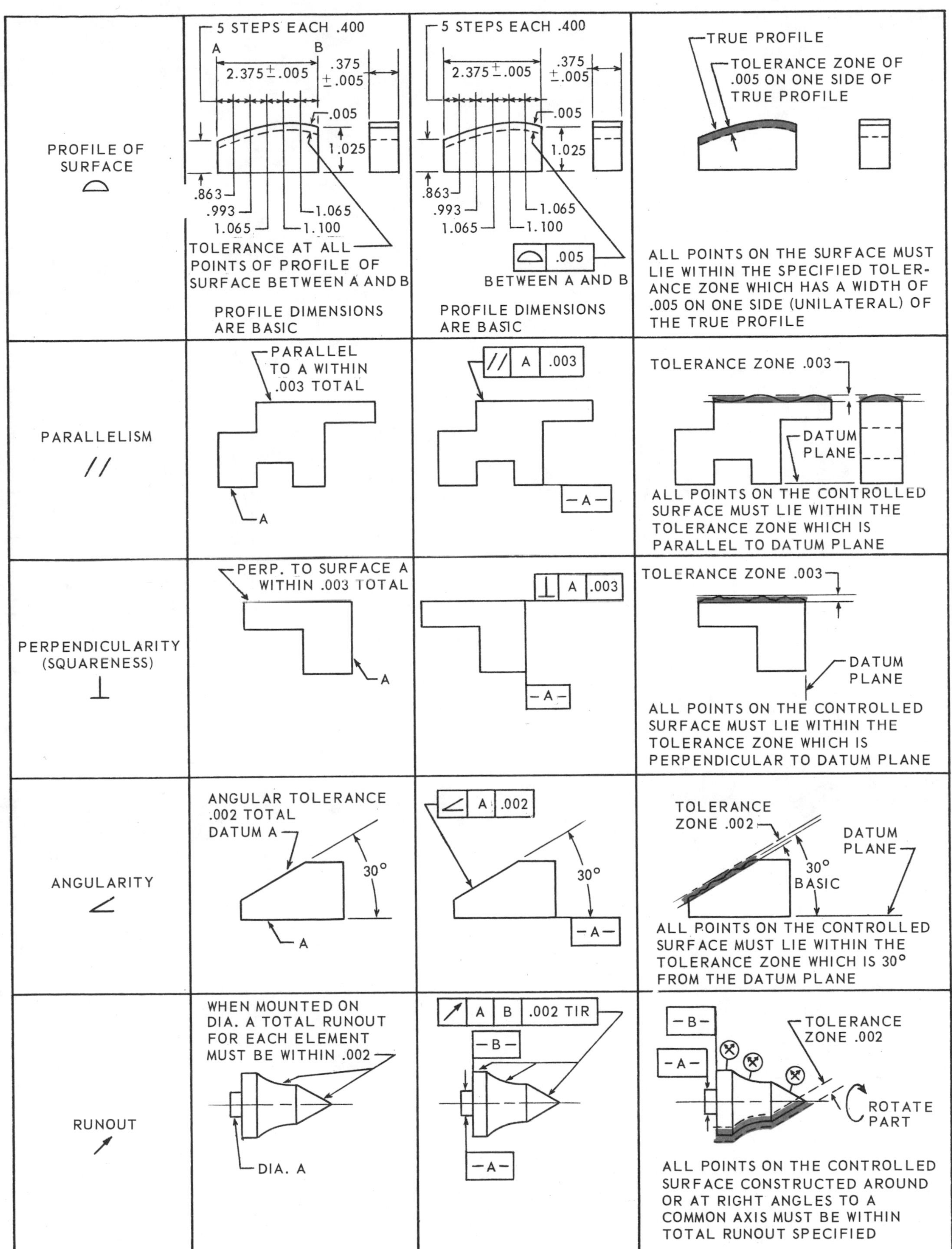

Fig. 20-8. (Continued)

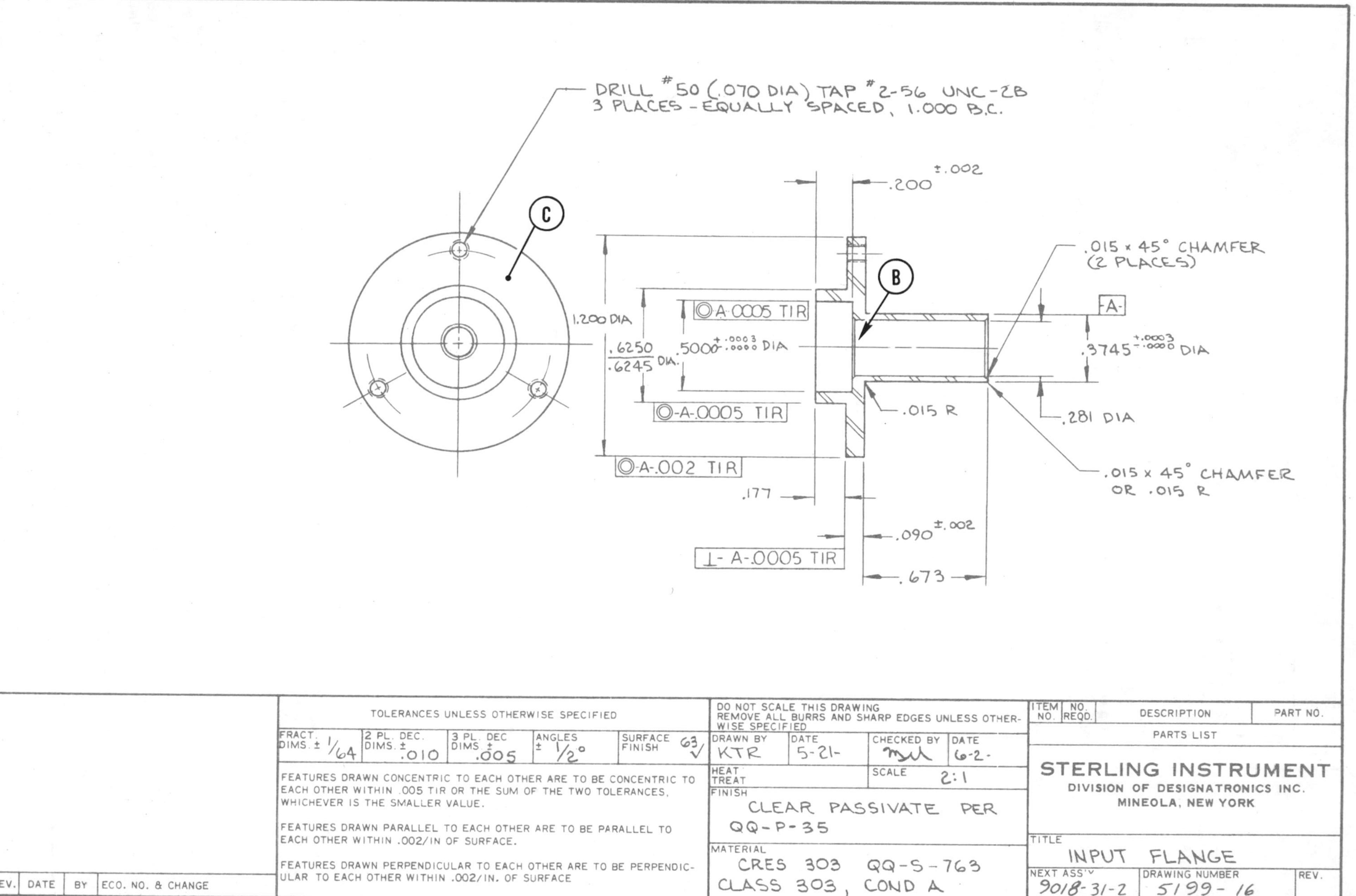

Fig. 20-BPR-1. Industry blueprint. *(Sterling Instrument Div., Designatronics Inc.)*

Blueprint Reading Activity 20-BPR-1
TOLERANCES OF FORM AND POSITION

Refer to the blueprint in Fig. 20-BPR-1 and answer the following questions.

1. What is the name of the part? 1. ______
2. What is the number of the print? 2. ______
3. What is the scale of the original plan? 3. ______
4. What is the material specification? 4. ______
5. Does the print show the drawing has been revised? 5. ______
6. What machine process is called out at B? 6. ______
7. Some holes are to be drilled and tapped in surface C. Interpret the callout. 7. ______
8. What surface texture is designated? 8. ______
9. What relationship is called for between surface C and Datum A? 9. ______
10. What relationship is called out between the major diameter of 1.200 and Datum A? 10. ______
11. Express the tolerance as limits for the .3745 hole. 11. ______
12. How are sharp edges and burrs to be treated? 12. ______

For additional Blueprint Activities for Unit 20 see pages 266 and 269.

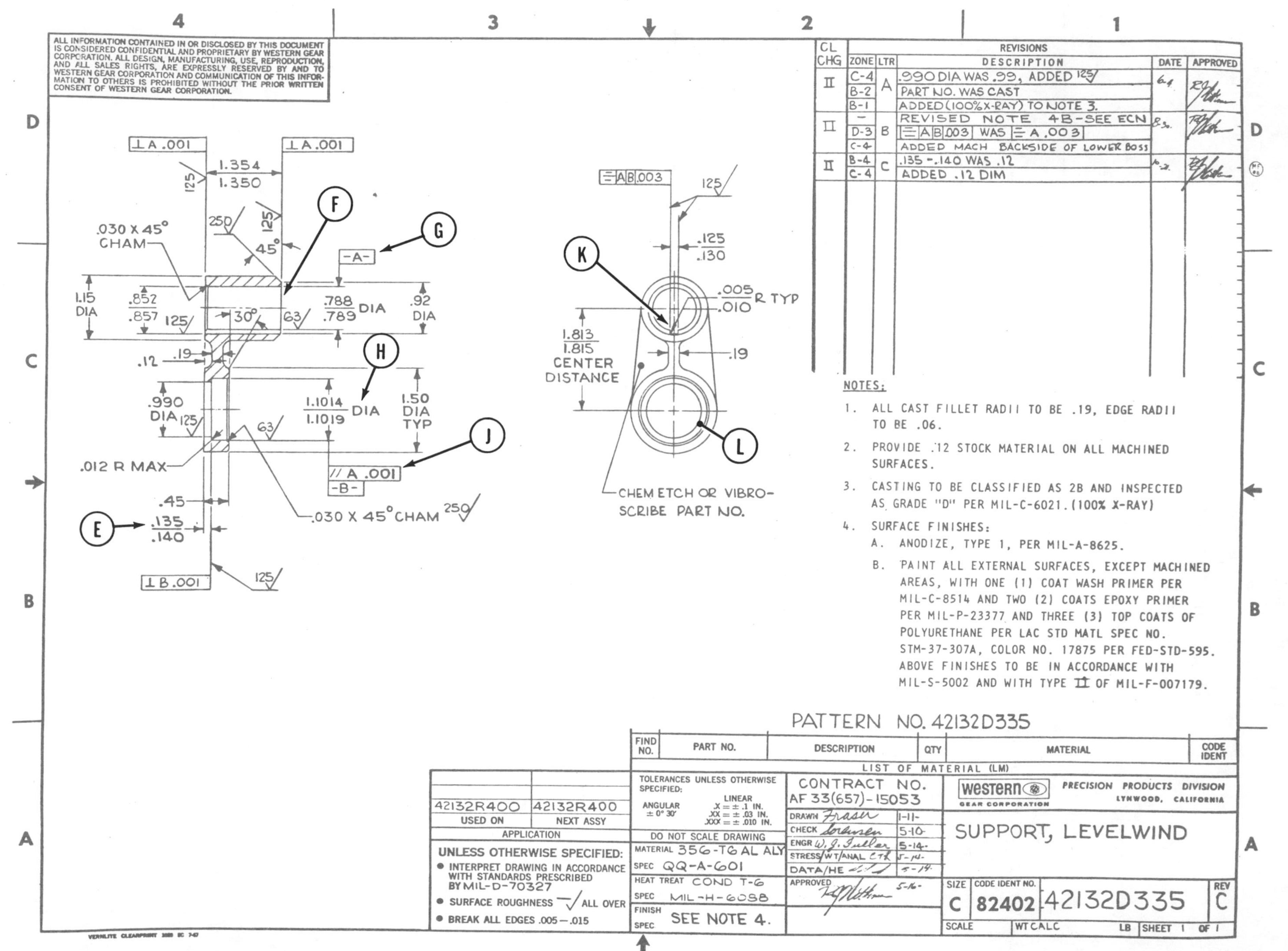

Fig. 20 BPR- Ind try b epri (W tern ar C .)

Blueprint Reading Activity 20-BPR-2
TOLERANCES OF FORM AND POSITION

Refer to the blueprint in Fig. 20-BPR-2 and answer the following questions.

1. What is the name of the part? 1. ____________

2. Give the print number. 2. ____________

3. Dimension E was what value before being changed? 3. ____________

4. What relationship is required between the end surface at F and hole G? 4. ____________

5. Give the dimension between the axes of the two holes as a limits dimension. 5. ____________

6. Interpret the feature control symbol at J. 6. ____________

7. What relationship is required between the keyway K and the axis of the hole? 7. ____________

8. Interpret the feature control symbol that conditions the relationship between surface L and diameter H. 8. ____________

9. What surface texture is called for at diameter H? At surface L? 9. H ____________ L ____________

10. What surface texture is required for the sides of the keyway? 10. ____________

Fig. 21-1. Roll checking a spiral bevel gear set.
(Western Gear Corp.)

Unit 21 Gears, Splines and Serrations

The purpose of this Unit is to acquaint you with gear terminology and how gears are represented and specified on blueprints.

There are many types and sizes of gears used in industry today, see Fig. 21-1. In this Unit we will consider three basic types of gears - - spur, bevel and worm.

Spur Gears

The spur gear is the most commonly used gear and resembles a wheel with a number of equally spaced teeth cut parallel to the axis, Fig. 21-2. Spur gears are used for drives on mechanisms such as machine lathes and mills where the axes of the gears are parallel and the gears are in the same plane with each other.

Fig. 21-2. Spur gears. (AiResearch Mfg. Co.)

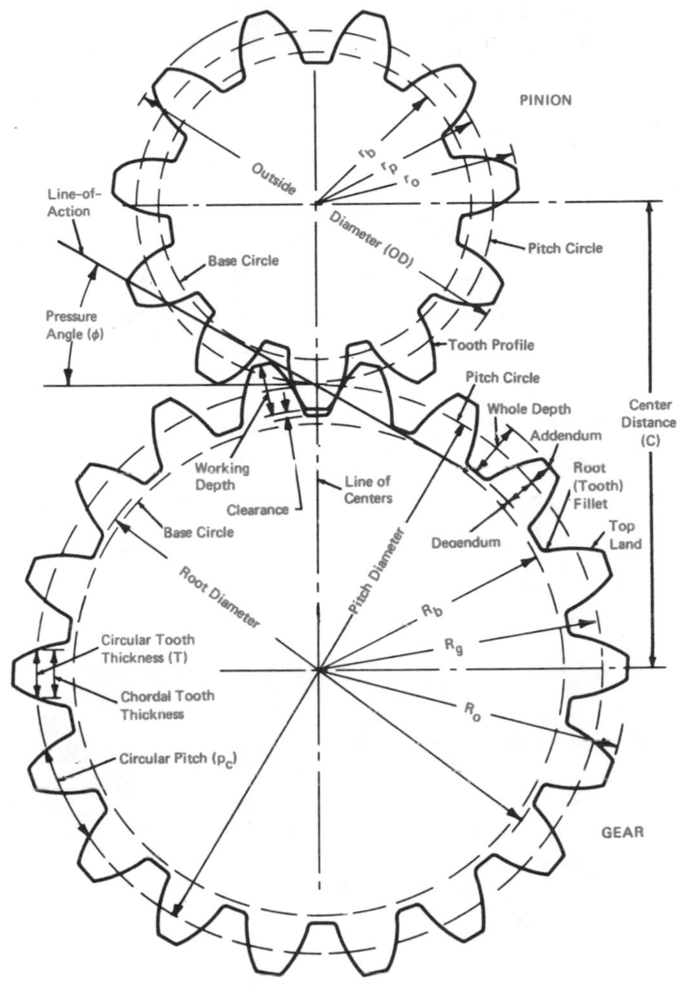

Fig. 21-3. Spur gear nomenclature.

Spur Gear Terminology

DIAMETRAL PITCH is the number of teeth in a gear per inch of pitch diameter - - that is, a gear having 48 teeth and a pitch diameter of 3 inches has a diametral pitch of 16. The diametral pitch would be indicated as:

16 DIAMETRAL PITCH
OR
DIAMETRAL PITCH 16

PITCH CIRCLE is an imaginary circle (located approximately half the distance from the roots and tops of the gear teeth) which is tangent with the pitch circle of a mating gear, Fig. 21-3.

PITCH DIAMETER is the diameter of the pitch circle.

ADDENDUM is the radial distance between the pitch circle and the top of the tooth.

DEDENDUM is the radial distance between the pitch circle and the bottom of the tooth.

ADDENDUM CIRCLE DIAMETER is equal to the pitch circle diameter plus twice the addendum (same as outside diameter).

DEDENDUM CIRCLE DIAMETER is equal to the pitch circle minus twice the dedendum (same as root diameter).

CLEARANCE is the radial distance between the top of a tooth and the bottom of the tooth space of a mating gear.

CIRCULAR PITCH is the length of the arc along the pitch circle between the center of one tooth to the center of the next.

TOOTH FACE is the curved surface of a tooth that lies beyond the pitch circle.

TOOTH FLANK is the curved surface of a tooth that lies inside the pitch circle.

TOOTH SPACE is the distance at the pitch circle between two adjacent teeth.

CIRCULAR THICKNESS is the length of the arc along the pitch circle between the two sides of the tooth.

CHORDAL ADDENDUM is the distance from the top of the tooth to the chord subtending the circular thickness arc. It is the height dimension used in setting gear-tooth calipers to measure tooth thickness, Fig. 21-4.

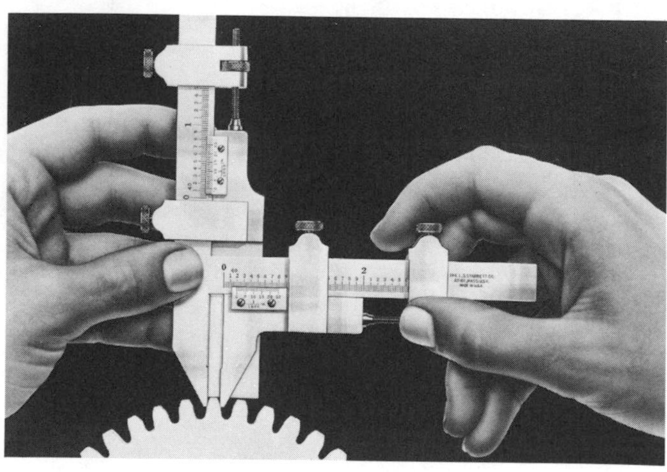

Fig. 21-4. Measuring tooth thickness with a gear Vernier Caliper. (L. S. Starrett Co.)

CHORDAL THICKNESS is the length of the chord along the pitch circle between the two sides of the tooth (the distance measured when a gear-tooth caliper is used to measure the tooth thickness at the pitch circle), Fig. 21-4.

WORKING DEPTH of two mating gears is the sum of their addendums.

WHOLE DEPTH is the total depth of a tooth space (addendum plus dedendum). Whole depth also equals the working space plus clearance.

PRESSURE ANGLE is the angle between the tooth profile and a radial line at their intersection on the pitch circle.

BACKLASH is the amount by which the width of a tooth space exceeds the thickness of the engaging tooth on the pitch circles.

INTERNAL DIAMETER is the diameter of a circle coinciding with the tops of the teeth of an internal gear.

OUTSIDE DIAMETER is the diameter of a circle coinciding with the tops of the teeth of an external gear (same as addendum circle).

RACK is a gear with the teeth spaced along a straight line.

CENTER DISTANCE is the distance between the axes of mating gears.

BASE CIRCLE is the diameter from which an involute tooth curve is generated or developed.

PIN DIAMETER is the diameter of the measuring pin or ball.

PIN MEASUREMENT is the dimension of the measurement over pins for an external gear; between pins for internal gear.

Spur Gear Representation and Specification

Most industries, rather than draw the tooth profile in gears, follow the practice of representing spur gears by showing the addendum circle (outside diameter) and the dedendum circle (root diameter) as a phantom line or a long dash line. A center line is used to show the pitch diameter (see blueprint page 174). Some prints will show a few teeth in the circle for clarity (see blueprint page 271).

Specifications necessary for the machining of the gear are usually given in a data block located on the drawing. This practice eliminates the possible error when the figuring is done in the shop. Occasionally it may be necessary to figure gear data and when the diametral pitch and number of teeth are known, most remaining data may be figured with the aid of one of the following formulae:

SYMBOLS		FORMULAE
Pitch Diameter	$= D$	$D = \frac{N P_c}{\pi}$
Circular Pitch	$= P_c$	$P_c = \frac{\pi D}{N}$
Diametral Pitch	$= P_d$	$P_d = \frac{N}{D}$
Outside Diameter	$= D_o$	$D_o = \frac{N + 2}{P_d}$
Number of Teeth	$= N$	

Other formulae may be found in a machinists' text or handbook. The American Gear Manufacturers Association (AGMA) has established a range of quality classes from 3 (crudest) to 15 (most precise) for coarse pitch gears. Fine pitch gears are separately standardized into classes numbered 5 (crudest) through 16 (most precise).

Bevel Gears

The bevel gear is another type of gear that is found in common use in industry. The spur gear, discussed in the previous section, is like a wheel with teeth in its edge. The bevel gear is more like a cone section with teeth on its conical side that transmits motion and power to a mating gear, Fig. 21-5. Most mating bevel gears have their shafts at right angles (90 deg.), Fig. 21-5, but the shaft angle may be other than 90 deg. The geometry and identification of bevel gear parts are given in Fig. 21-6.

Fig. 21-5. Bevel gears. (Western Gear Corp.)

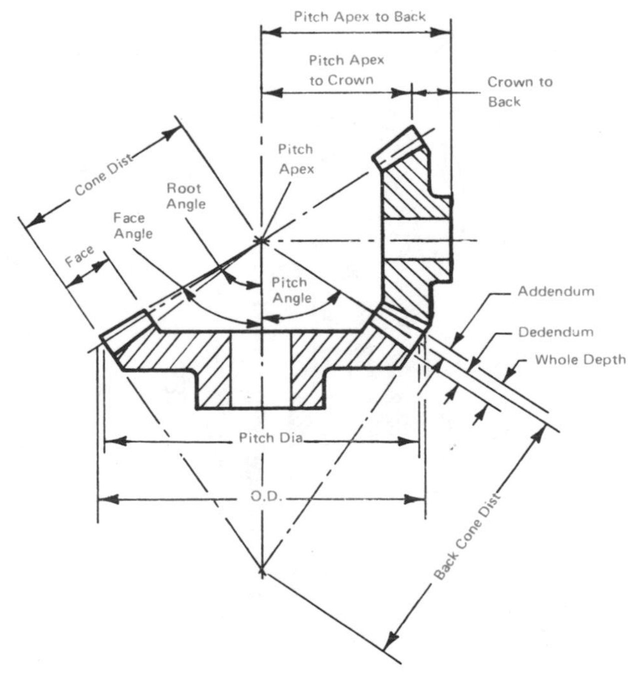

Fig. 21-6. Bevel gear nomenclature. (Sterling Instrument Div. of Designatronics, Inc.)

The specifications necessary for cutting a bevel gear are usually given in the gear data block on the drawing. Formulas for cutting bevel gears vary with the type of gear and are too numerous to list in this blueprint reading text. If a certain formula is needed, it may be found in a machinists' handbook.

Worm Gears

The worm gear is another gear type in common use, Fig. 21-7. It is used for transmitting motion and power usually at 90 deg. angles. The worm gear (wheel) and the worm (screw) are also used for speed reduction since one revolution of a single thread worm is required to advance one tooth on the gear.

Fig. 21-7. Worm gear and worm. (Western Gear Corp.)

To increase the length-of-action, the worm gear is made of a throated (concave) shape to wrap around the worm. On double-enveloping worms, the worm is a "hour-glass" shape.

Fig. 21-8. Worm gear nomenclature.

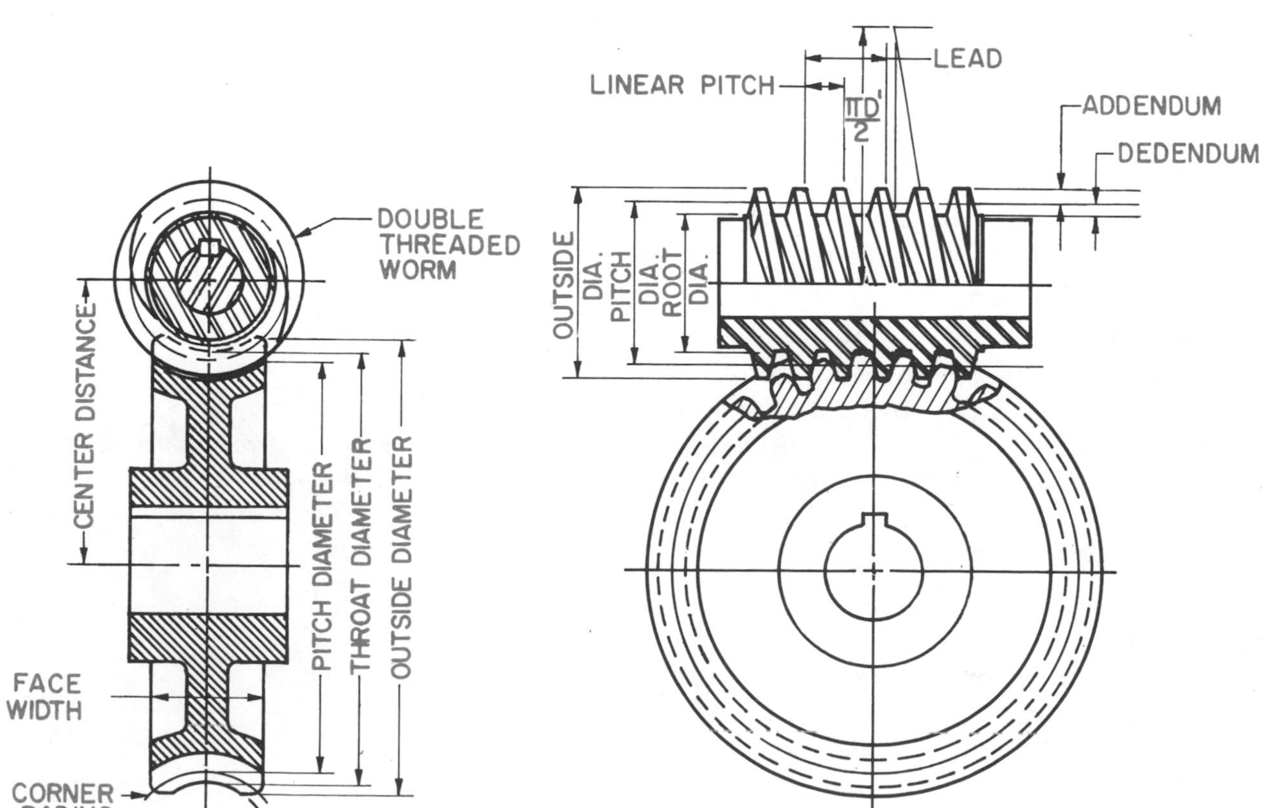

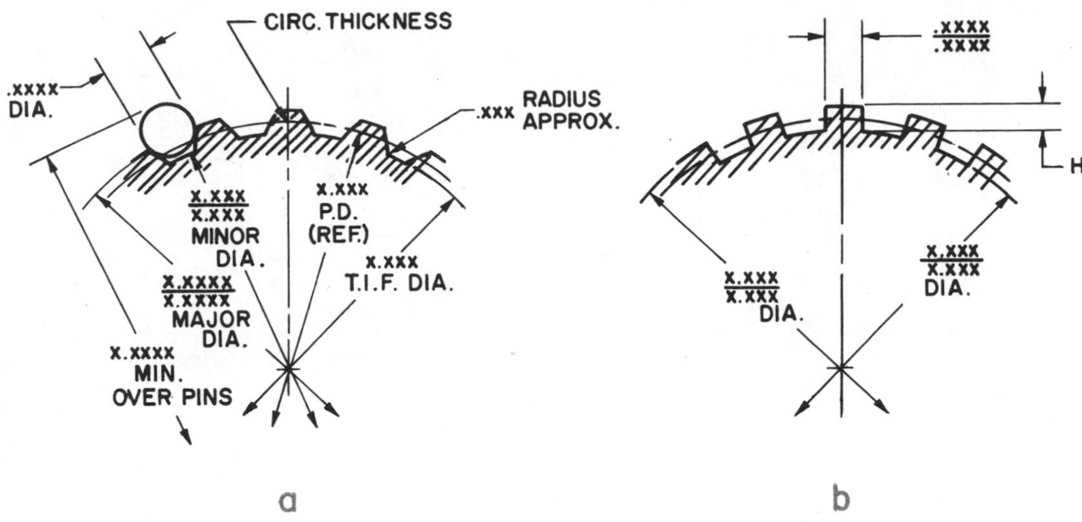

Fig. 21-9. Splines (a) involute and (b) parallel.

Worm Gear Terminology

The terminology for worm gearing, Fig. 21-8, is much the same as for spur and bevel gearing except for a few terms:

LINEAR PITCH is distance from given point on one worm thread to the next. It is equal to circular pitch of worm gear.

LEAD is the distance any one thread advances in one revolution. For a single-threaded worm, the linear pitch and the lead are the same. For a double-threaded (two threads) worm, the lead is twice the linear pitch; for a four-threaded worm the lead is four times the linear pitch, etc.

LEAD ANGLE is the angle the lead makes with a perpendicular line to the worm axis.

THROAT DIAMETER is the diameter of a circle coinciding with the tops of the worm gear teeth at their center plane.

The specifications necessary for cutting a worm gear and worm are usually given in a data block on the drawing. If additional data are needed, check the required formula in a machinists' handbook.

Splines and Serrations

Splines and serrations are like multiple keys on a shaft which prevent rotation between the shaft and its related member.

SPLINES may have parallel sided teeth but the use of splines with involute teeth are increasing in use today, Fig. 21-9. Involute splines are produced with the same technique and equipment as is used for gears. There are two types of fits for splines: (1) on the major diameter and (2) on the sides of the teeth. For each type of fit there are three classes of fits: (1) sliding (class 1 side fit), (2) close (class 2 side fit) and (3) press.

SERRATIONS are primarily intended for parts fitted permanently together and have different tooth proportions and higher pressure angles than splines. They are well adapted for use on thin wall tubing.

Terms for involute splines and serrations are the same as for spur gears. Usually, data for producing splines or serrations are given on the blueprint. When formulae are needed, check a machinists' handbook.

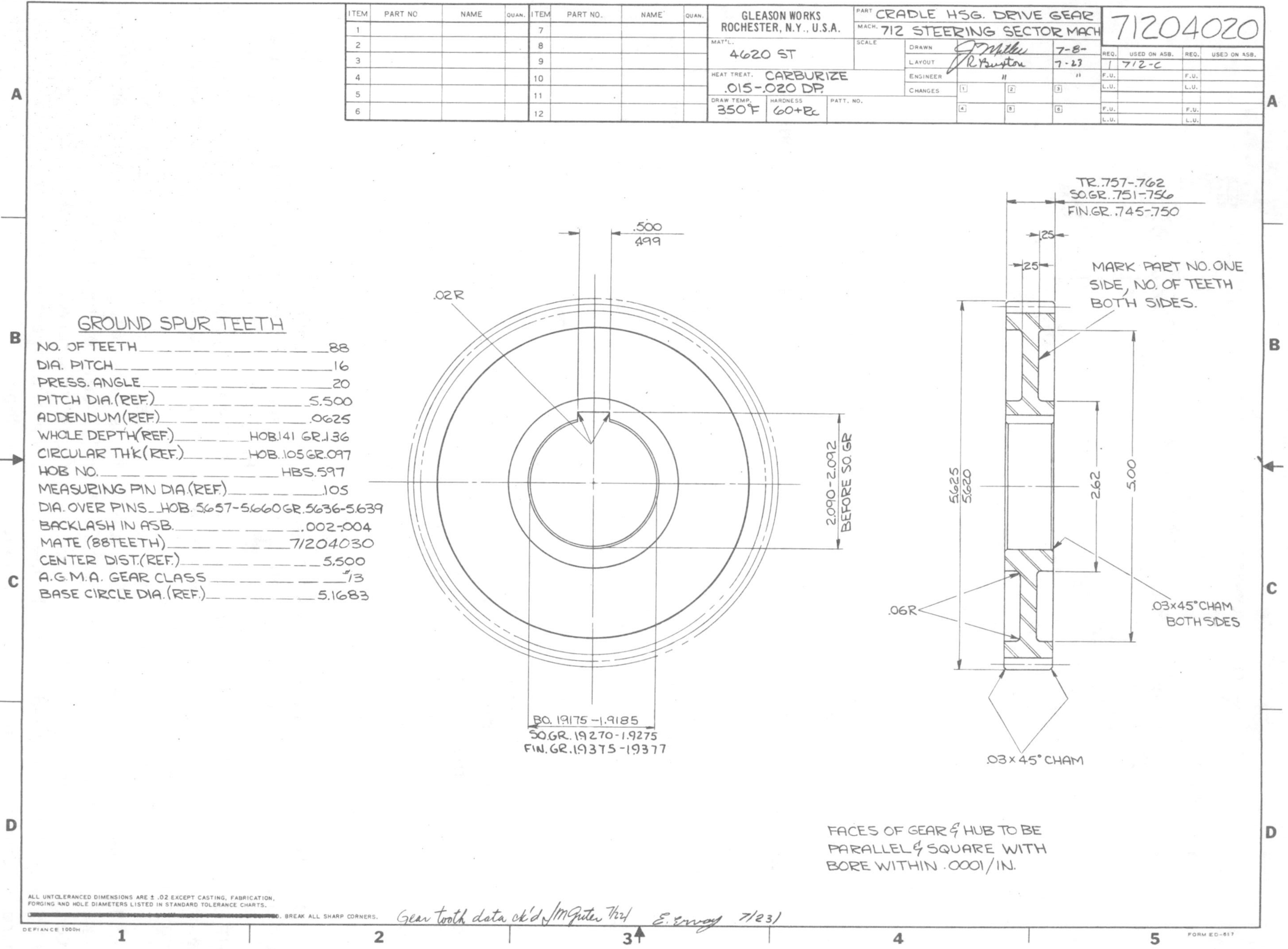

Fig. 21-BPR-1. Industry blueprint. (Gleason Works)

Gears, Splines and Serrations

Blueprint Reading Activity 21-BPR-1
SPUR GEAR

Refer to the blueprint in Fig. 21-BPR-1 and answer the following questions.

1. Give the name of the part. 1. ______
2. What is the number of the print? 2. ______
3. What material is to be used? 3. ______
4. What heat treatment is the part to receive? 4. ______
5. The finished gear must meet what hardness test? 5. ______
6. How many are required and on what assembly is this part to be used? 6. ______
7. How many teeth is the gear to have? 7. ______
8. Give the pressure angle. 8. ______
9. What is the meaning of the abbreviation (REF.) following the addendum? 9. ______
10. What is the number of the mating gear and how many teeth does it have? 10. ______
11. Give the measurement over pins for the gear after hobbing. 11. ______
12. Give the tolerance limits for the hole after the finish grinding operation. 12. ______
13. List the dimensions for the keyway before grinding. 13. ______
14. What relationship must exist between faces of gear and hub with the bore? 14. ______
15. What chamfer is called for on the hole diameter? 15. ______
16. Give the width of the face of the gear after finish grinding. 16. ______

For additional Blueprint Activities for Unit 21 see pages 270, 273, 274, and 277.

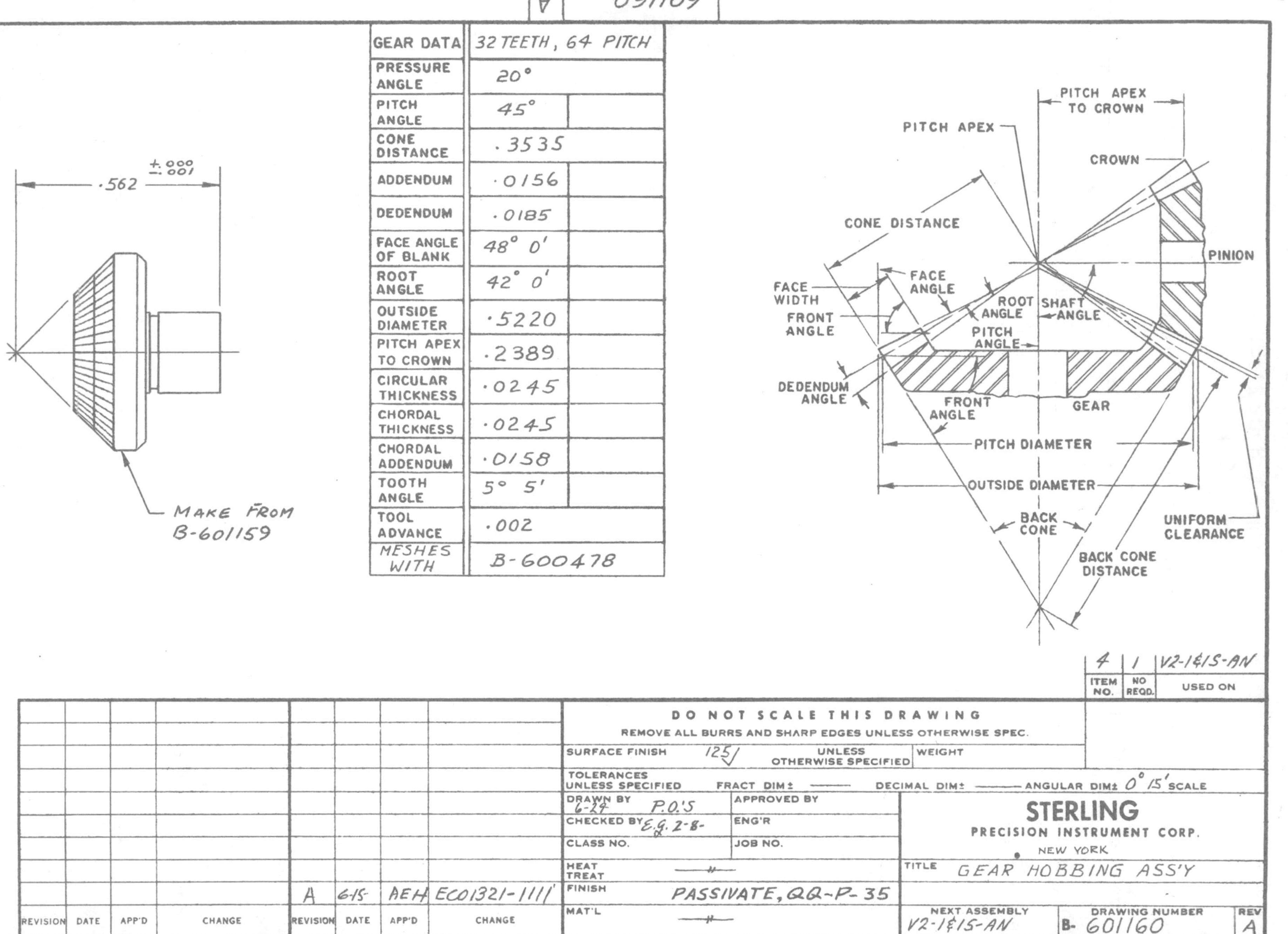

Fig. 21-BPR-2. Industry blueprint. (Sterling Instrument Div., Designatronics Inc.)

Blueprint Reading Activity 21-BPR-2
BEVEL GEAR

Refer to the blueprint in Fig. 21-BPR-2 and answer the followings questions.

1. What is the name of the part and drawing number?	1. ____________________
2. Give the number of the part from which the gear is to be cut.	2. ____________________
3. This print indicates a change order. How do you know the print has been revised to reflect the change?	3. ____________________
4. What surface texture is specified?	4. ____________________
5. What finish is specified?	5. ____________________
6. How many are required for the next assembly? Give assembly number.	6. ____________________
7. How many teeth is the gear to have?	7. ____________________
8. What is the size of the pressure angle?	8. ____________________
9. Give the circular thickness of the tooth.	9. ____________________
10. Give the pitch angle of the gear.	10. ____________________
11. How much is the tool to be advanced on each cut?	11. ____________________
12. This gear meshes with what gear?	12. ____________________

Worm Wheel Data	
Pitch (Diam)	48
Thread	Four (R.H.)
Lead of Worm	.2618
Pressure ∡	25°
Lead Angle	14°-2'
Whole Depth	.0407
AGMA	Prec 1
Testing Pressure	20 Oz.
Tooth Form	Involute

Section E-E

10° (TYP) ⊥-A-.001 .240 1/32 Saw Slot +.000 -.001 P.D. T.D. +.000 -.002 O.D. 63/ "B" DIA "C" DIA "A" DIA -A- .250 ⊙-A-.002 T.I.R 3/16 ⊥-A-.001 T.I.R ⊥-A-.001 T.I.R. .460

Worm Wheel Data (Mates with W11)

Part No	No of Teeth	Pitch Dia	Throat Dia	O.D.	"A" Dia	"B" Dia	"C" Dia
W5-1S	30	.6250	.6624	.6832	+.0005 -.0000 .1248 Dia	+.000 -.003 .1875 Dia	±.005 .250 Dia
-2S	40	.8333	.8707	.8915			
-3S	50	1.0417	1.0791	1.0999			
-4S	60	1.2500	1.2874	1.3082			
-5S	70	1.4583	1.4957	1.5165			
-6S	80	1.6667	1.7041	1.7249			
-7S	90	1.8750	1.9124	1.9332			
-8S	100	2.0833	2.1208	2.1415			
W5-9S	120	2.5000	2.5374	2.5582			
W5-10S	30	.6250	.6624	.6832	+.0005 -.0000 .1873" Dia	+.000 -.003 .2500 Dia	±.005 .312 Dia
-11S	40	.8333	.8707	.8919			
-12S	50	1.0417	1.0791	1.0999			
-13S	60	1.2500	1.2874	1.3082			
-14S	70	1.4583	1.4957	1.5165			
-15S	80	1.6667	1.7041	1.7249			
-16S	90	1.8750	1.9124	1.9332			
-17S	100	2.0833	2.1208	2.1415			
W5-18S	120	2.5000	2.5374	2.5582			
W5-19-S	30	.6250	.6624	.6832	+.0005 -.0000 .2498 Dia	+.000 -.003 .3125 Dia	±.005 .375 Dia
-20S	40	.8333	.8707	.8919			
-21S	50	1.0417	1.0791	1.0999			
-22S	60	1.2500	1.2874	1.3082			
-23S	70	1.4583	1.4957	1.5165			
-24S	80	1.6667	1.7041	1.7249			
-25S	90	1.8750	1.9124	1.9332			
-26S	100	2.0833	2.1208	2.1415			
-27S	120	2.5000	2.5374	2.5582			
W5-28S	180	3.7500	3.7874	3.8082			

Item No	Symbol	No Req'd	Description	Part or Catalog No	Spec No or Mfg

LIST OF MATERIAL

Revision	Date	App'd	Change
F	12/4/	GUY	1321-2314 O.D. CONC. .002 WAS .0005
E	1/11/	NU	ECO1321-2119
D	10/4	AEH	ECO1321-959
C	9/18	AEH	ECO1321-941

Item No | No Req'd | Used On | Revision | Date | App'd | Change

DO NOT SCALE THIS DRAWING

REMOVE ALL BURRS AND SHARP EDGES UNLESS OTHERWISE SPEC

SURFACE FINISH 125/ UNLESS OTHERWISE SPECIFIED | WEIGHT

TOLERANCES UNLESS SPECIFIED FRACT DIM ± 1/64 DECIMAL DIM ±.005 ANGULAR DIM ± ∽ SCALE NONE

DRAWN BY 9-17- A.E.H | APPROVED BY

CHECKED BY 9-19- A.J.C. | ENG'R

CLASS NO | JOB NO

HEAT TREAT ∽

FINISH ∽

MAT'L BRONZE AS PER QQ-B-637, COMP. 1

STERLING PRECISION CORPORATION INSTRUMENT DIVISION NEW YORK

TITLE WORM WHEEL (CLAMP TYPE) 48 PITCH R.H. 3/16 FACE

NEXT ASSEMBLY — | DRAWING NUMBER B.W5-1S TO W5-28S | REV F

DRAWING NO W5-1S TO W5-28S | REV F

Fig. 21-BPR-3. Industry blueprint. (Sterling Instrument Div., Designatronics Inc.)

Blueprint Reading Activity 21-BPR-3
WORM GEARS

Refer to the blueprint in Fig. 21-BPR-3 and answer the following questions.

1. Give the name and number of the blueprint. 1. ____________________ ____________________

2. What is the material specification? 2. ____________________ ____________________

3. What general surface texture is required? 3. ____________________

For Gear No. W5-10S:

4. How many teeth does the gear have? 4. ____________________

5. What is its: 5. Pitch diameter ____________________
Throat diameter ____________________
Outside diameter ____________________

6. What is the diameter of its center bore? 6. ____________________

7. What surface texture is specified for this hole? 7. ____________________

8. How many threads are on the mating worm? 8. ____________________

9. Give the lead angle. 9. ____________________

10. This worm wheel mates with what worm? 10. ____________________

11. What relationship must exist between datum A and the outside diameter? 11. ____________________ ____________________ ____________________

12. State the relationship which must exist between the end surface of the gear hub and the center bore. 12. ____________________ ____________________ ____________________

PART NO. PT 8584TTI

1/2 -13 UNC 2A TH'D.
P.D. .4485-.4435

8 1/8

1 3/8

6 11/32

1 1/4

31/32

7/32 DIA. - 5/32 DEEP

7/16

.4985 TH'D.
.4876 O.D.

1/16 X .010 RELIEF

32

1/8

2

.1865
.1880

.517
.512

SECTION A-A

A

A

.750
.749 DIA.

23/32 DIA.

1/32 X 45° CHAM.
BOTH ENDS

.6695
.6692 DIA.

.624
.623 DIA.

3/8 DIA.

EXTERNAL INVOLUTE SPLINE DATA		
A. S. A. STANDARD INVOLUTE SPLINE WITH 30° PRESSURE ANGLE CLASS 2 FIT		
NUMBER OF TEETH		9
PITCH		10 / 20
PITCH DIAMETER		.9000
MAJOR DIAMETER		.9760
FORM DIAMETER		.8091
MINOR DIAMETER		.7760
CIR. TOOTH THICKNESS	MAX EFF.	.1571
	MIN ACTUAL	.1544
PIN DIAMETER		.1920
PIN MEASUREMENT		1.1717 - 1.1754
REMARKS:		

UNLESS OTHERWISE SPECIFIED:

TOLERANCES: MICROINCH FINISH ON ALL MACHINED SURFACES 63 MAX. FRACTIONAL DIMENSIONS ± .010". ALL OTHER DIMENSIONS TO LIMITS GIVEN.

V THREADS: COUNTERSINK TAPPED HOLES TO MAJOR DIAMETER 120° INCLUDED ANGLE. CHAMFER MALE THREADED PARTS TO MINOR DIAMETER 120° INCLUDED ANGLE. THREAD LENGTH OR DEPTH DIMENSION INDICATES FULL THREADS.

ANGLES: ± 1/2 DEGREE. COUNTERSINKS: 82°. REMOVE ALL BURRS

SYM.	BY	CHANGE	DATE

SOUTH BEND LATHE — ONE OF THE Amsted INDUSTRIES
SOUTH BEND, IND. U.S.A.

PART OR UNIT	SHAFT HEADSTOCK	
SUPERSEDED NUMBER		DATE
MAT'L.	C 1213 C.D.S.	
D'WN. BY SIGNATURE	Barry L. Stull	DATE 6·5·
C'KD. BY SIGNATURE	J.R. Fultz	DATE 6·10·
SCALE	SHT. NO. 1 OF 1	CODE NO.
PART NO.	PT 8584TTI	B

Fig. 21-BPR-4. Industry blueprint. (South Bend Lathe)

Blueprint Reading Activity 21-BPR-4
SPLINES

Refer to the blueprint in Fig. 21-BPR-4 and answer the following questions.

1. What is the name and number of the part?	1. ______________________
2. What material is specified?	2. ______________________
3. Has a change order been issued against the print?	3. ______________________
4. What surface texture is required?	4. ______________________
5. What type spline is to be machined?	5. ______________________
6. Give the specified pressure angle.	6. ______________________
7. Give the class of fit. Is this a sliding, close or press fit?	7. ______________________
8. How many teeth are to be cut?	8. ______________________
9. What is the major diameter of the spline? Length?	9. Diameter ______________ Length ______________
10. Give the pin diameter and the measurement over pins.	10. Pin Dia. ______________ Measurement over Pins ______________
11. Give the keyway dimensions.	11. ______________________
12. Interpret the thread specification.	12. 1/2 ______________ 13 ______________ UNC ______________ 2A ______________ P.D. ______________

PART 5
READING BLUEPRINTS IN SPECIALIZED AREAS

Unit 22
Reading Numerical Control Documents

Today, products are being designed and blueprints are being prepared specifically for numerically controlled (N/C) machining, Fig. 22-1. N/C programs, written in a manner that electronic devices can "read," are prepared by individuals called PROGRAMMERS (persons who understand the N/C language and have a technical background in machine shop procedures, machine tool capabilities and blueprint reading).

Fig. 22-1. Numerically controlled drill showing tape reader and control panel. (American Standard, Inc.)

N/C Programming Documents

Machinists or technicians who operate N/C equipment must know how to read and interpret N/C documents which accompany the production job, to maintain a high level of performance of numerically controlled equipment and to give helpful information to the programmer for the purpose of improving programmed instructions.

N/C PRODUCTION BLUEPRINTS. The blueprint may or may not be labeled for numerically controlled machining, but it is likely that the dimensioning was done from datum points, edges, surfaces or planes (see blueprint on page 188). Features will be located by datum (fixed or zero point) dimensioning or tabular dimensioning to facilitate the "writing" of the N/C program.

PROGRAM MANUSCRIPT SHEET indicates the name of the part, part number, tape number, operator number, programmer's name, date, set points for location of part on machine table, operator's instructions and a program manuscript (in a numerical language) for each sequence (see the program sheet on page 186). Each sequence number or line is called a BLOCK and controls a specific action of the machine tool.

MACHINE SETUP SHEET details the tool and machine setup and gives the operator instructions for locating the part on the machine table (see the sheet on page 187). The sheet also indicates the total number of tools used and total number of HITS or times the tools are used.

PROGRAM MANUSCRIPT PRINTOUT is a typed copy of the program for easy referral and verification of the punched or magnetic tape program, Fig. 22-2. It may be produced at the same time the tape is prepared or it may be run later from the tape.

```
001  01437  02889  t11
002  02312
003  00312  00375  t13
004  03437
```

Fig. 22-2. Typed copy of N/C program manuscript.

PERFORATED TAPE is used to control the operation of most N/C machines, Fig. 22-3. Small holes are punched in the tape forming the NUMERICAL LANGUAGE that the electronic reader and control panel on the N/C machine can "read and understand."

Reference Point Systems

The two systems of programming N/C machines are the INCREMENTAL and the ABSOLUTE.

INCREMENTAL SYSTEM refers to the preceding point when making the next movement. Each move is described as a distance and direction from the preceding point - - it does not refer to a fixed zero or datum point. The incremental system is adaptable to the POINT-TO-POINT or POSITIONING method of N/C machining (Example: drilling a number of holes) as well as the CONTINUOUS PATH or CONTOURING method (Example: milling an irregular form).

ABSOLUTE SYSTEM refers to a zero point which is fixed or a datum point. Each move is described as a distance and direction from the point of origin or zero point. The absolute system is adaptable only to the point-to-point method of N/C machining. The program sheet on page 186 is an example of the absolute system of programming.

ZERO SET POINTS. An N/C machine is wired either for a FIXED ZERO set point or a FLOATING ZERO set point. In the fixed zero system the machine refers to this point as zero and parts to be machined are located with reference to this point. On the program manuscript sheet for the REAR COVER BRACKET, page 186, the set point X0 is to be 1.000 inch from machine zero. The Y0 is to be 3.000 inches from machine zero.

In the floating zero system, the N/C programmer may establish zero at any

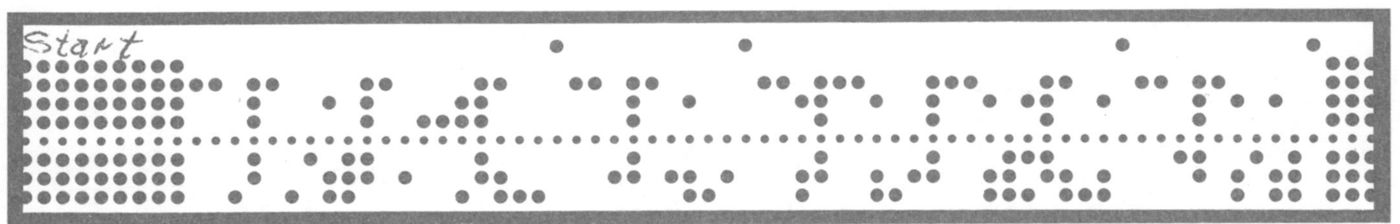

Fig. 22-3. Perforated tape which contains "numerical language" to control machines.

convenient point by loading it into the electronic control system.

Interpreting an N/C Program

Dimensional instructions for an N/C program are given on a two or three dimensional coordinate plane, Fig. 22-4. When the operator is facing a vertical spindle machine, table movement to the left or right is the X direction or axis. Table movement away from the operator or toward the operator is called the Y direction or axis. Vertical movement of the working tool is called the Z direction or axis. The machining of some parts requires the use of only the X and Y coordinates; others require X, Y and Z.

It is easier to understand the direction of movement if you assume the table remains stationary and the tool moves over the work. When the work is located on the table with the datum zero point at the zero point of the machine, movement of the tool along the X axis to the right is in the +X direction and to the left is in the -X direction. Movement of the tool into the work away from the operator is known as +Y. Movement of the tool toward the operator is -Y. Tool movement down into the work is known as -Z, and up from the work is known as +Z. Tool movements are assumed to be in the PLUS direction unless marked MINUS.

Refer to the program sheet on page 186. Sequence number 001 calls for an X dimension of 01.437 and a Y dimension of 02.889 (dimensions are given in thousandths). Station number 11 of the turret punch press is used for this sequence or operation. This station is equipped with a rectangular tool .187 x .375 x 1/32R mounted in the direction of the Y axis (see Turret Setup sheet, page 187). The photo, Fig. 22-5, shows an N/C punch press similar to the one used for this job.

A study of the X and Y dimensions on the blueprint, page 188, indicates that sequence 001 centers the punch in station 11 directly over the upper left-hand rectangular slot B and will punch the slot during the sequence operation.

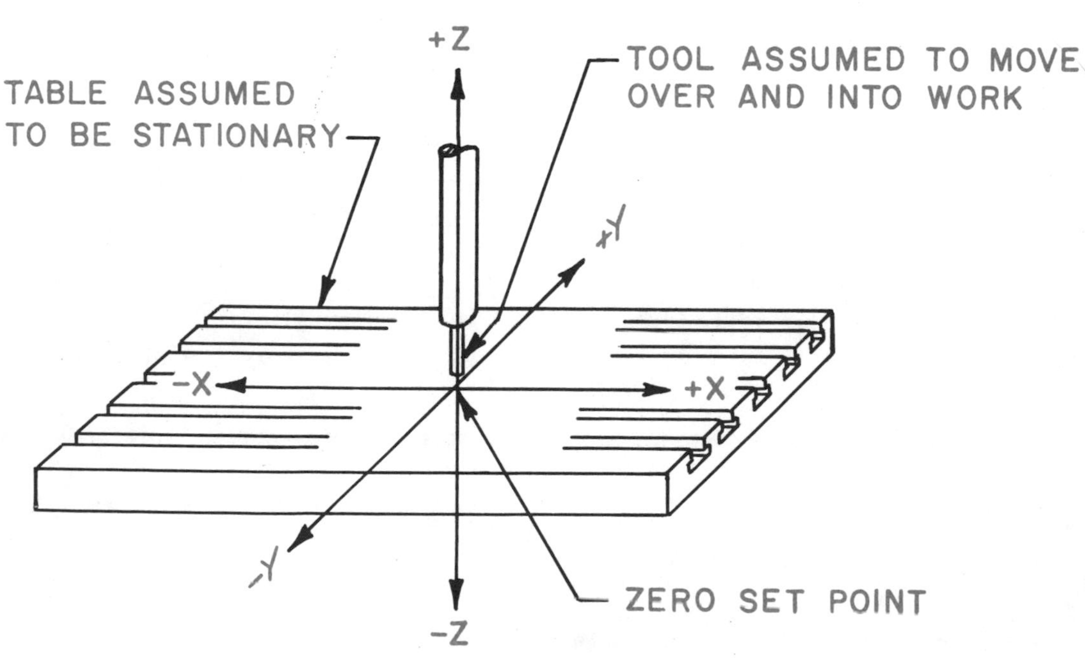

Fig. 22-4. Numerical control coordinates.

Fig. 22-5. A numerically controlled punch press in operation. (Perkin-ElmerCorp.)

Sequence 002 changes the X axis dimension to 02.312 and the Y axis remains unchanged. A study of the blueprint indicates the machine is positioned and will punch the other rectangular slot.

Sequence 003 changes the X axis to 00.312, the Y axis to 00.375 and the station to 13. A study of the blueprint reveals that this sequence will punch the lower left-hand hole A. Notice that the punch at station 13 is a .375 round and that the size of hole A specified in the data block on the blueprint is .375.

Let us now look at sequence 005. This contains a minus X positioning. This means that the -X dimension is centered .250 to the left of the datum. The Y axis is plus and located 1.084 above the datum. Sequence 005 is using station 1 and a study of the machine setup sheet indicates this station is equipped with a rectangular punch .500 x 2.250 mounted in the direction of the Y axis. When this tool is centered at -X .250 and Y 1.084, it is in exact position to shear along side F.

You will have an opportunity in the blueprint reading activities which follow to try your skill at interpreting N/C blueprints and related documents.

WIEDEMATIC PROGRAM SHEET NO. 1 OF 1
G.E. MARK CENTURY CONTROL

PART NAME BRACKET-REAR COVER
PART NO. 219-1850 DET. NO. 1 REV. A
TAPE NO. 1 OPER. NO.
PROG. BY: P. TISANO DATE 6-5-
SET WORKHOLDERS AT: 6 36
66

OPERATOR'S INSTRUCTIONS:

NO.	AUXILIARY FUNCTIONS
T	Turrets Left - increase no's.
T-	Turrets Right - decrease no's
M00	Program Stop
M02	Information Delete
M03	Position - Don't punch
M06	Operator's instructions

	(M)		NO. X	ΔX	X0	No. Y	ΔY	Y0
PATTERN	1	GRID	010	05130	01000	008	03830	03000

PATTERN

XP	YP

M 1 Complete one pattern
M 2 Complete one seq. all patterns
No. X Number of patterns in "x" axis
Δ X Distance Between Patterns in "x" axis
X0 From mach. zero (edge of sheet) to pattern "0"
No. Y Number of patterns in "Y" axis
Δ Y Distance between patterns in "Y" axis
Y0 From Mach. zero (edge of sheet) to pattern "0"

SEQ. NO.	TAB		MINUS	X-DIMEN.	TAB		MINUS	Y-DIMEN.	TAB	TURR. DIR.	STA. NO.	AUX. FUNCTIONS
001		X		01437		Y		02889		T	11	
002		X		02312		Y						
003		X		00312		Y		00375		T	13	
004		X		03437		Y						
005		X	—	00250		Y		01084		T-	01	
006		X		04000		Y						
007		X		00687		Y		02668		T-	20	
008		X				Y		02918				
009		X		00438		Y		02668				
010		X		03062		Y						
011		X				Y		02918				
012		X		03312		Y		02668				
013		X		02625		Y		00480				
014		X		01125		Y						
015		X		01875		Y						
		X				Y						
		X				Y						
		X				Y						

FORM 1-0444-00

Program sheet for Fig. 22-BPR-1.

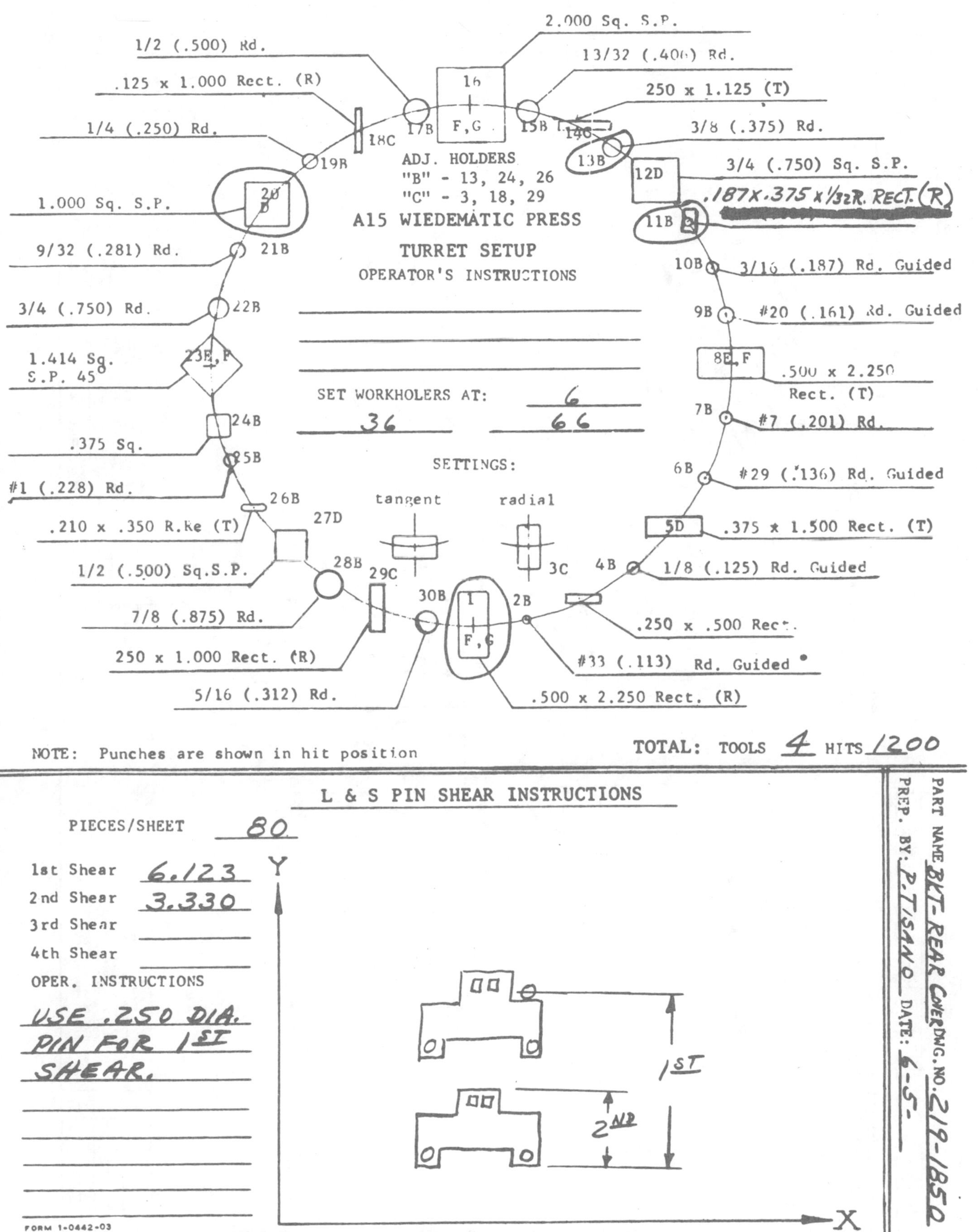

Machine setup sheet for Fig. 22-BPR-1.

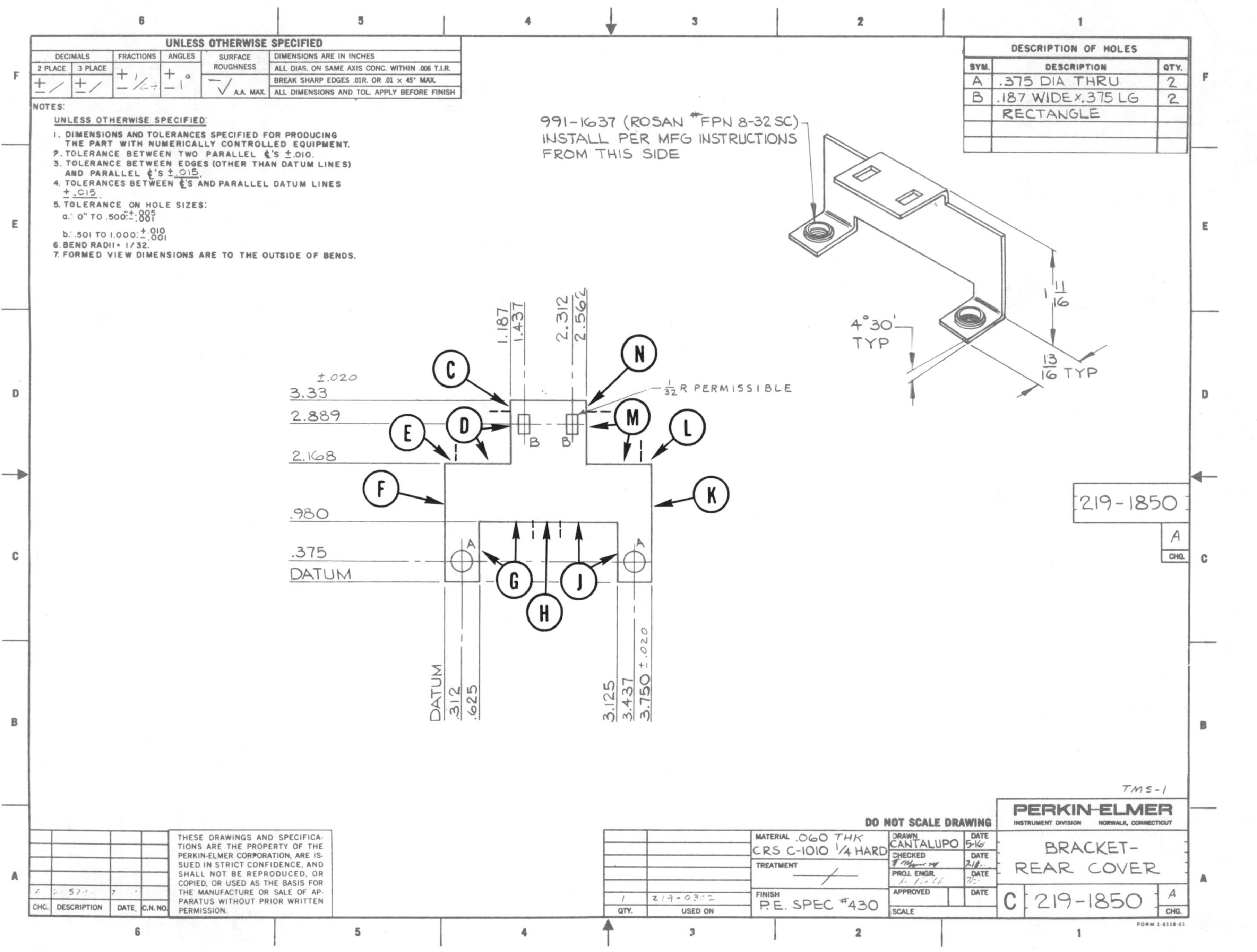

Fig. [illegible]-BPR 1. Industry bluepr[illegible] ([illegible]kin-Elmer Corp.)

Blueprint Reading Activity 22-BPR-1
NUMERICAL CONTROL

Refer to the blueprint in Fig. 22-BPR-1 and related documents on pages 186 and 187 and answer the following questions.

1. What is the name and number of the part? 1. ______________________________
2. What material is specified? 2. ______________________________
3. Give the overall size of the part in the flat. 3. ______________________________
4. Give the size of the holes A. 4. ______________________________
5. Give the dimensions for slot B. 5. ______________________________
6. What tolerance is permitted between the two holes marked A? 6. ______________________________
7. What tolerance is permitted on the dimension 2.889? 7. ______________________________
8. Describe the work being performed during sequence number 4 including the type of tool used. 8. ______________________________
9. Describe the work being performed during sequence numbers 6 through 15. 9. Seq. 6 ____________
   Seq. 7 ____________
   Seq. 8 ____________
   Seq. 9 ____________
   Seq. 10 ____________
   Seq. 11 ____________
   Seq. 12 ____________
   Seq. 13 ____________
   Seq. 14 ____________
   Seq. 15 ____________
10. Give the overall dimensions of the folded part. 10. ______________________________

For additional Blueprint Activities for Unit 22 see pages 278 and 286.

WIEDEMATIC PROGRAM SHEET NO. 1 OF 2
G.E. MARK CENTURY CONTROL

PART NAME BRACKET-FLAME IGNITER

PART NO. 303-1928 DET. NO. 1 REV. A

TAPE NO. 1 OPER. NO. ________

PROG. BY: P. TISANO DATE 10-14-

SET WORKHOLDERS AT: 6 36 66

OPERATOR'S INSTRUCTIONS: ________

NO.	AUXILIARY FUNCTIONS
T	Turrets Left - increase no's.
T-	Turrets Right - decrease no's
M00	Program Stop
M02	Information Delete
M03	Position - Don't punch
M06	Operator's instructions

	(M)		NO. X	ΔX	XO		No. Y	ΔY	YO
PATTERN	1	GRID	005	12500	01000	/	007	04620	00000

PATTERN	
XP	YP

M 1	Complete one pattern
M 2	Complete one seq. all patterns
No. X	Number of patterns in "x" axis
Δ X	Distance Between Patterns in "x" axis
XO	From mach. zero (edge of sheet) to pattern "0"
No. Y	Number of patterns in "Y" axis
Δ Y	Distance between patterns in "Y" axis
YO	From Mach. zero (edge of sheet) to pattern "0"

SEQ. NO.	TAB		MINUS	X-DIMEN.	TAB		MINUS	Y-DIMEN.	TAB	TURR. DIR.	STA. NO.	AUX. FUNCTIONS
001		X		02330		Y		00721		T	03	
002		X		01954		Y		00069				
003		X		02142		Y		00395				
004		X		08095		Y		03157		T	09	
005		X		07657		Y		02562				
006		X		10438		Y		02093				
007		X		01220		Y		01500		T	10	
008		X		02595		Y						
009		X		09220		Y		02332				
010		X				Y		03366				
011		X		11036		Y		02064		T	12	
012		X		11023		Y		00923				
013		X		01220		Y		00562		T	17	
014		X		10507		Y		03275		T	18	
015		X		01938		Y		02407		T	19	
016		X		02438		Y						
017		X		10440		Y		02480				
018		X		10628		Y		01500				

FORM 1-0444-00

Program sheet for Fig. 22-BPR-2.

WIEDEMATIC PROGRAM SHEET

G. E. MARK CENTURY CONTROL

DRAWING NO: 303-1928 REV. A SHEET NO: 2 OF 2 SHEETS

DET. NO. 1 TAPE NO. 1 PROG BY. P. TISANO DATE: 10-14-

SEQ. NO.			MINUS	X-DIMEN.			MINUS	Y-DIMEN.	MINUS	TURR. DIR.	STA. NO.	AUX. FUNCTIONS
019		X		00345		Y		02437		T	20	
020		X		11130		Y		02562				
021		X		06903		Y		03140		T	29	
022		X		07717		Y		03610				
023		X		05782		Y		02970		T	05	
024	TAB	X		02906	TAB	Y			TAB			
025		X		01468		Y						
026		X		04344		Y						
027		X		00438		Y	–	00188				
028		X		01875		Y						
029		X		03097		Y		00812				
030		X		05972		Y						
031		X		08848		Y						
032		X		10130		Y						
033		X		07410		Y						
034		X		04534		Y						
		X				Y						
		X				Y						
		X				Y						
		X				Y						
		X				Y						
		X				Y						
	TAB	X			TAB	Y			TAB			
		X				Y						
		X				Y						
		X				Y						
		X				Y						
		X				Y						
		X				Y						
		X				Y						
		X				Y						
		X				Y						
		X				Y						
		X				Y						
		X				Y						

FORM 1-C443-01

Program sheet for Fig. 22-BPR-2 (Continued).

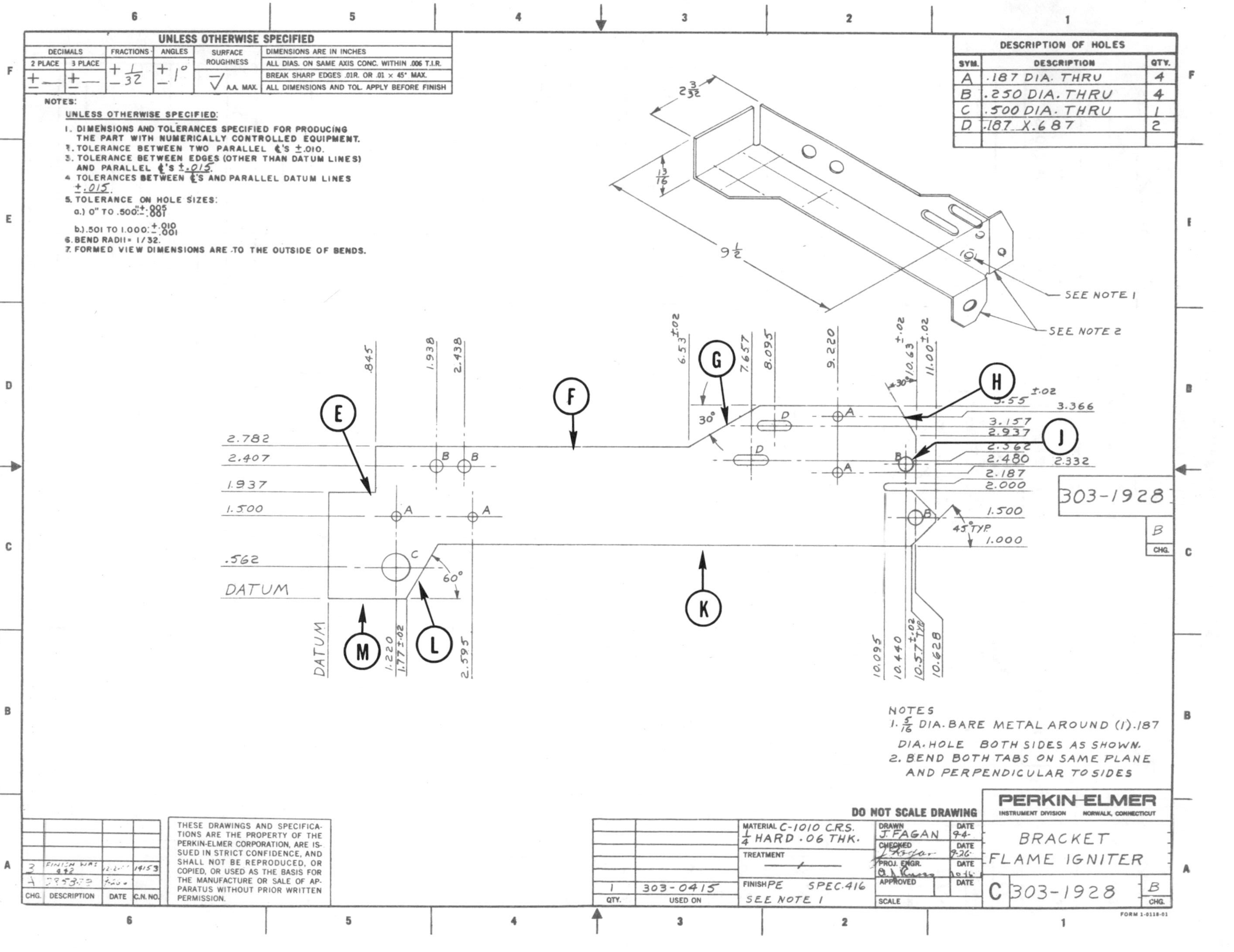

Fig. 22-BPR-2. Industry blueprint. (Perkin–Elmer Corp.)

Blueprint Reading Activity 22-BPR-2
NUMERICAL CONTROL

Refer to the blueprint in Fig. 22-BPR-2 and related documents on pages 190, 191, and 194 and answer the following questions.

1. What is the name of the part and print number? 1. ____________ ____________

2. Give the material specification. 2. ____________

3. What is the overall size of the flat stretchout? 3. ____________

4. What tolerances are specified between holes? 4. ____________

5. What change was made in the latest revision of the print? 5. ____________ ____________

6. On what assembly is this part used and in what quantity? 6. ____________

7. Give the size for the following features: 7. A ____________ B ____________ C ____________ D ____________

8. Interpret sequences 4 through 6. 8. Seq. 4 ____________ Seq. 5 ____________ Seq. 6 ____________

9. Interpret sequences 7 through 10. 9. Seq. 7-10 ____________

10. Interpret sequences 11 and 12. 10. Seq. 11-12 ____________

11. What sequences are used to punch the four holes marked B? 11. ____________

12. The C hole is punched in what sequence? 12. ____________

13. List the sequence(s) used in machining the following features: 13. E ______ F ______ G ______ H ______ J ______ K ______ L ______ M ______

14. At what angle are the end tabs to be bent? 14. ____________

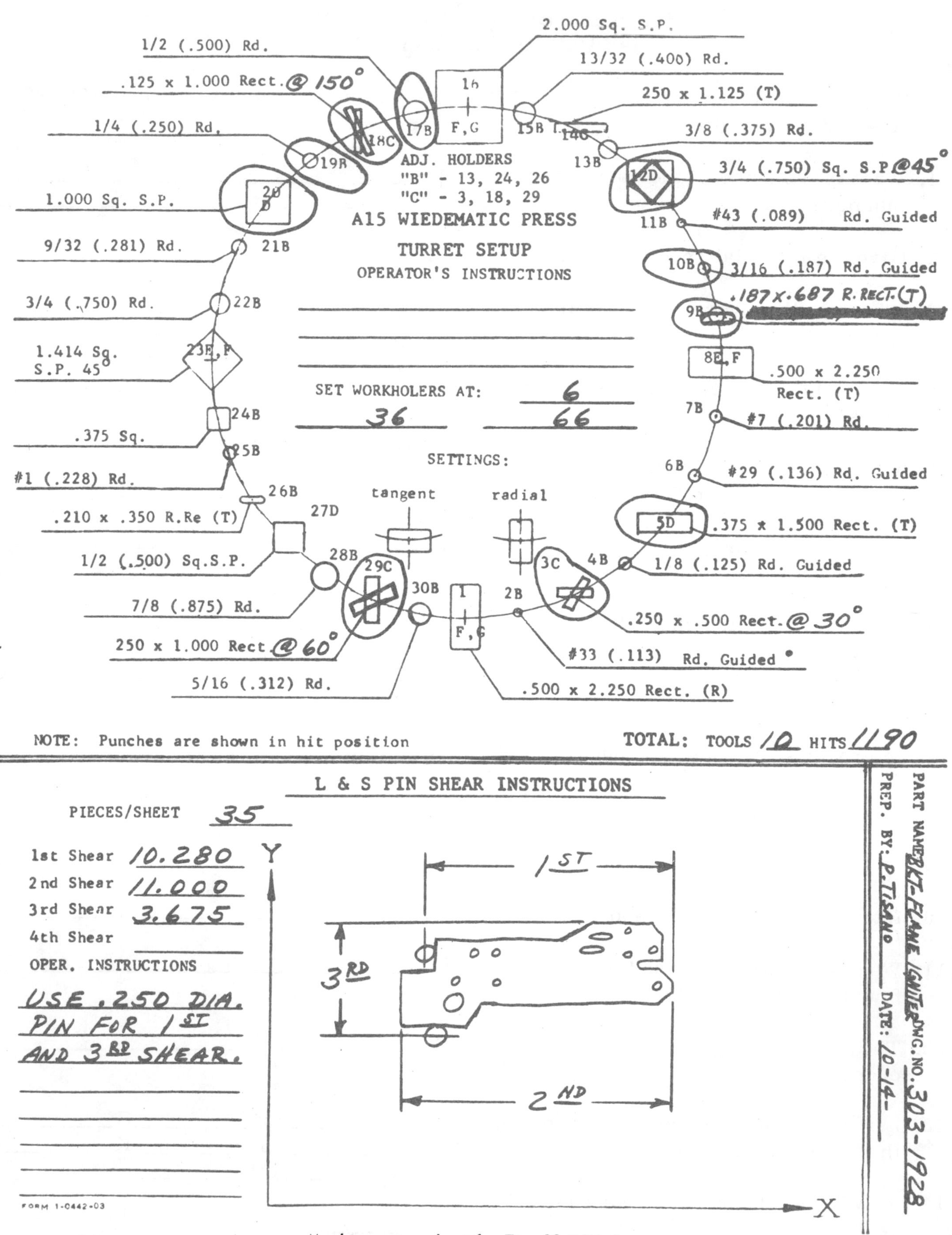

Machine setup sheet for Fig. 22-BPR-2.

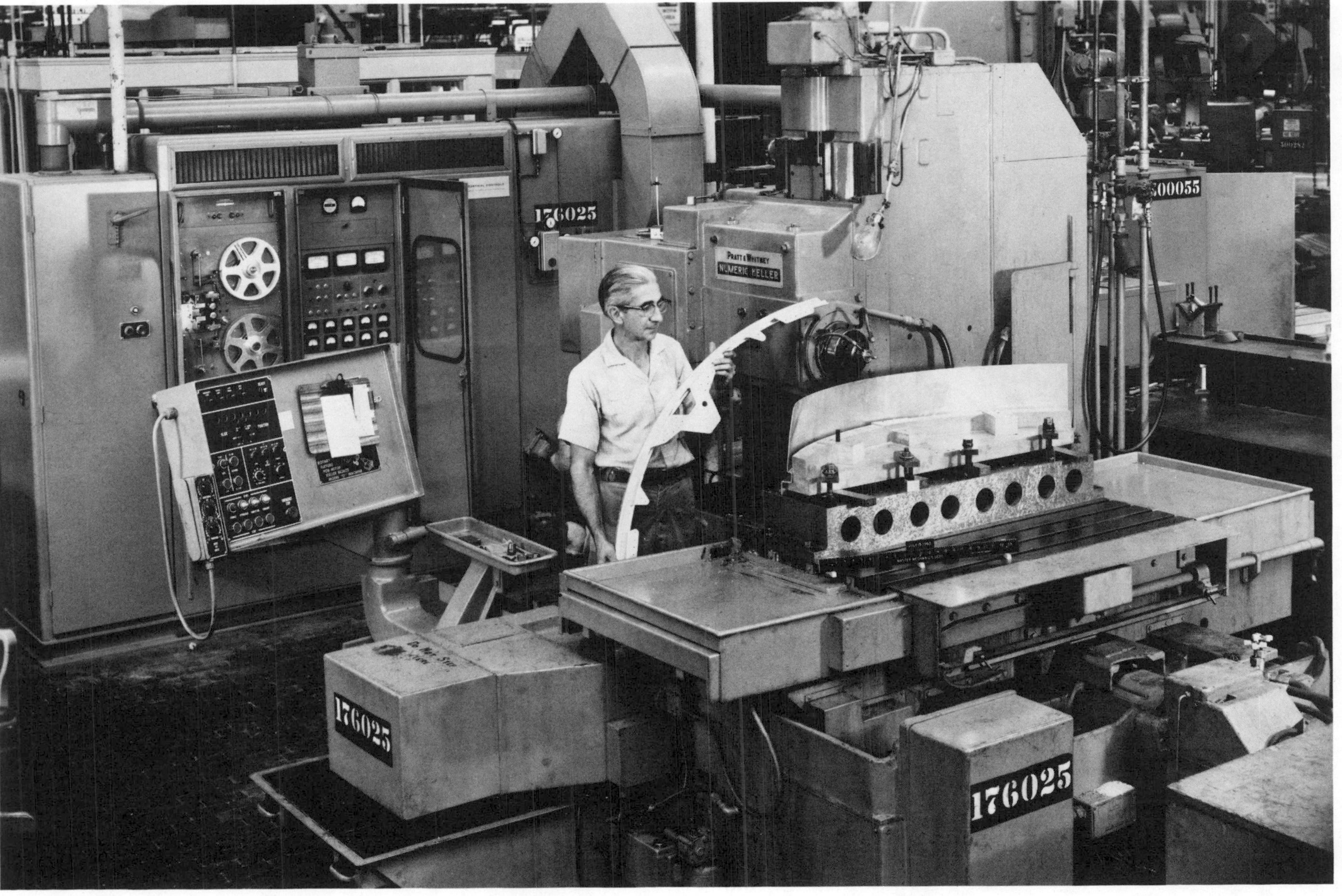

Industry photo. An automated, continuous-path numerically controlled machine capable of high-accuracy milling, drilling, reaming, tapping, and boring. (Pratt and Whitney Aircraft)

Unit 23
Precision Sheet Metal Blueprints

Growth in the instrumentation, electronic, and similar industries has brought about the need for a special type of work in thin-gage metals. This field of work is known as PRECISION SHEET METAL and may be defined as "working thin-gaged metals to machine shop tolerances." The calculations and layout are so exact that when the metal is machined in the flat position and then folded or assembled, the relationship and location of the resulting planes and features are within their specified tolerances, Fig. 23-1.

Calculations and Layouts

Usually blueprints will be dimensioned in the flat layout form by the design-draftsman to achieve the desired dimension of the part and its features after it is folded or assembled. However, it is helpful to know how the dimensions are calculated in the event this must be done in the shop or field.

The calculations for allowances in bend radii may be obtained directly from a set-

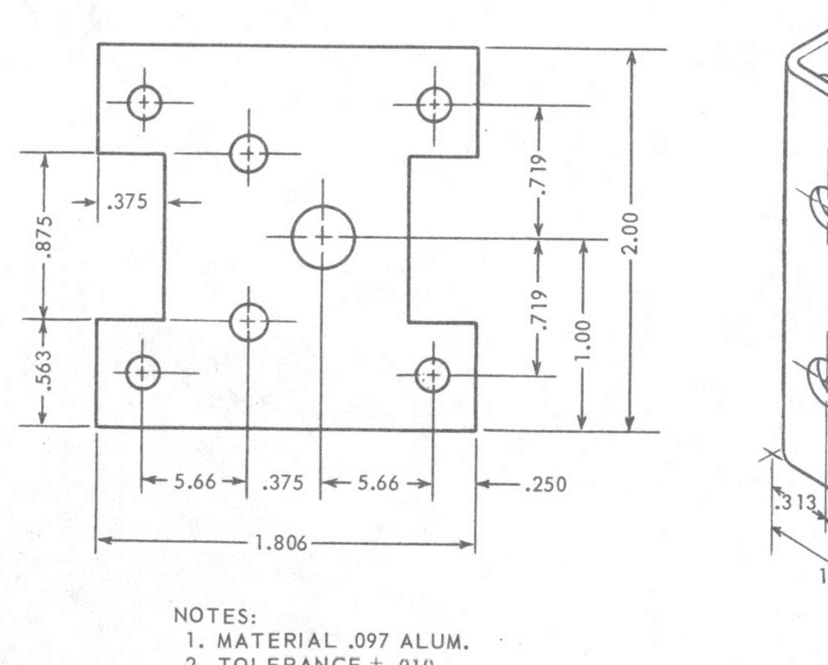

Fig. 23-1. Precision sheet metal part.

back chart or by use of the formulae by which the charts were developed.

Set-Back Charts

Set-back charts are available in most industries where precision sheet metal work is done. A portion of a chart for 90 deg. bends is shown in Fig. 23-2. A more complete chart is shown on page 333. These charts will save time and errors in calculations in the shop.

The set-back figure is found by following across the row representing the bend radius until it meets the vertical column representing the thickness of the sheet metal to be bent. For example, a bend radius of 1/8 in. on metal .040 thickness

PRECISION SHEET METAL
SET–BACK CHART

MATERIAL THICKNESS

90 DEG. BEND RADIUS	.016	.020	.025	.032	.040	.051	.064
1/32	.034	.039	.046	.055	.065	.081	.097
3/64	.041	.046	.053	.062	.072	.086	.104
1/16	.048	.053	.059	.068	.079	.093	.110
5/64	.054	.060	.066	.075	.086	.100	.117
3/32	.061	.066	.073	.082	.092	.107	.124
7/64	.068	.073	.080	.089	.099	.113	.130
1/8	.075	.080	.086	.095	.106	.120	.137
9/64	.081	.087	.093	.102	.113	.127	.144
5/32	.088	.093	.100	.109	.119	.134	.150
11/64	.095	.100	.107	.116	.126	.140	.157
3/16	.102	.107	.113	.122	.133	.147	.164

Fig. 23-2. Precision sheet metal set-back chart for 90 deg. bends.

would require a set-back figure of .106, Fig. 23-2. The diagram in Fig. 23-3 and the formula which follows show the application of the set-back figure in calculating the DEVELOPED LENGTH (length in the flat to produce desired folded size) of a precision sheet metal part.

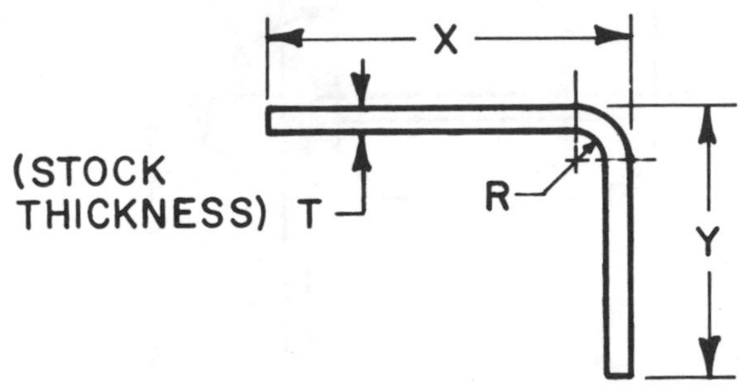

Fig. 23-3. Diagram for calculating developed length from set-back chart.

Using the set-back chart:

Developed Length = X + Y - Z

Where: X = outside distance of one side
Y = outside distance of other side
Z = set-back allowance for the 90 deg. bend (from chart, Fig. 23-2)

Example:

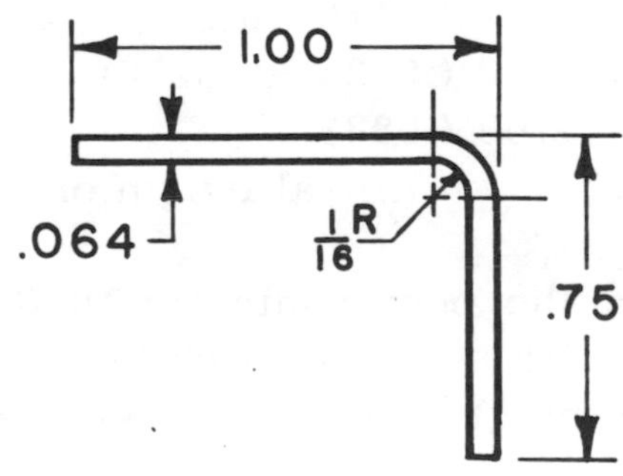

Developed Length = X + Y - Z
= 1.00 + .75 - .110
= 1.75 - .110
Developed Length = 1.64

Formula

One of two formulae is used for calculating the lineal length of the bend on precision sheet metal parts, depending on the size of the bend radius and the thickness of the metal. After finding the lineal length, the developed length (length in the flat) can be calculated.

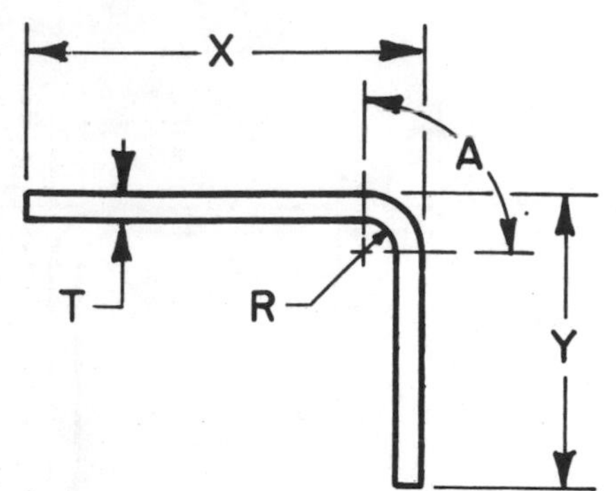

Fig. 23-4. Diagram for calculating developed length using formula.

To find A (the lineal length of a 90 deg. bend), Fig. 23-4, when R (inside radius) is LESS THAN TWICE THE STOCK THICKNESS use the formula:

$$A = 1/2\,\pi\ (R + .4T)$$

To find A, where R = 1/16 or .0625 and T = .064 (inside radius is less than twice the stock thickness):

$$\begin{aligned} A &= 1/2\,\pi\ (R + .4T) \\ &= 1/2 \times 3.1416\ (.0625 + .4 \times .064) \\ &= 1.5708\ (.0625 + .0256) \\ &= 1.5708\ (.0881) \\ A &= .1384 = \text{Lineal length of bend} \end{aligned}$$

When the bend radius is MORE THAN TWICE THE STOCK THICKNESS the following formula should be used to find A (LINEAL length of bend):

$$A = 1/2\,\pi\ (R + .5T)$$

To find the developed length of the part after the lineal bend length (A) has been found, refer to the diagram in Fig. 23-4 and use formula which follows:

$$\text{Developed Length} = X + Y + A - (2R + 2T)$$

Where: X = outside distance of one side
Y = outside distance of other side
A = lineal length of bend
R = the bend radius
T = the material thickness

Example:

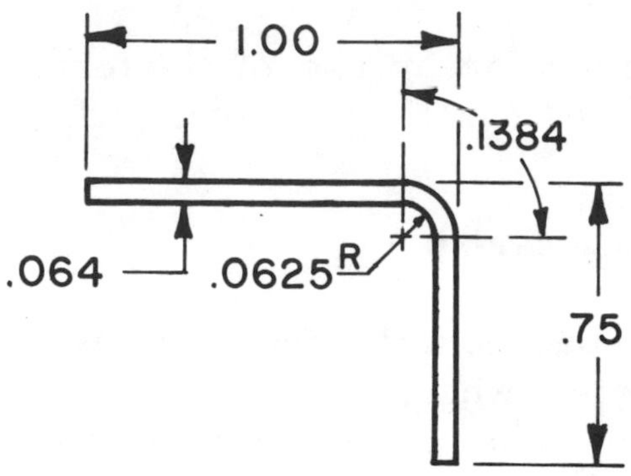

$$\begin{aligned} \text{Developed Length} &= 1.00 + .75 + .1384 - (.125 + .128) \\ &= 1.8884 - .253 \\ &= 1.6354 \text{ or } 1.64 \end{aligned}$$

Note that the two examples used the same part and dimensions. The same answer was obtained using the chart as was obtained using the formulae. However, the process is much shorter using the set-back chart.

Layout and Fold Procedure

1. Calculate (by formula or from chart) developed length and feature locations, and lay out.
2. Make two identical pieces to overall dimensions.
3. With forming die, form one piece and check measurements for length.
4. If tolerances for length are off, make adjustments and make two more identical blanks.
5. Lay out features and form one piece and check length and location of features.
6. If tolerances are off, make adjustments and repeat steps 4 and 5 until a part within tolerances is obtained.
7. When a satisfactory folded part has been obtained, its identical mating piece becomes the blank die pattern for building the blank die or making subsequent parts for the form die.

The formulae as well as the set-back charts for precision sheet metal work provide data that are approximate due to the differences in the coefficients of expansion in bending of the various blank metals used. It is, therefore, advisable to have a duplicate blank in the "flat" to refer to in making adjustments until overall measurements and feature locations meet the tolerances specified.

Reading Precision Sheet Metal Blueprints

Blueprints for precision sheet metal parts are read in the same manner as other blueprints in the metal machining and fabrication industries. Some prints will be fully dimensioned as the blueprint on page 200. Other prints will contain few, if any, dimensions but will be accompanied by a polyester film print that is dimensionally stable, from which the toolmaker will work directly in making a "flat pattern tool," (see blueprint on page 202). Parts will then be made with the use of this flat pattern.

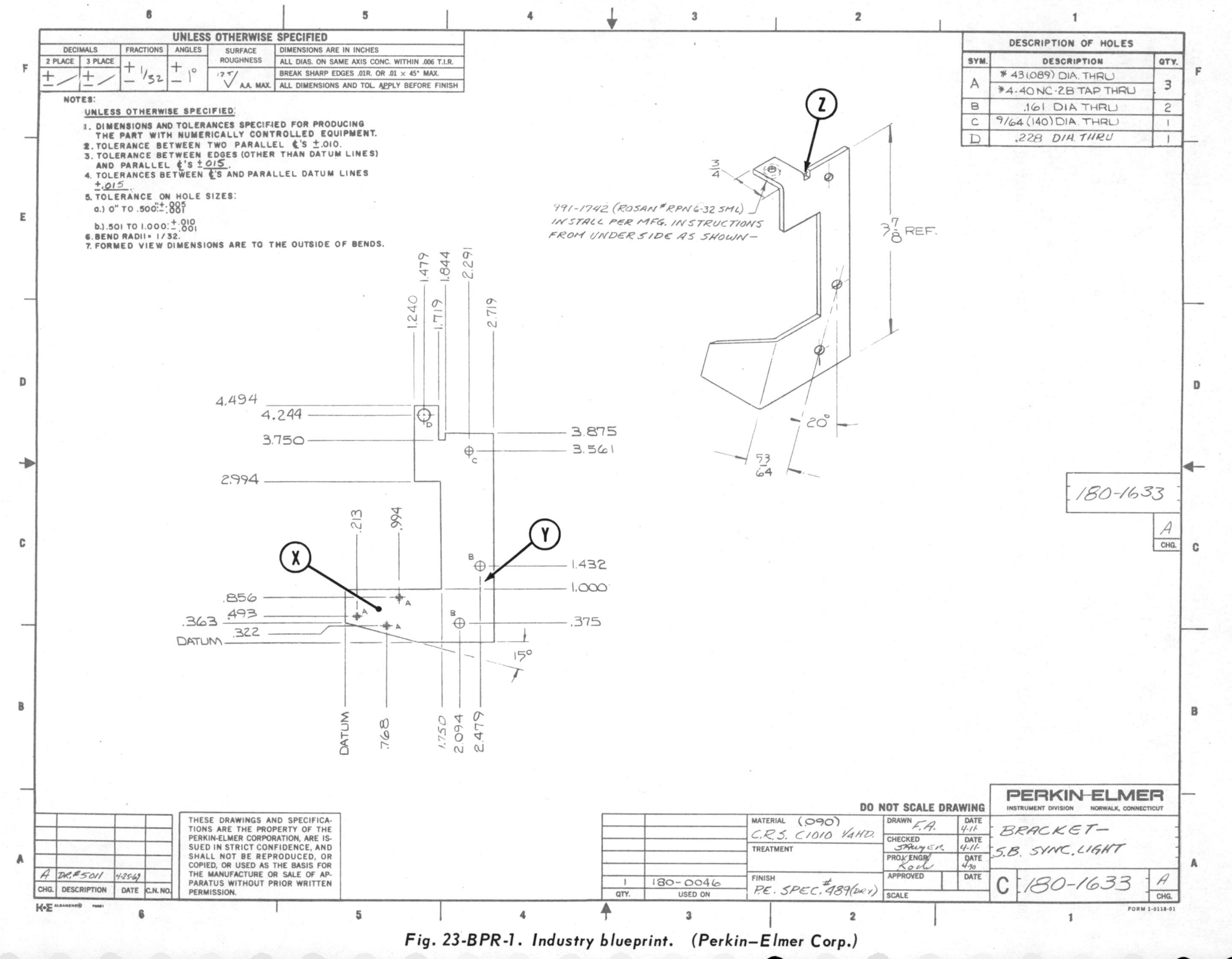

Fig. 23-BPR-1. Industry blueprint. (Perkin–Elmer Corp.)

Blueprint Reading Activity 23-BPR-1
PRECISION SHEET METAL

Refer to the blueprint in Fig. 23-BPR-1 and answer the following questions.

1. Give the name of the part and drawing number. — 1. ________________________________ ________________________________

2. What material is specified? — 2. ________________________________

3. What tolerances are to be held between holes? Between holes and edges? — 3. ________________________________ ________________________________

4. Give the overall size of the part in the flat position. — 4. ________________________________

5. Interpret the final machine process for the three holes marked A. — 5. ________________________________ ________________________________ ________________________________ ________________________________

6. Give the size of hole D as limit dimensions. — 6. ________________________________

7. What angle must surface X make with the vertical center line Y after the part is formed? — 7. ________________________________

8. What is the bend radii for the part? — 8. ________________________________

9. Give the width of the slot at Z. — 9. ________________________________

10. Give the number of the assembly on which this part is used and the quantity. — 10. ________________________________

For additional Blueprint Activities for Unit 23 see pages 292 and 295.

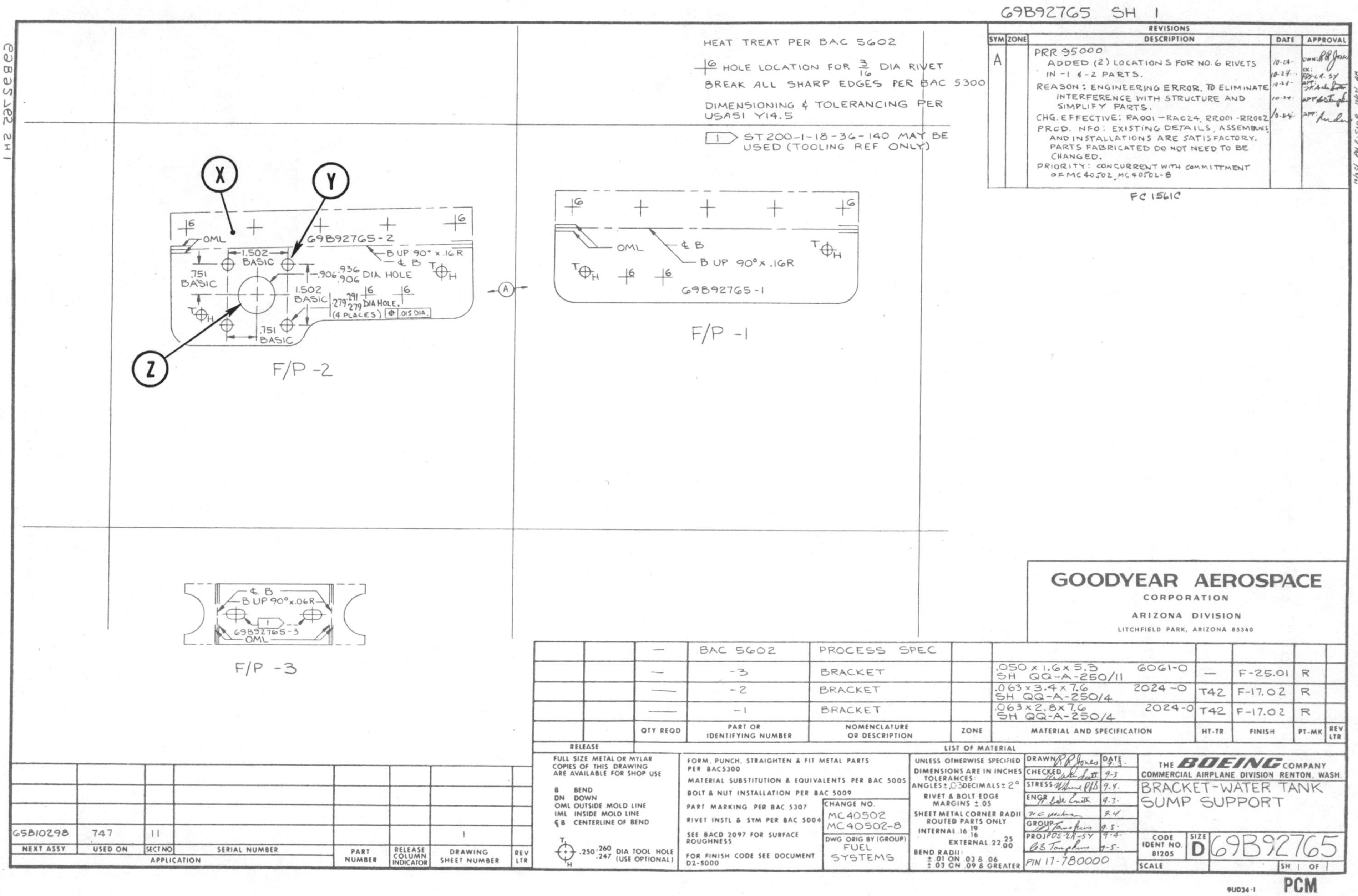

69B92765 SH 1

HEAT TREAT PER BAC 5602

16 HOLE LOCATION FOR $\frac{3}{16}$ DIA RIVET

BREAK ALL SHARP EDGES PER BAC 5300

DIMENSIONING & TOLERANCING PER USASI Y14.5

1 ST200-1-18-36-140 MAY BE USED (TOOLING REF ONLY)

SYM	ZONE	REVISIONS DESCRIPTION	DATE	APPROVAL
A		PRR 95000 ADDED (2) LOCATIONS FOR NO. 6 RIVETS IN -1 & -2 PARTS. REASON: ENGINEERING ERROR. TO ELIMINATE INTERFERENCE WITH STRUCTURE AND SIMPLIFY PARTS. CHG. EFFECTIVE: RA001 - RAC24, RR001 - RR002 PROD. NFO: EXISTING DETAILS, ASSEMBLIES AND INSTALLATIONS ARE SATISFACTORY. PARTS FABRICATED DO NOT NEED TO BE CHANGED. PRIORITY: CONCURRENT WITH COMMITTMENT OF MC40502, MC40502-8		

FC 1561C

X Y Z

69B92765-2 OML 1.502 BASIC .751 BASIC B UP 90° x .16R ℄ B .906/.936 DIA HOLE 1.502 BASIC .279/.291 DIA HOLE (4 PLACES) .015 DIA. .751 BASIC

F/P -2

OML ℄ B B UP 90° x .16R 69B92765-1

F/P -1

℄ B B UP 90° x .06R 69B92765-3 OML

F/P -3

GOODYEAR AEROSPACE CORPORATION
ARIZONA DIVISION
LITCHFIELD PARK, ARIZONA 85340

QTY REQD	PART OR IDENTIFYING NUMBER	NOMENCLATURE OR DESCRIPTION	ZONE	MATERIAL AND SPECIFICATION	HT-TR	FINISH	PT-MK	REV LTR
—	BAC 5602	PROCESS SPEC						
—	-3	BRACKET		.050 x 1.6 x 5.3 SH QQ-A-250/11 6061-0	—	F-25.01	R	
—	-2	BRACKET		.063 x 3.4 x 7.6 SH QQ-A-250/4 2024-0	T42	F-17.02	R	
—	-1	BRACKET		.063 x 2.8 x 7.6 SH QQ-A-250/4 2024-0	T42	F-17.02	R	

RELEASE — LIST OF MATERIAL

FULL SIZE METAL OR MYLAR COPIES OF THIS DRAWING ARE AVAILABLE FOR SHOP USE

B BEND
DN DOWN
OML OUTSIDE MOLD LINE
IML INSIDE MOLD LINE
℄B CENTERLINE OF BEND

.250/.260 .247 DIA TOOL HOLE (USE OPTIONAL)

FORM, PUNCH, STRAIGHTEN & FIT METAL PARTS PER BAC5300
MATERIAL SUBSTITUTION & EQUIVALENTS PER BAC 5005
BOLT & NUT INSTALLATION PER BAC 5009
PART MARKING PER BAC 5307
RIVET INSTL & SYM PER BAC 5004
SEE BACD 2097 FOR SURFACE ROUGHNESS
FOR FINISH CODE SEE DOCUMENT D2-5000

CHANGE NO. MC40502 MC40502-8
DWG ORIG BY (GROUP) FUEL SYSTEMS

UNLESS OTHERWISE SPECIFIED DIMENSIONS ARE IN INCHES
TOLERANCES: ANGLES ± .030 DECIMALS ± 2°
RIVET & BOLT EDGE MARGINS ± .05
SHEET METAL CORNER RADII ROUTED PARTS ONLY
INTERNAL .16 19/16 EXTERNAL 22.25/.00
BEND RADII: ±.01 ON .03 & .06; ±.03 ON .09 & GREATER

DRAWN — CHECKED — STRESS — ENGR — GROUP — PROJ — PIN 17-780000

THE BOEING COMPANY
COMMERCIAL AIRPLANE DIVISION RENTON, WASH.
BRACKET-WATER TANK SUMP SUPPORT
CODE IDENT NO. 81205 SIZE D 69B92765
SCALE SH 1 OF 1

NEXT ASSY	USED ON	SECT NO	SERIAL NUMBER	PART NUMBER	RELEASE COLUMN INDICATOR	DRAWING SHEET NUMBER	REV LTR
65B10298	747	11				1	

APPLICATION

9U034-1 PCM

Fig. 23-BPR-2. Industry blueprint. (Goodyear Aerospace)

Blueprint Reading Activity 23-BPR-2
PRECISION SHEET METAL

Refer to the blueprint in Fig. 23-BPR-2 and answer the following questions.

1. What is the name of the part and print number? 1. ____________

2. Give the material size and specification for the -1 part. 2. ____________

3. How are the holes to be located for the rivets? 3. ____________

4. What is the diameter of the rivet holes? 4. ____________

5. Give the diameter of holes Y as limit dimensions. 5. ____________

6. How many holes Y are to be machined, and what relationship must exist between the holes? 6. ____________

7. State the diameter of hole Z as a plus or minus tolerance. 7. ____________

8. What angle does surface X make with the rest of the part after bending? 8. ____________

9. Give the bend radius of -3 part. 9. ____________

10. What does T ⊕ H mean on the print? 10. ____________

11. What finish is specified for -2 part? 11. ____________

Unit 24
Welding Blueprints

Reading welding drawings is quite similar to reading other drawings except for the symbols involved. The American Welding Society has developed and adopted standard procedures for using symbols to indicate the exact location, size, strength, geometry and other information necessary to describe the weld required. The welding symbols included in this Unit will assist you in reading and interpreting drawings involving welding processes.

Elements of Welding Symbols

It is important to distinguish between the WELD SYMBOL and the WELDING SYMBOL. The weld symbol indicates the type of weld only, and the welding symbol consists of the following elements:

REFERENCE LINE is the horizontal line (may appear vertically on some prints) portion of a welding symbol which is joined by an arrow and a tail, Fig. 24-1.

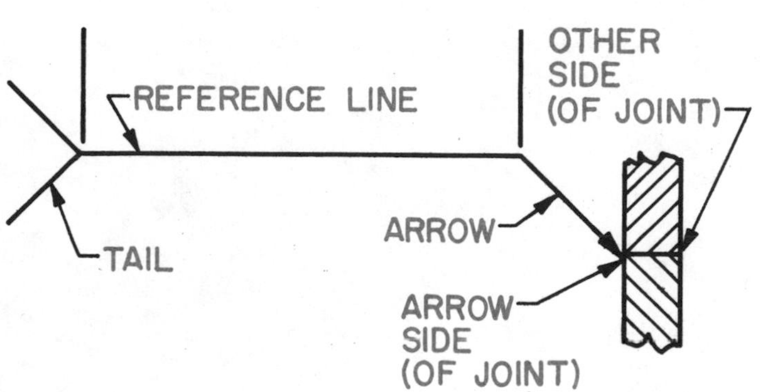

Fig. 24-1. Basic welding symbol.

An ARROW is used to connect the welding symbol reference line to one side of the joint. This is considered the ARROW SIDE of the joint. The side opposite the arrow is considered the OTHER SIDE of the joint, Fig. 24-1.

The TAIL SECTION shown in Figs. 24-1 and 24-2 is used for designating the welding specification, process (for abbreviations see page 335), or other reference such as an industry specification.

Fig. 24-2. Tail section designating Laser Beam Welding.

BASIC WELD SYMBOLS for various types of welds are shown in Fig. 24-3. A more comprehensive list of weld symbols and their applications are shown on page 334.

LOCATION OF WELDS are indicated by their placement on the reference line. Weld symbols placed on the side of the reference line nearest the reader indicate welds on the ARROW SIDE of the joint, Fig. 24-4. Weld symbols on the reference

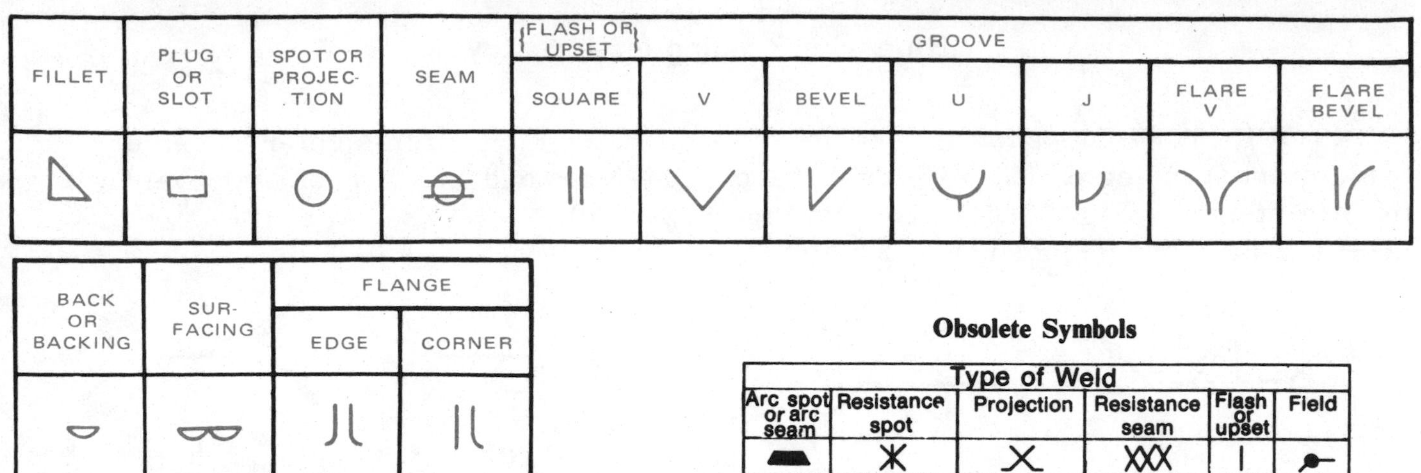

Fig. 24-3. Basic weld symbols. (With permission of The American Welding Society)

line side away from the reader indicate welds on the OTHER SIDE of the joint. Weld symbols on both sides of the reference line indicate welds on BOTH SIDES of the joint.

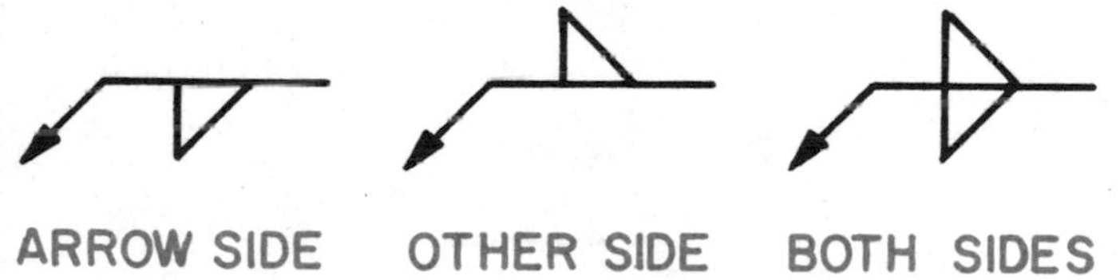

Fig. 24-4. Location of welds.

DIMENSIONS OF WELDS are shown on the same side of the reference line as the weld symbol, Fig. 24-5(a). When the dimensions are covered by a general note, such as "ALL FILLET WELDS 3/8" IN SIZE UNLESS OTHERWISE NOTED," the welding symbol need not be dimensioned, Fig. 24-5(b). When both welds have the same dimensions, one or both may be dimensioned, Fig. 24-5(c). The pitch of staggered intermittent welds is shown to the right of the length of the weld, Fig. 24-5(d).

SUPPLEMENTARY AND OTHER SYMBOLS used with the welding symbol to further specify the type of weld are discussed under the following:

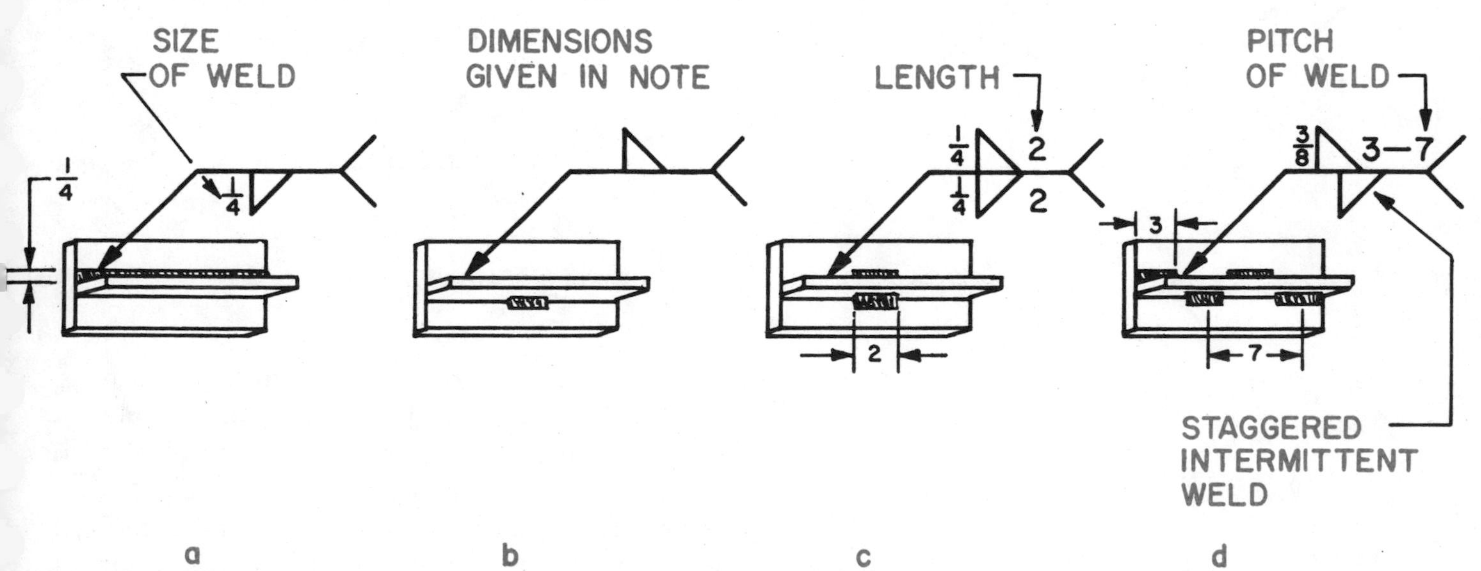

Fig. 24-5. Dimensions of welds.

CONTOUR SYMBOL shown next to the weld symbol indicates fillet welds that are to be flat-faced, Fig. 24-6(a); convex, Fig. 24-6(b); or concave-faced, Fig. 24-6(c).

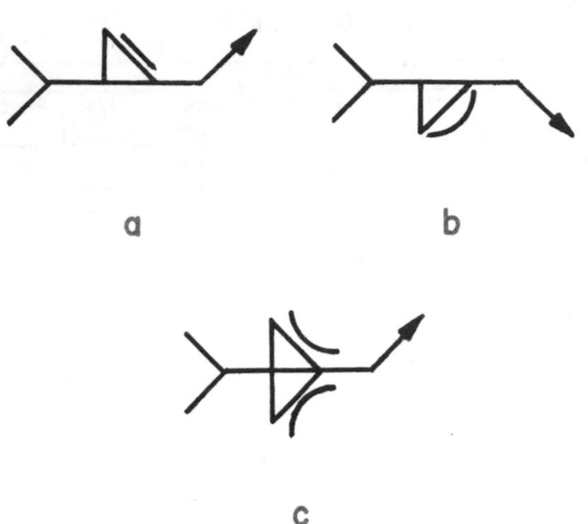

Fig. 24-6. Contour symbols.

GROOVE ANGLE, Fig. 24-7(a), is shown on the same side of the reference line as the weld symbol. The size (depth) of groove welds is shown to the left of the weld symbol, see Fig. 24-7(b). The root opening of groove welds, Fig. 24-7(c), is shown inside the weld symbol.

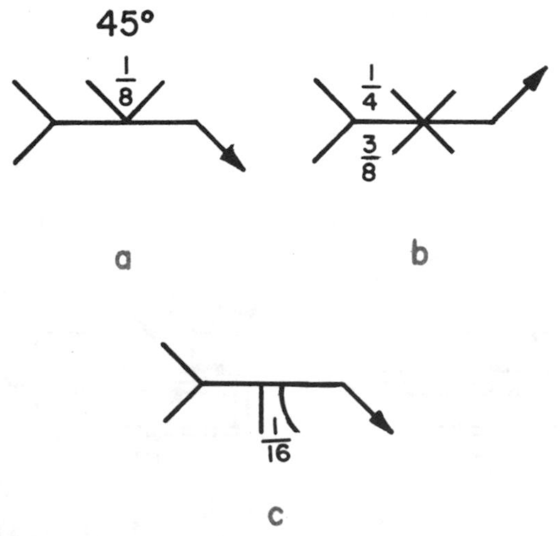

Fig. 24-7. Groove symbols.

SPOT WELDS are called out for size in diameter, Fig. 24-8(a), strength in pounds, Fig. 24-8(b); pitch (center to center), Fig. 24-8(c); and number of spot welds, Fig. 24-8(d).

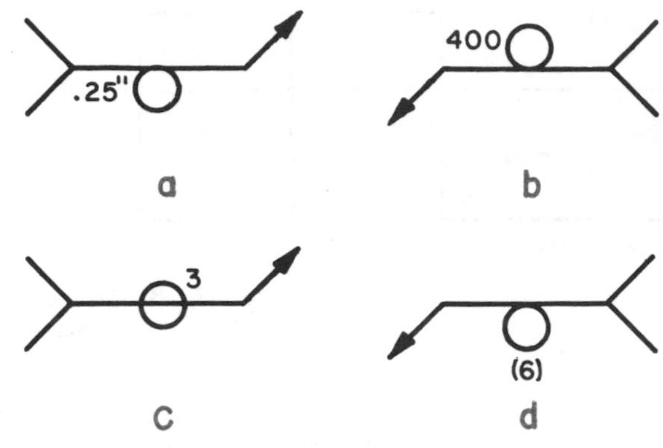

Fig. 24-8. Spot weld symbols.

WELD-ALL-AROUND symbol, Fig. 24-9, indicates the welds extending completely around a joint.

Fig. 24-9. Weld-all-around symbol.

FIELD WELD symbol indicates the welds which are not made in the shop or place of initial construction, Fig. 24-10.

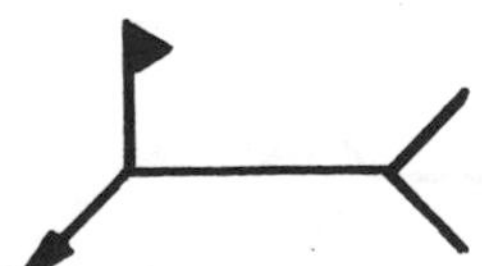

Fig. 24-10. Field weld symbol.

MELT-THRU symbol indicates the welds where 100 percent joint or member penetration plus reinforcement is required in welds made from one side, Fig. 24-11(a). When melt-thru welds are to be finished by machining or some other process, a contour symbol is added, Fig. 24-11(b).

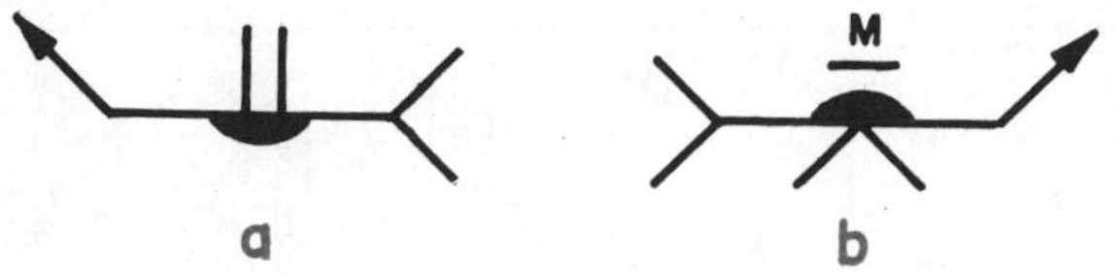

Fig. 24-11. Melt-Thru symbols.

FINISH SYMBOL indicates method of finishing (C = chipping; G = grinding; M = machining; R = rolling; H = hammering) and not the surface texture (see Melt-Thru above).

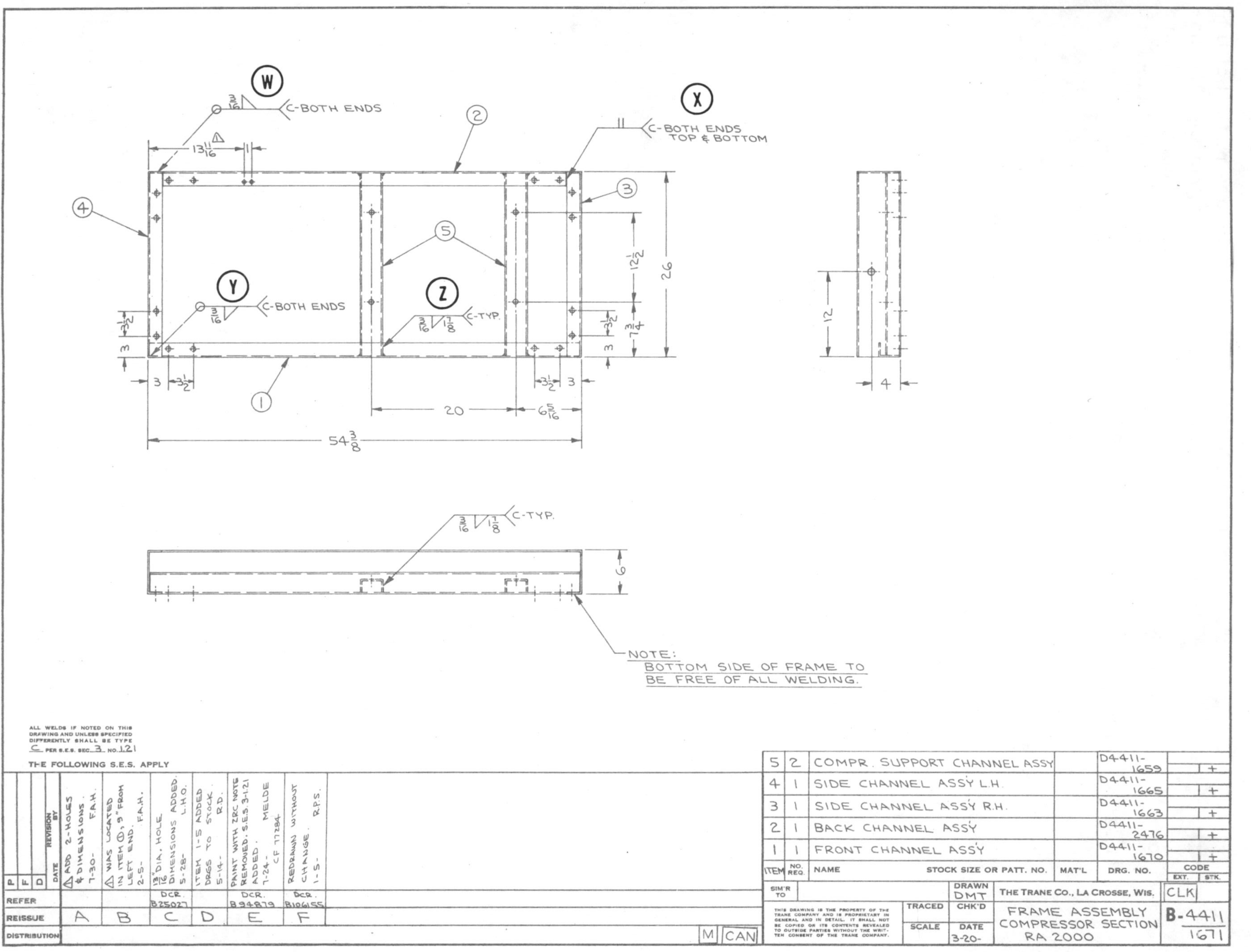

Fig. 24-BPR-1. Industry blueprint. (Ross Heat Exchanger Division)

Blueprint Reading Activity 24-BPR-1
WELDING

Refer to the blueprint in Fig. 24-BPR-1 and answer the following questions.

1. Give the name and drawing number of the assembly.

   1. ______________________

2. Have changes been issued against the drawing? If so, how many?

   2. ______________________

3. How many separate parts are required in the assembly?

   3. ______________________

4. What effect does the local note have on the welding to be done?

   4. ______________________

5. What process is specified for all of the welds?

   5. ______________________

6. Refer to the symbols and interpretations in this Unit and to the AWS Standard Welding Symbols on page 334 and interpret the following symbols on the blueprint:

   6. W ______________________

   X ______________________

   Y ______________________

   Z ______________________

For additional Blueprint Activities for Unit 24 see pages 296 and 299.

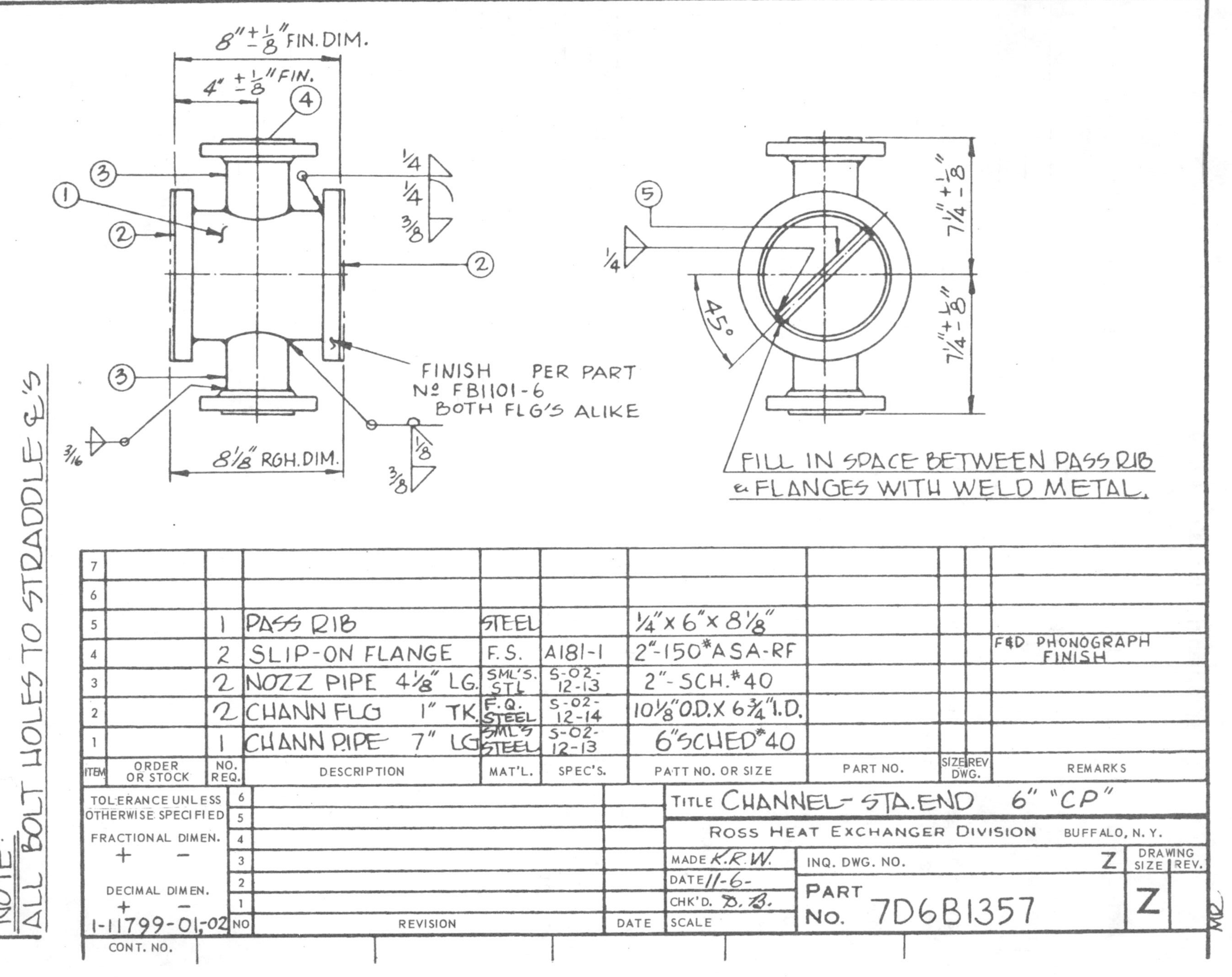

ITEM	ORDER OR STOCK	NO. REQ.	DESCRIPTION	MAT'L.	SPEC'S.	PATT NO. OR SIZE	PART NO.	SIZE DWG.	REV	REMARKS
7										
6										
5		1	PASS RIB	STEEL		1/4" x 6" x 8 1/8"				
4		2	SLIP-ON FLANGE	F.S.	A181-1	2"-150# ASA-RF				F&D PHONOGRAPH FINISH
3		2	NOZZ PIPE 4 1/8" LG.	SML'S. STL	S-02-12-13	2"-SCH. #40				
2		2	CHANN FLG 1" TK.	F.Q. STEEL	S-02-12-14	10 1/8" O.D. x 6 3/4" I.D.				
1		1	CHANN PIPE 7" LG.	SML'S STEEL	S-02-12-13	6" SCHED #40				

Fig. 24-BPR-2. Industry blueprint. (The Trane Co.)

Blueprint Reading Activity 24-BPR-2
WELDING

Refer to the blueprint in Fig. 24-BPR-2 and answer the following questions:

1. What is the name and number of the part?

   1. ______________________________
   ____

2. What is the total number of parts required to make the weldment (the welded assembly)?

   2. ______________________________

3. Give the description, material and size of item 5.

   3. ______________________________
   ____

4. How are bolt holes to be aligned?

   4. ______________________________

5. Interpret the type of welds required to join parts:

   5. 1 and 2 ______________________________
   ____
   ____
   ____

   1 and 3 ______________________________
   ____
   ____

   3 and 4 ______________________________
   ____

   1 and 5 ______________________________
   ____

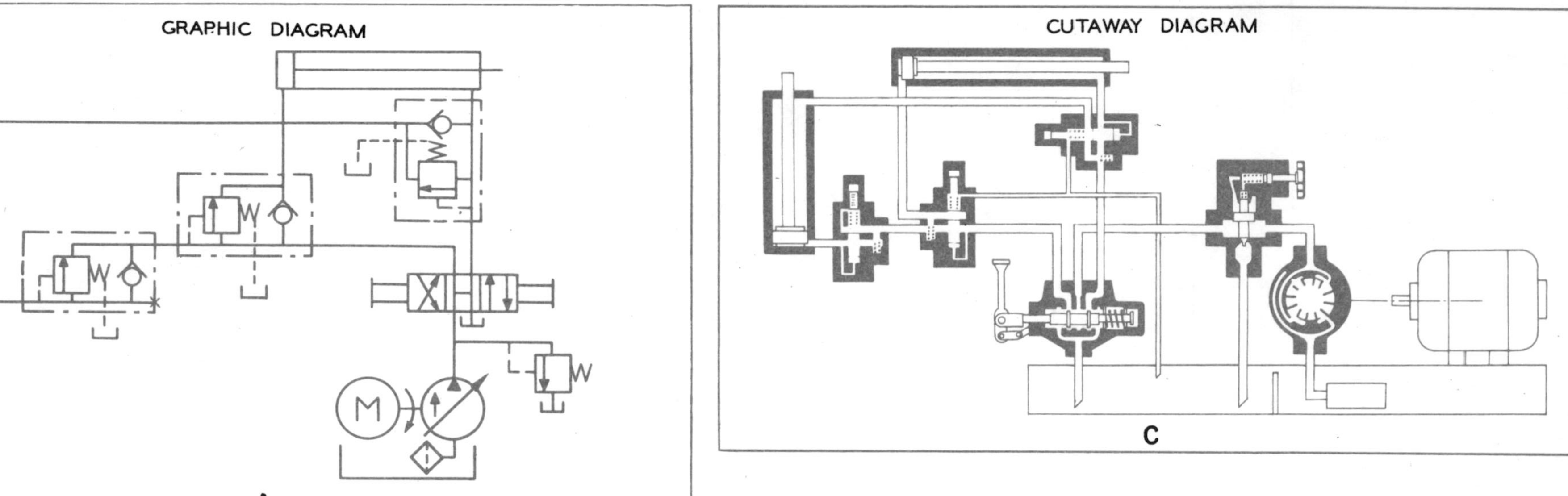

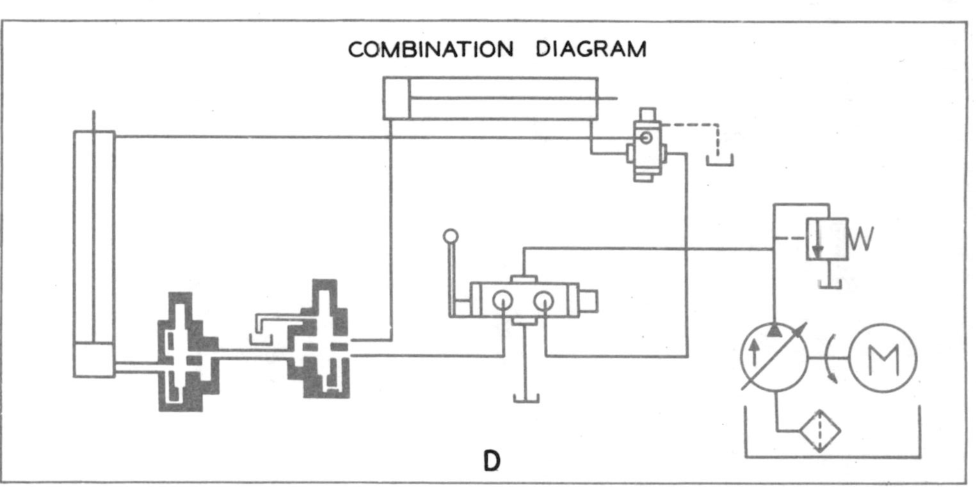

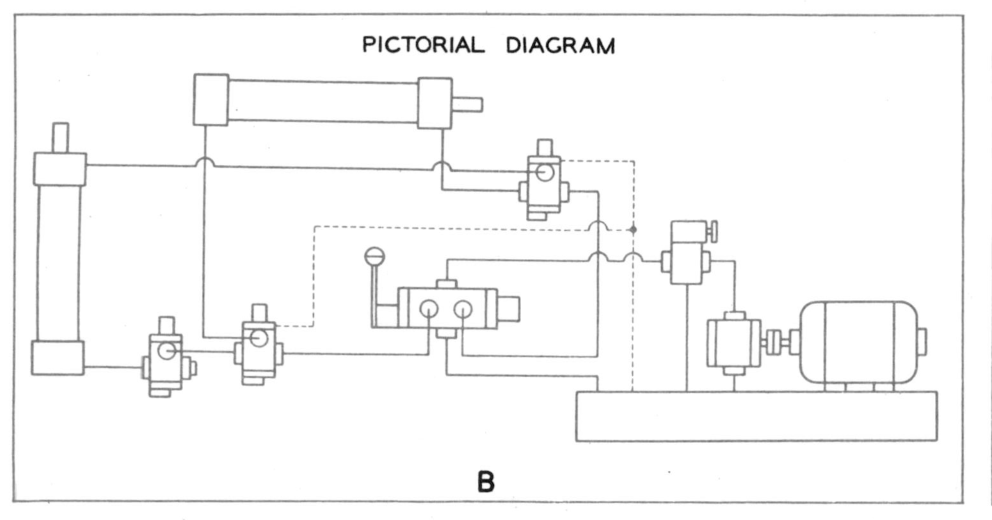

Fig. 25-1. Examples of graphic, pictorial, cutaway and combination diagrams for fluid power circuits.[1]

Unit 25
Instrumentation and Control Diagrams

This Unit will introduce you to the basic symbols and diagrams used in instrumentation and control circuits and assist you in reading and interpreting graphic diagrams.

Types of Diagrams

Several types of diagrams or drawings are used in showing instrumentation and control circuits. These include GRAPHIC, PICTORIAL, CUTAWAY and COMBINATION DIAGRAMS, Fig. 25-1. The emphasis in this Unit will be graphic diagrams since these are the most widely used in industry, and since the graphic symbols have been standardized.

GRAPHIC DIAGRAMS consist of graphic symbols joined by lines that provide an easy method of emphasizing functions of the circuit and its components, see Fig. 25-1(A).

PICTORIAL DIAGRAMS are used when piping is to be shown between components, Fig. 25-1(B).

CUTAWAY DIAGRAMS consist of cutaway symbols of components and emphasize component function and piping between components, Fig. 25-1(C).

COMBINATION DIAGRAMS utilize, in one diagram, the type of component illustration that best suits the purpose of the diagram, Fig. 25-1(D).

After you have learned the basic symbols and sequence of operations used in instrumentation and control circuits, you will have little difficulty in reading any of these four types of diagrams.

Terms Used in Instrumentation and Control Work

ACCUMULATOR: A container in which fluid is stored under pressure as a source of fluid power.

ACTUATE: To put devices and circuits into action.

ACTUATOR: A device for converting hydraulic energy into mechanical energy. A motor or cylinder.

CIRCUIT: The complete path of flow in a fluid power system including the flow-generating device.

COMPONENT: A single hydraulic or pneumatic unit.

CONTROL: A device used to regulate the function of a component or system.

ENCLOSURE: A phantom line rectangle drawn around components to indicate the limits of an assembly on a diagram. A housing for components.

FLUID: A liquid or gas.

PORT: An internal or external terminus of a passage in a component.

RESERVOIR: A container for storage of liquid in a fluid power system.

RESTRICTION: A reduced cross-sectional area in a line or passage which produces a pressure drop.

Graphic Symbols for Fluid Power Diagrams

Many instrumentation and control circuits for machine tools involve fluid power (liquid or gas) and the symbols in Fig. 25-2 are those commonly found on blueprints for these circuits.

Supplementary Information to Accompany Circuit Diagrams

Once you have become familiar with the symbols, graphic diagrams are relatively easy to read and understand. In addition to the graphic diagram, blueprints of circuit diagrams usually include a listing of the sequence of operations, solenoid chart, and components used to facilitate understanding of the function and purpose of the circuit and its components.

SEQUENCE OF OPERATIONS is an explanation of the various functions of the circuit explained in order of occurrence. Each phase of the operation is numbered or lettered and a brief description is given of the initiating and resulting action. This information is usually given in the upper part of the sheet or on an attached sheet.

SOLENOID CHART. If solenoids are used in the circuit, a chart is normally located in the lower left corner of the blueprint to help explain the operation of the electrically controlled circuit. Solenoids are given a letter on the diagrams and the chart shows whether the solenoids are energized (+) or de-energized (-) at each phase of system operation.

COMPONENT LIST, sometimes called a Bill of Materials, includes the name, model number, quantity, and supplier or manufacturer of components in the circuit. Each component in the diagram is numbered and keyed to the component list. Where two or more identical components are used, the same key number is used, followed by a letter. The list usually appears in the upper right corner of the blueprint.

Reading a Graphic Diagram

Let's try reading the pneumatic graphic circuit diagram shown in Fig. 25-3.

A. Familiarize yourself with the diagram by matching the numbers in the component list with their graphic symbols in the diagram. Refer to the list of graphic symbols in Fig. 25-2 for further clarification.

B. Follow the sequence of operations:
   1. Valve (5) is depressed (manual--see symbols) to start cycle.
   2. After valve (5) is shifted, follow the air flow through valve (7A) on to where it shifts valve (4) to direct air to head end of cylinder (10).
   3. The piston in cylinder (10) extends to contact work (riveting operation)

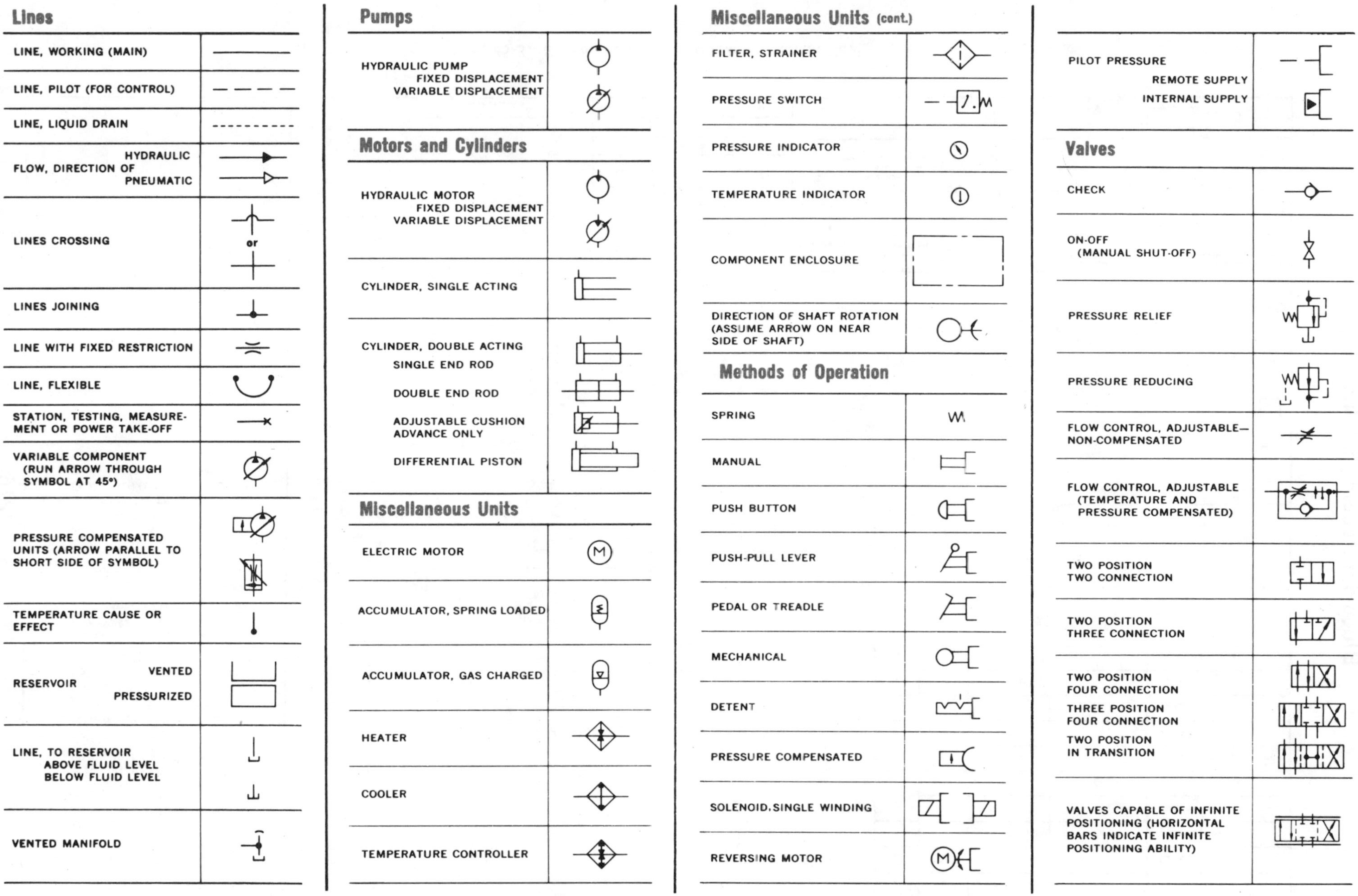

Lines

LINE, WORKING (MAIN)

LINE, PILOT (FOR CONTROL)

LINE, LIQUID DRAIN

FLOW, DIRECTION OF
HYDRAULIC
PNEUMATIC

LINES CROSSING (symbol shown two ways, marked "or")

LINES JOINING

LINE WITH FIXED RESTRICTION

LINE, FLEXIBLE

STATION, TESTING, MEASUREMENT OR POWER TAKE-OFF

VARIABLE COMPONENT (RUN ARROW THROUGH SYMBOL AT 45°)

PRESSURE COMPENSATED UNITS (ARROW PARALLEL TO SHORT SIDE OF SYMBOL)

TEMPERATURE CAUSE OR EFFECT

RESERVOIR
VENTED
PRESSURIZED

LINE, TO RESERVOIR
ABOVE FLUID LEVEL
BELOW FLUID LEVEL

VENTED MANIFOLD

Pumps

HYDRAULIC PUMP
FIXED DISPLACEMENT
VARIABLE DISPLACEMENT

Motors and Cylinders

HYDRAULIC MOTOR
FIXED DISPLACEMENT
VARIABLE DISPLACEMENT

CYLINDER, SINGLE ACTING

CYLINDER, DOUBLE ACTING
SINGLE END ROD
DOUBLE END ROD
ADJUSTABLE CUSHION ADVANCE ONLY
DIFFERENTIAL PISTON

Miscellaneous Units

ELECTRIC MOTOR

ACCUMULATOR, SPRING LOADED

ACCUMULATOR, GAS CHARGED

HEATER

COOLER

TEMPERATURE CONTROLLER

Miscellaneous Units (cont.)

FILTER, STRAINER

PRESSURE SWITCH

PRESSURE INDICATOR

TEMPERATURE INDICATOR

COMPONENT ENCLOSURE

DIRECTION OF SHAFT ROTATION (ASSUME ARROW ON NEAR SIDE OF SHAFT)

Methods of Operation

SPRING

MANUAL

PUSH BUTTON

PUSH-PULL LEVER

PEDAL OR TREADLE

MECHANICAL

DETENT

PRESSURE COMPENSATED

SOLENOID, SINGLE WINDING

REVERSING MOTOR

PILOT PRESSURE
REMOTE SUPPLY
INTERNAL SUPPLY

Valves

CHECK

ON-OFF (MANUAL SHUT-OFF)

PRESSURE RELIEF

PRESSURE REDUCING

FLOW CONTROL, ADJUSTABLE—NON-COMPENSATED

FLOW CONTROL, ADJUSTABLE (TEMPERATURE AND PRESSURE COMPENSATED)

TWO POSITION
TWO CONNECTION

TWO POSITION
THREE CONNECTION

TWO POSITION
FOUR CONNECTION

THREE POSITION
FOUR CONNECTION

TWO POSITION
IN TRANSITION

VALVES CAPABLE OF INFINITE POSITIONING (HORIZONTAL BARS INDICATE INFINITE POSITIONING ABILITY)

Fig. 25-2. Graphic symbols for fluid power diagrams. (Vickers Industrial Div.)

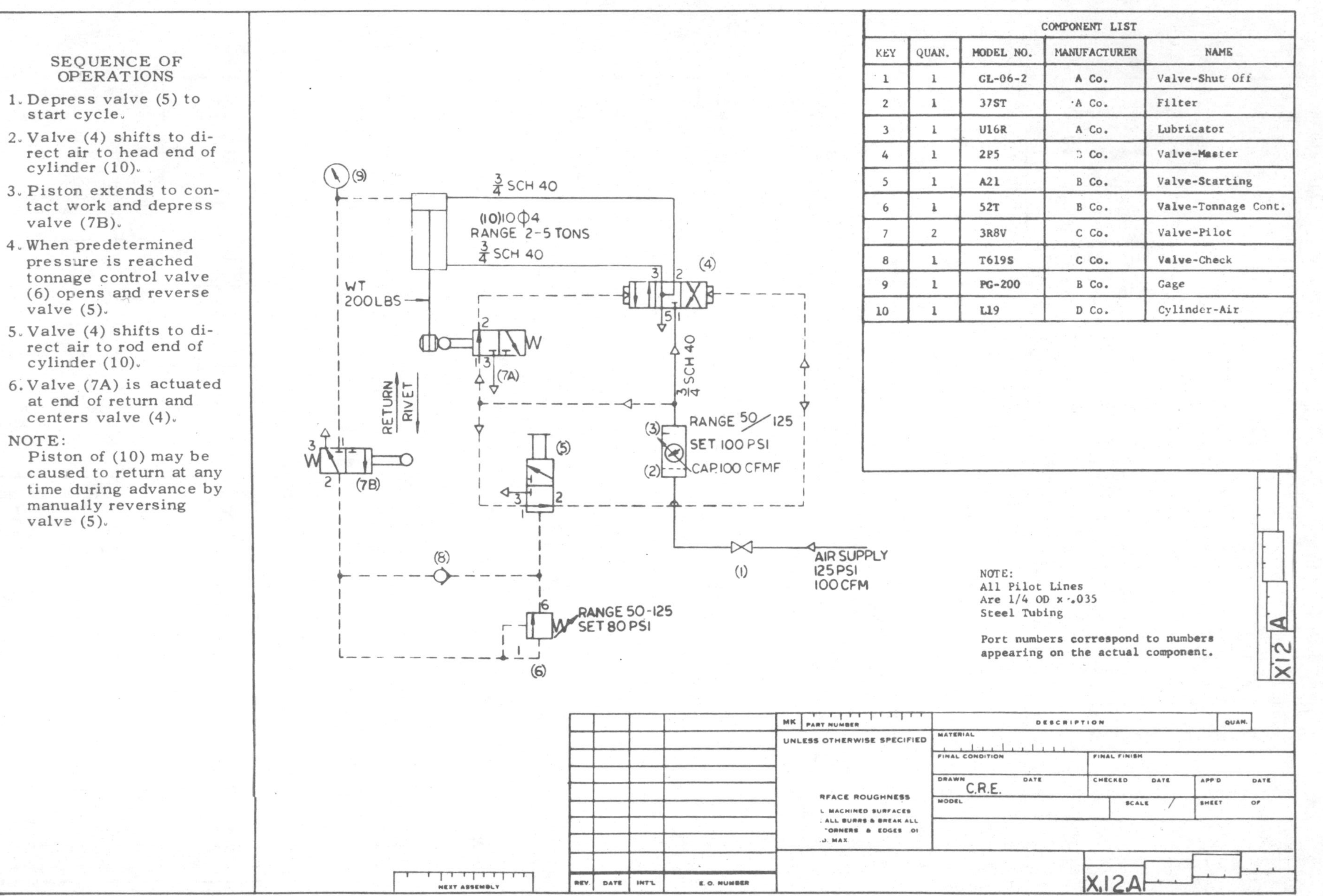

COMPONENT LIST

KEY	QUAN.	MODEL NO.	MANUFACTURER	NAME
1	1	GL-06-2	A Co.	Valve-Shut Off
2	1	37ST	A Co.	Filter
3	1	U16R	A Co.	Lubricator
4	1	2P5	B Co.	Valve-Master
5	1	A21	B Co.	Valve-Starting
6	1	52T	B Co.	Valve-Tonnage Cont.
7	2	3R8V	C Co.	Valve-Pilot
8	1	T619S	C Co.	Valve-Check
9	1	PG-200	B Co.	Gage
10	1	L19	D Co.	Cylinder-Air

Fig. 25-3. Graphic circuit diagram – Pneumatic.[1]

and also depresses valve (7B).

4. When a predetermined pressure is reached, tonnage control valve (6) (set at 80 pounds per square inch) opens and reverses valve (5).
5. The reversing of valve (5) now shifts the air flow and shifts valve (4) to direct the air flow to the rod end of cylinder (10). Air pressure that was on the head of the cylinder is exhausted to the atmosphere by this same shift of valve (4).
6. Valve (7A) is actuated at the end of the piston rod return which permits the air flow to flow through valve (7A) and centers valve (4), neutralizing the pressure on the rod end and head end of cylinder (10).

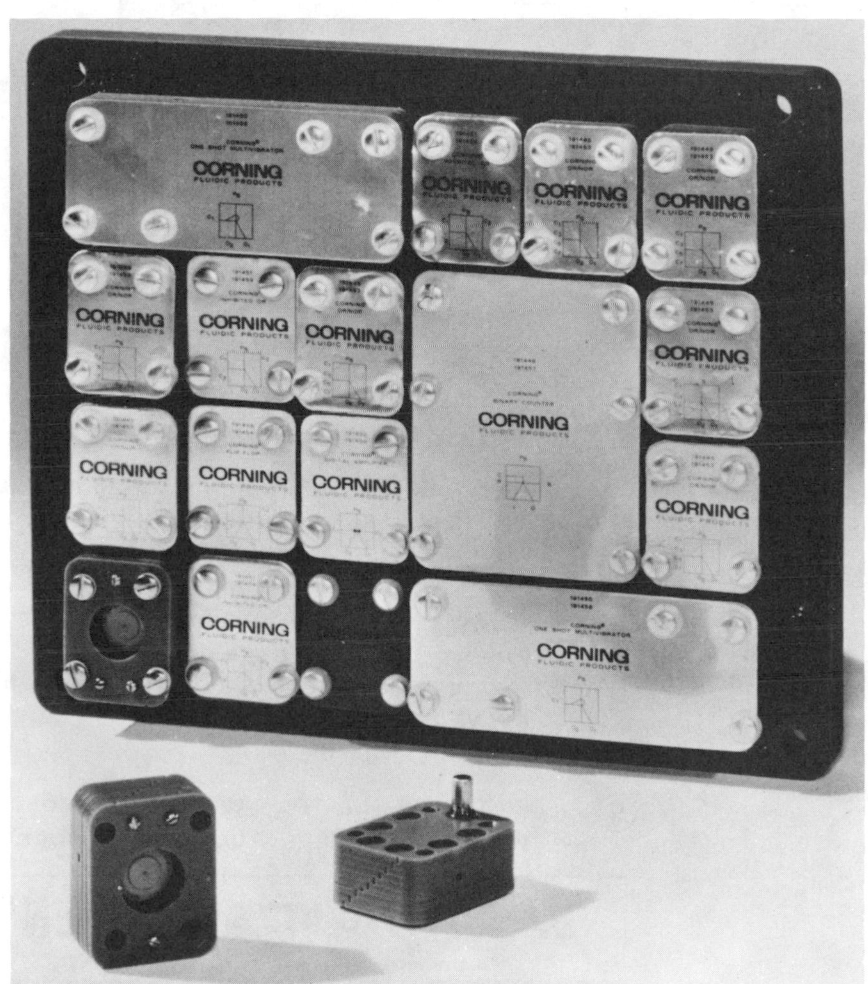

A 20-gate fluidic manifold containing various devices. (Corning Glass Works)

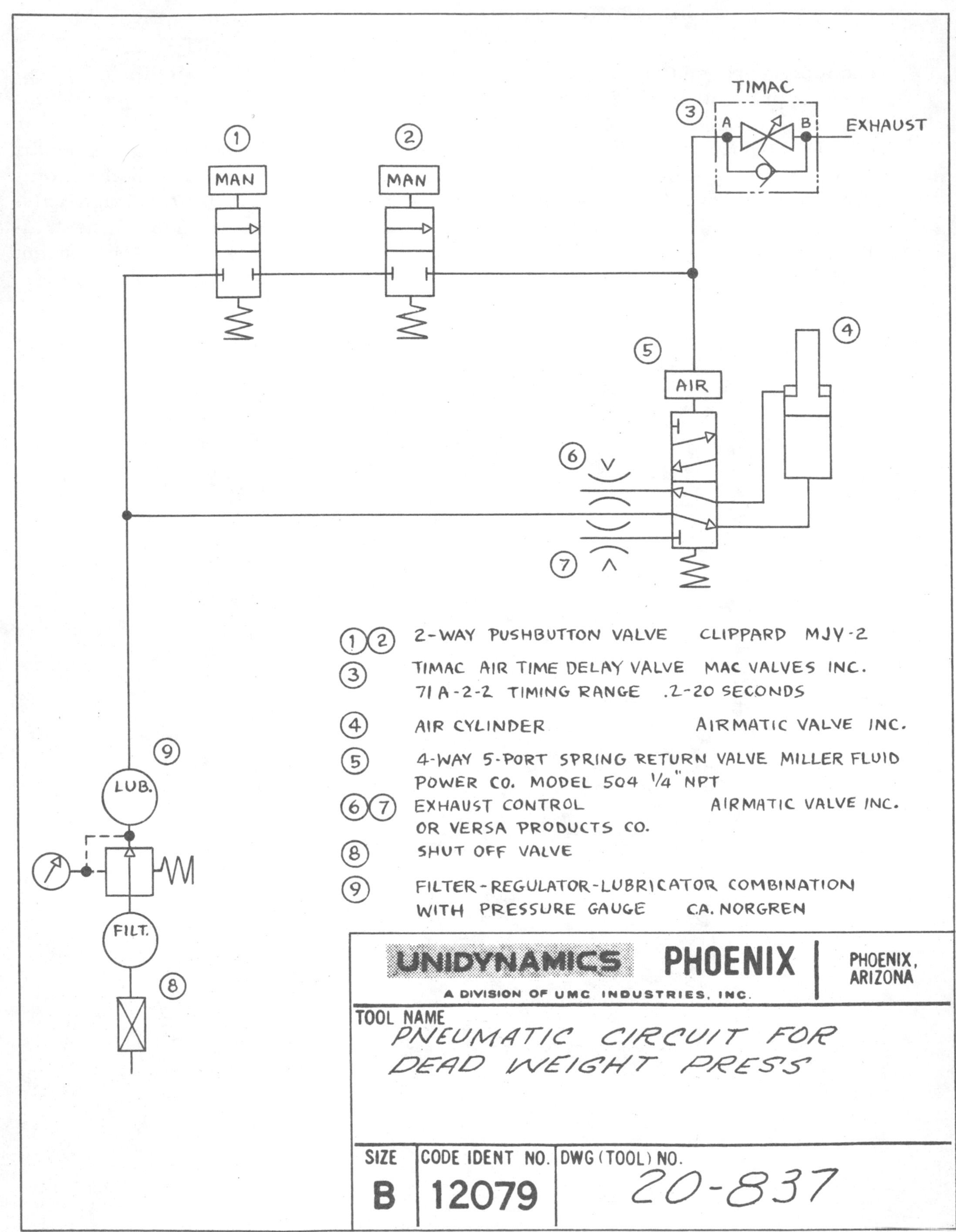

Fig. 25-BPR-1. Industry blueprint. (Unidynamics)

Sequence of Operation After First Start-Up

(Refer to Blueprint Reading Activity 25-BPR-1.)

1. Operator actuates the two pushbutton valves (1) and (2), until weight starts to descend.
2. Air passes through (1) and (2) to (3) and pilot of (5).
3. Four-way valve (5) is shifted, exhausting head end of cylinder (4) and pressurizing rod end of cylinder.
4. Weight of ram descends.
5. Speed of descend is controlled by exhaust control (7).
6. Timac air timing valve (3) is timing until time interval has elapsed.
7. Timac vents air line from (2) and (5).
8. With pilot exhausted, spring return of four-way valve (5) will shift valve to normal position.
9. Rod end of cylinder (4) is exhausting and head end is pressurized.
10. Weight rises.
11. Speed of ascend is controlled by exhaust control (6).

Blueprint Reading Activity 25-BPR-1
INSTRUMENTATION AND CONTROL DIAGRAMS - PNEUMATIC

Refer to the sequence of operations above and to the blueprint in Fig. 25-BPR-1, and answer the following questions.

1. What is the name of the circuit?
2. Give the drawing number.
3. How many different component parts are listed in the list of components?
4. What starts the sequence of operations?
5. What causes valve 5 to shift?
6. When valve 5 shifts, what happens to cylinder 4?
7. What part controls the time of the cylinder decent?
8. What causes cylinder 4 to reverse and rise?
9. What does the phantom line around valve 3 indicate? What components are included?
10. Is it necessary to depress valves 1 and 2 during entire cycle of press operation? Why?

1. ________________

2. ________________

3. ________________

4. ________________

5. ________________

6. ________________

7. ________________

8. ________________

9. ________________

10. ________________

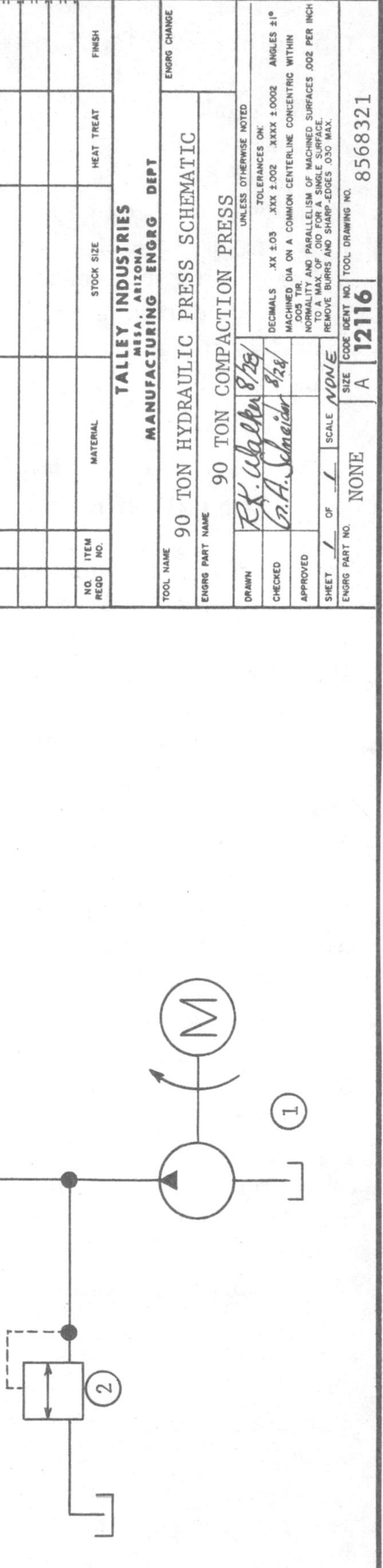

Fig. 25-BPR-2. Industry blueprint. (Talley Industries)

90 Ton Hydraulic Press Schematic

(Refer to Blueprint Reading Activity 25-BPR-2.)

Sequence of Operations

Description: A system to control a 90 ton hydraulic "explosive compaction" press. The press ram decends at a controlled rate, compacting the explosive charge slowly, to avoid the possibility of detonation due to a sudden shock.

1. Cycle is initiated by energizing solenoid A which shifts control valve (3) and supplies pressure to the head end of the hydraulic cylinder (7) which operates the ram of the press.
2. The ram cylinder (7) extends at a slow rate controlled by the fluid in the cylinder rod end which must pass thru the flow control valve (5).
3. When the cylinder is fully extended and the pressing pressure is reached, pressure switch (6) de-energizes solenoid A and energizes solenoid B. This shifts control valve (3) to the extreme opposite position and retracts ram cylinder (7). Control valve (5) has an integral check valve which permits the free flow of fluid in the opposite direction, causing cylinder (7) to retract rapidly.
4. Retraction of cylinder (7) actuates limit switch (8) which de-energizes solenoid B returning the control valve (3) to the neutral position, completing the cycle.
5. Two relief valves in the circuit provide two working pressures. On the upstroke of the cylinder (7) the low-pressure relief valve (4) controls the pressure. On the down-stroke the high-pressure relief valve (2) limits the maximum press tonnage.

The circuit has a safety feature in the spring centered, double solenoid operated, 3-position, 4-way valve (3). If an electrical power failure should occur, both solenoids would de-energize and valve (3) would return to the center position, with the center position ports blocked, thus stopping the ram cylinder (7) travel immediately.

Blueprint Reading Activity 25-BPR-2
INSTRUMENTATION AND CONTROL DIAGRAMS - HYDRAULIC

Refer to the sequence of operations above and to the blueprint in Fig. 25-BPR-2, and answer the following questions.

1. Give the name of the circuit. 1. ______________________
____
2. What is the drawing number? 2. ______________________
3. What starts the sequence of operations? 3. ______________________
____
4. The shifting of valve 3 supplies pressure to which end of cylinder 7? 4. ______________________
5. What causes the cylinder to extend at a slow controlled rate? 5. ______________________

6. What may occur in the sequence to cause cylinder 7 to retract?

6. ______________________

7. Cylinder 7 retracts rapidly. What is the cause of this?

7. ______________________

8. What completes the cycle?

8. ______________________

9. In case of an electrical power failure, what would happen to the system? Why?

9. ______________________

10. May the system be stopped at any time? How?

10. ______________________

Fluidic Control Systems

Another type of control circuitry that is finding many applications in industry because of its flexibility and reliability is FLUIDICS - - a low-pressure air system (3-10 psig). Fluidics are used as a means of sensing and controlling fluid power circuits.

Terms Used in Fluidic Control Systems

It is necessary to become familiar with the basic terminology and abbreviations used in fluidic control systems in order to read and interpret symbols and circuit diagrams. Those terms common to other instrumentation and control systems, and included earlier in this Unit, are not repeated here.

FLUIDICS: A technology concerned with logical control functions and uses fluid interactions to produce control signals.

DEVICE (GATE): Another name for a fluidic component.

CONTROL SIGNAL: A pressure signal used to operate the device.

OUTPUT SIGNAL: Air pressure at an output port sufficient to operate other fluidic or interface devices.

MONOSTABLE: A device that is stable only on the O_2 port and will switch back to the O_2 port when the control signal is not maintained.

BISTABLE: Device with two stable states and when switched from one output port to the other will remain at that port even though control signal is not maintained.

INTERFACE DEVICE: A device which is controlled by a fluidic device and which actuates various pneumatic and hydraulic components such as cylinders and rotary actuators that perform the actual work required.

P_S: Pressure in the form of air supply.

C_1, C_2, C_3 . . . : Control ports.

O_1, O_2: Output ports.

Fluidic Control Devices

There are numerous standardized components for fluidic control circuits but only those most commonly used are discussed here. A more complete listing is available in the American Standards prepared by the National Fluid Power Association.

FLIP-FLOP is a bistable device which can be made to switch its output from one port to the other. This is accomplished by

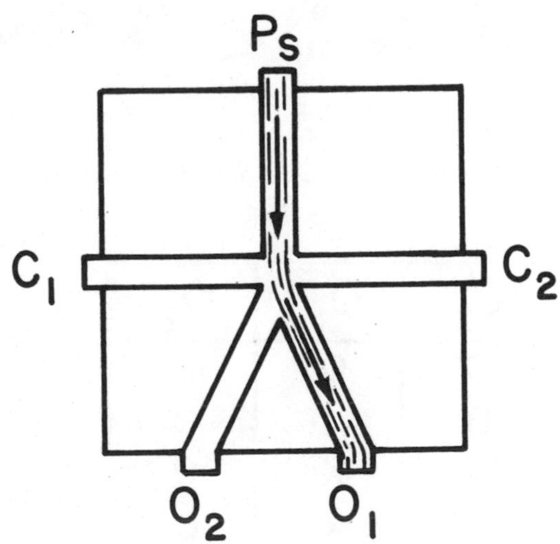

Fig. 25-4. Schematic of air flow in a Flip-Flop.

providing the proper control signal at one of the control ports, say C_1 to produce an output pressure at O_1, Fig. 25-4.

The output will continue at this port (pressure at P_S is assumed in all of discussions in this section) even though the control signal C_1 is removed. A control signal at C_2 will cause the output pressure to shift to O_2 where it will continue until a control pressure is applied at C_1.

The Flip-Flop derives its name from the two stable states, O_1 and O_2, and can be made to Flip-Flop from one to the other.

The graphic symbol for the Flip-Flop as recommended by the National Fluid Power Association is shown in Fig. 25-5. A TRUTH TABLE is also shown which depicts the operation of the Flip-Flop. The symbol X represents the presence of pressure while the symbol O represents the absence of pressure.

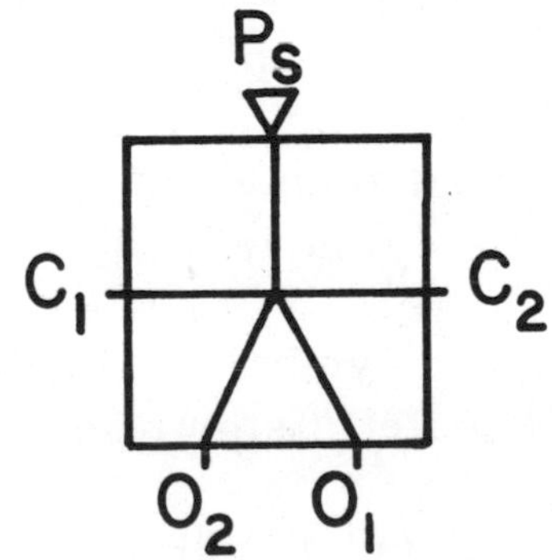

C_1	C_2	O_1	O_2
X	O	X	O
O	O	X	O
O	X	O	X
O	O	O	X

Fig. 25-5. Graphic symbol and truth table for fluidic Flip-Flop device.

OR/NOR GATE is a monostable device designed to provide an output at the O_2 (NOR) port when a control signal is not present. When pressure is present at one

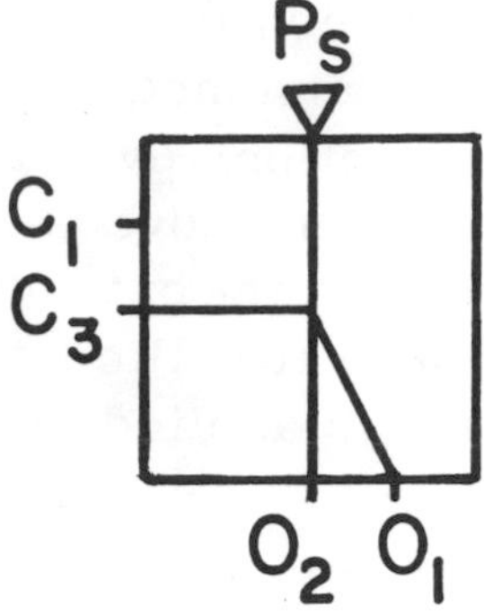

C_1	C_3	O_2(NOR)	O_1 (OR)
O	O	X	O
O	X	O	X
O	O	X	O
X	O	O	X
O	O	X	O
X	X	O	X

Fig. 25-6. Graphic symbol and truth table for OR/NOR gate.

or any combination of the control ports, the output will switch to the O_1 (OR port). This device is used in the controlling of components such as a spring-loaded hydraulic valve in a fluid power circuit to actuate a cylinder. The valve would be actuated as long as there was a signal present at C_1 or C_3 (or both). When the signal is removed from the control port(s), the output signal would be shifted to the O_2 port, permitting the spring-loaded hydraulic valve to return to neutral position, thus retracting the cylinder.

The graphic symbol and truth table for the OR/NOR gate is shown in Fig. 25-6.

INHIBITED OR GATE is very similar to the OR/NOR gate except that it provides for the O_2 and output to be overridden when desired by the presence of an inhibiting factor at C_2. For example, in a drilling operation it may be desirable to actuate the drill motors and move them into the workpiece, but not if another signal is present indicating the workpiece is not in position. This later signal (C_2) inhibits the O_1 (OR) gate and prevents the signal from moving to the O_1 port until the inhibited signal is removed.

The symbol and truth table for the Inhibited OR gate are shown in Fig. 25-7. Note that the output is at O_2 if either C_2 is present or C_1 and C_3 are absent. Output will be at O_1 (OR) only if C_1 or C_3 are present and C_2 is absent.

AND GATE is a monostable device used to determine when two control signals are

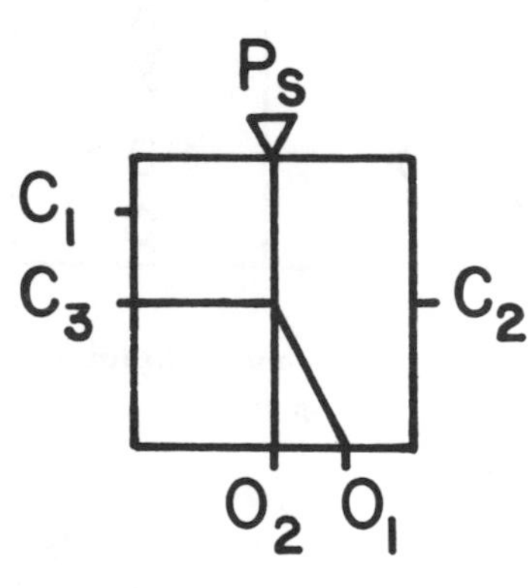

C_1	C_3	C_2	O_1	O_2
O	O	O	O	X
X	O	O	X	O
O	X	O	X	O
X	X	O	X	O
X	O	X	O	X
O	X	X	O	X
X	X	X	O	X

Fig. 25-7. Graphic symbol and truth table for Inhibited OR gate.

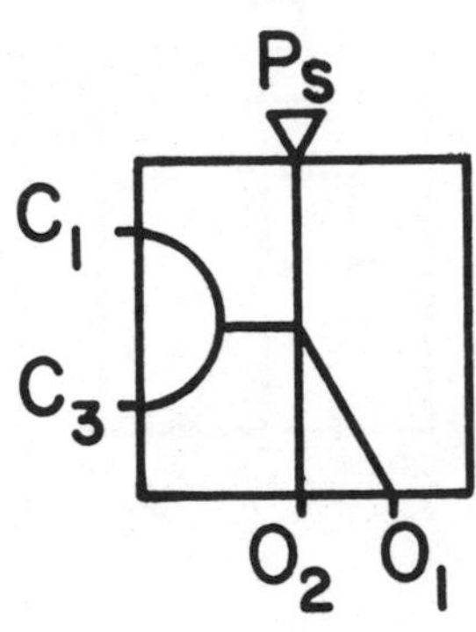

C_1	C_3	O_1	O_2
O	O	O	X
X	O	O	X
O	O	O	X
O	X	O	X
O	O	O	X
X	X	X	O
O	O	O	X

Fig. 25-8. Graphic symbol and truth table for AND gate.

present and may be contrasted to the NOR gate which is used to determine when none of the signals is present. A typical use of the AND gate would be when two signals must be present before another function is permitted to occur. For example, before a stamping operation can be performed, the correct tool must be in place and the work positioned. An AND gate could be used to signal this condition and initiate the stamping operation. Fig. 25-8 shows the standard graphic symbol and the truth table for the AND gate.

DIGITAL AMPLIFIER is similar to the Flip-Flop in that it is possible to switch the output from one port to the other. However, the Digital Amplifier is monostable and it is necessary to keep a signal present at the proper control port in order to maintain the desired output.

The advantage of the Digital Amplifier is that it has greater output pressure than other fluidic devices, approximately 60 percent of supply pressure, and can be used to control interface devices where other fluidic devices may not have sufficient pressure, particularly when several devices are used in series (one after the other) or when long control lines are necessary.

The standard symbol and the truth table for the Digital Amplifier are shown in Fig. 25-9.

ONE-SHOT is a device used to convert a sustained signal into a 10 millisecond pulse of fixed duration. A control signal at C_1 will cause the output to shift to O_1 for a period of 10 milliseconds. Then, even though the signal at C_1 is maintained, the

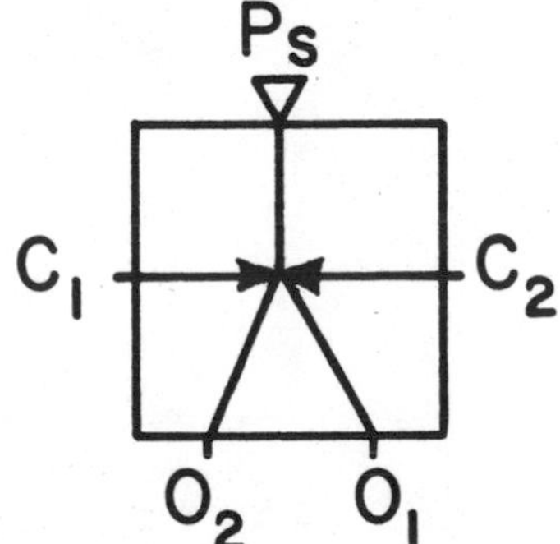

C_1	C_2	O_1	O_2
X	O	X	O
O	X	O	X
* O	O	—	—
* X	X	—	—

* Invalid States – in that a control signal must be present always at one, and only one, control port.

Fig. 25-9. Graphic symbol and truth table for the Digital Amplifier.

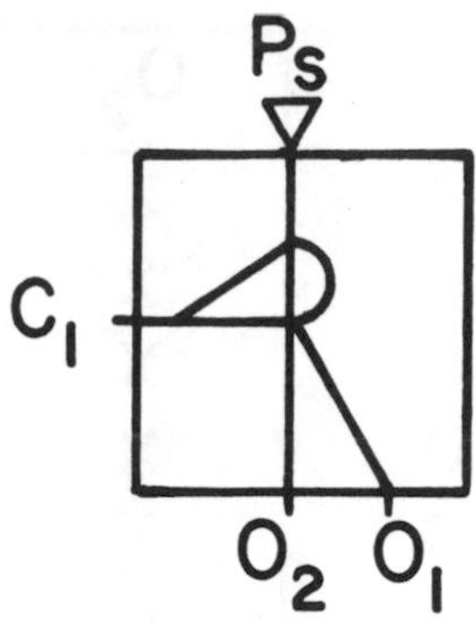

Fig. 25-10. Graphic symbol for One-Shot.

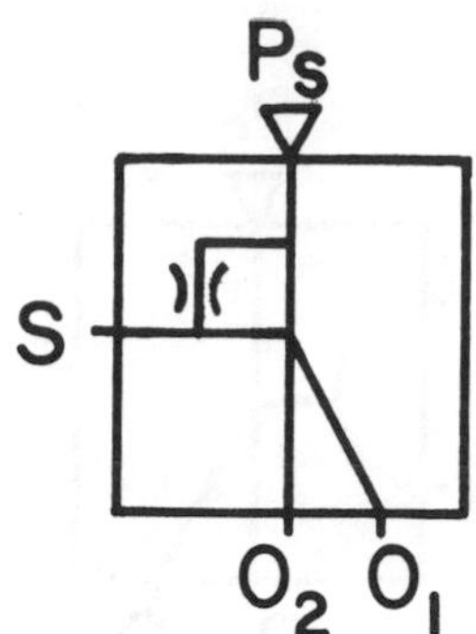

Fig. 25-11. Graphic symbol for Back Pressure Switch.

output shifts back to O_2. A typical application of the One-Shot might be when both pilots of a spool valve (double pilot operated valve) are pressurized at the same time and the valve will not move; therefore, one pilot must be exhausted before the opposite can be pressurized. This is accomplished by sending the shifting signal through a One-Shot before it actuates the valve pilot. Exhausting one pilot allows the valve spool to move when it receives pressure at the opposite pilot.

The symbol for the One-Shot is shown in Fig. 25-10.

BACK PRESSURE SWITCH (opposite of the OR gate) is used to detect the closure of a sensor. The output is normally from O_2. When S is blocked, see Fig. 25-11, the output will switch to O_1. The Back Pressure Switch is used in conjunction with mechanical components such as a pneumatic push button to initiate various sequences. See blueprint on page 228.

---

1. Extracted from USA Standard Drafting Practices Fluid Power Diagrams (USASI Y14.7–1966), with permission of the publisher, The American Society of Mechanical Engineers, United Engineering Center, 345 E. 47th Street, New York, N. Y. 10017.

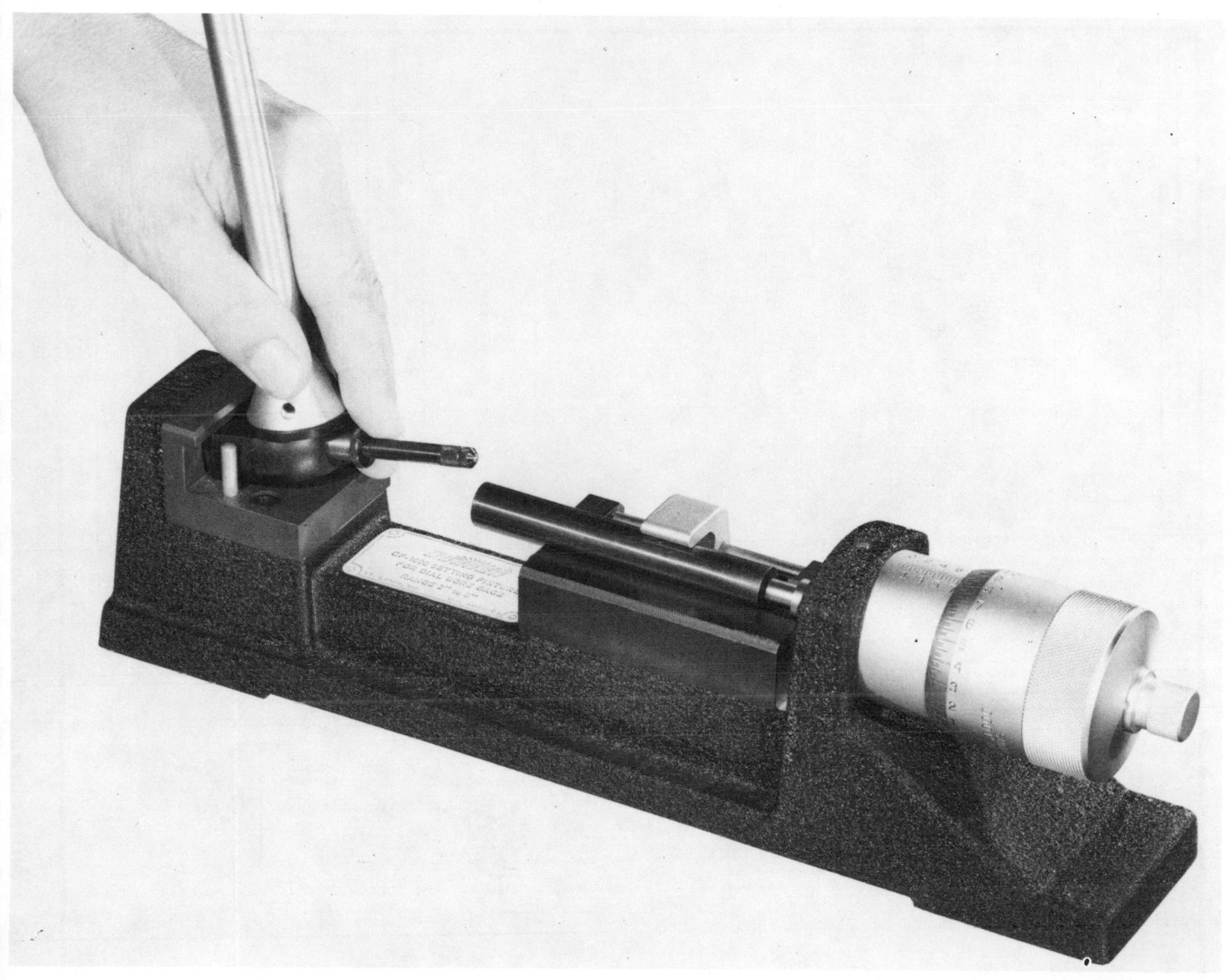

Industry photo, showing the use of the Gage Nest which is drawn in Fig. A-8-BPR-1 on page 231.

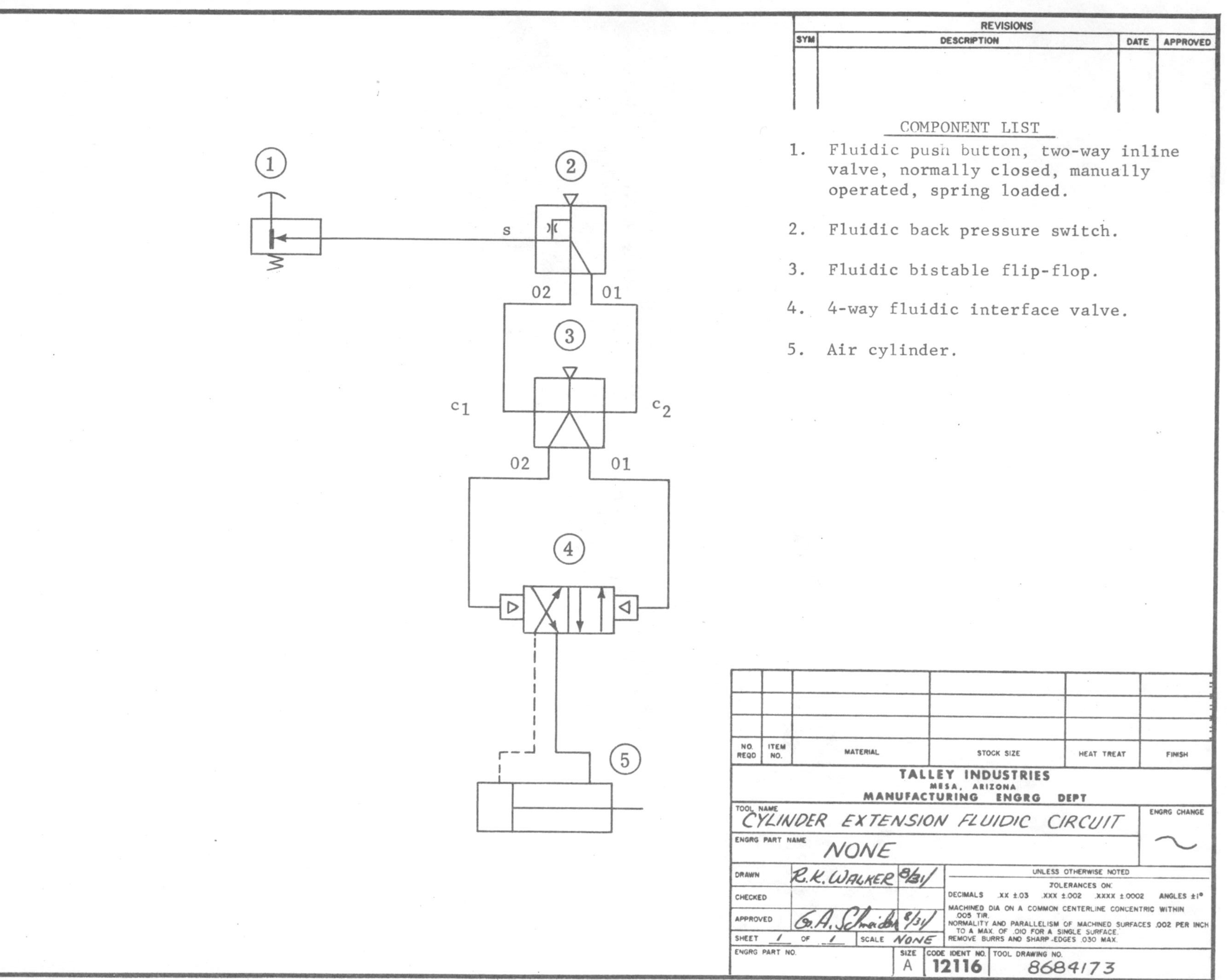

Fig. 25-BPR-3. Industry blueprint. (Talley Industries)

Cylinder Extension Fluidic Circuit

(Refer to Blueprint Reading Activity 25-BPR-3.)

Sequence of Operations

Description: A fluidic control system to extend and retract a pneumatic cylinder which in turn performs a clamping operation. The sequence of operations is as follows:

1. The normal position of the cylinder is in the retracted position. This is effected by the use of a push button, two-way inline, normally closed, manually operated, spring-loaded valve (1).
2. Initially, when the control air is turned on, pressure from the Back Pressure Switch (2) is blocked, by valve (1), causing an output at O_1.
3. This O_1 output signals the C_2 control port of Flip-Flop (3) which switches its output to the O_2 side, thus actuating the 4-way fluidic interface valve (4) and holds cylinder (5) in the retracted position.
4. When the operator desires to extend cylinder (5), thereby clamping the workpiece, he depresses push button of valve (1).
5. Depressing of the push button of valve (1) exhausts the Back Pressure S port and permits an output at O_2.
6. The O_2 output of the Back Pressure Switch signals C_1 of Flip-Flop (3) which switches its output to the O_1 port.
7. The O_1 output of Flip-Flop (3) actuates control valve (4), extending cylinder (5) to a clamping position.
8. Release of the push button on valve (1) completes the cycle by retracting the cylinder, releasing the workpiece and readying the system for the next piece.

Blueprints Reading Activity 25-BPR-3
INSTRUMENTATION AND CONTROL DIAGRAM - FLUIDICS

Refer to the sequence of operations above and to the blueprint in Fig. 25-BPR-3, and answer the following questions.

1. What is the name of the circuit diagram? The drawing number?
2. What is the normal position of the cylinder?
3. What actuates the system and clamps the workpiece?
4. When the workpiece is in a clamped position, valve 4 is controlled by which output port of Flip-Flop 3?
5. What is the normal position for valve 4 as shown?
6. Must the push button for valve 1 be depressed during entire clamping period? Why?

1. ______________________________
____

2. ______________________________

3. ______________________________
____

4. ______________________________

5. ______________________________
____
____

6. ______________________________
____
____
____

For additional Blueprint Activities for Unit 25 see pages 300, 304, and 307.

PART 6
ADVANCED AND ACHIEVEMENT BLUEPRINTS

Advanced Blueprint Activities

Blueprint Reading Activity A-8-BPR-1
DETAIL AND ASSEMBLY DRAWINGS

Refer to the blueprint in Fig. A-8-BPR-1 and answer the following questions.

1. Is this a detail or assembly drawing? — 1. ____________

2. What views are shown? — 2. ____________

3. Give the overall size of the part. — 3. ____________

4-12. Identify the following features in the other views.

	Top	Front	Rt. Side
4.	A	- -	
5.		- -	X
6.	C		
7.	D	- -	
8.		K	- -
9.			Z
10.	E	- -	
11.		L	- -
12.	- -		Y

13. Give the diameter of P. — 13. ____________

14. Give the reamed diameter of R. — 14. ____________

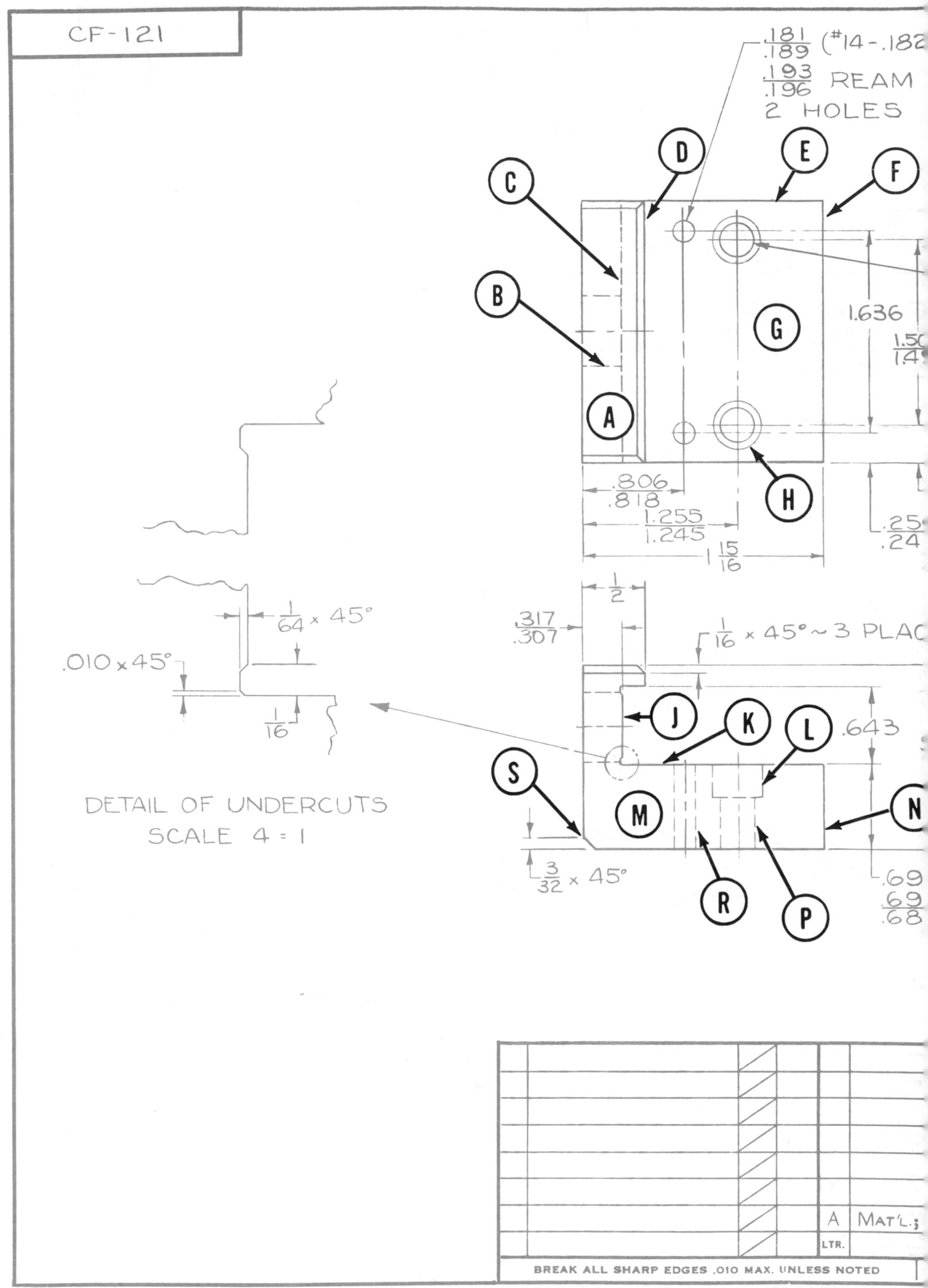

Fig. A-8-BPR-1. Industry blueprint. (Sunnen Products)

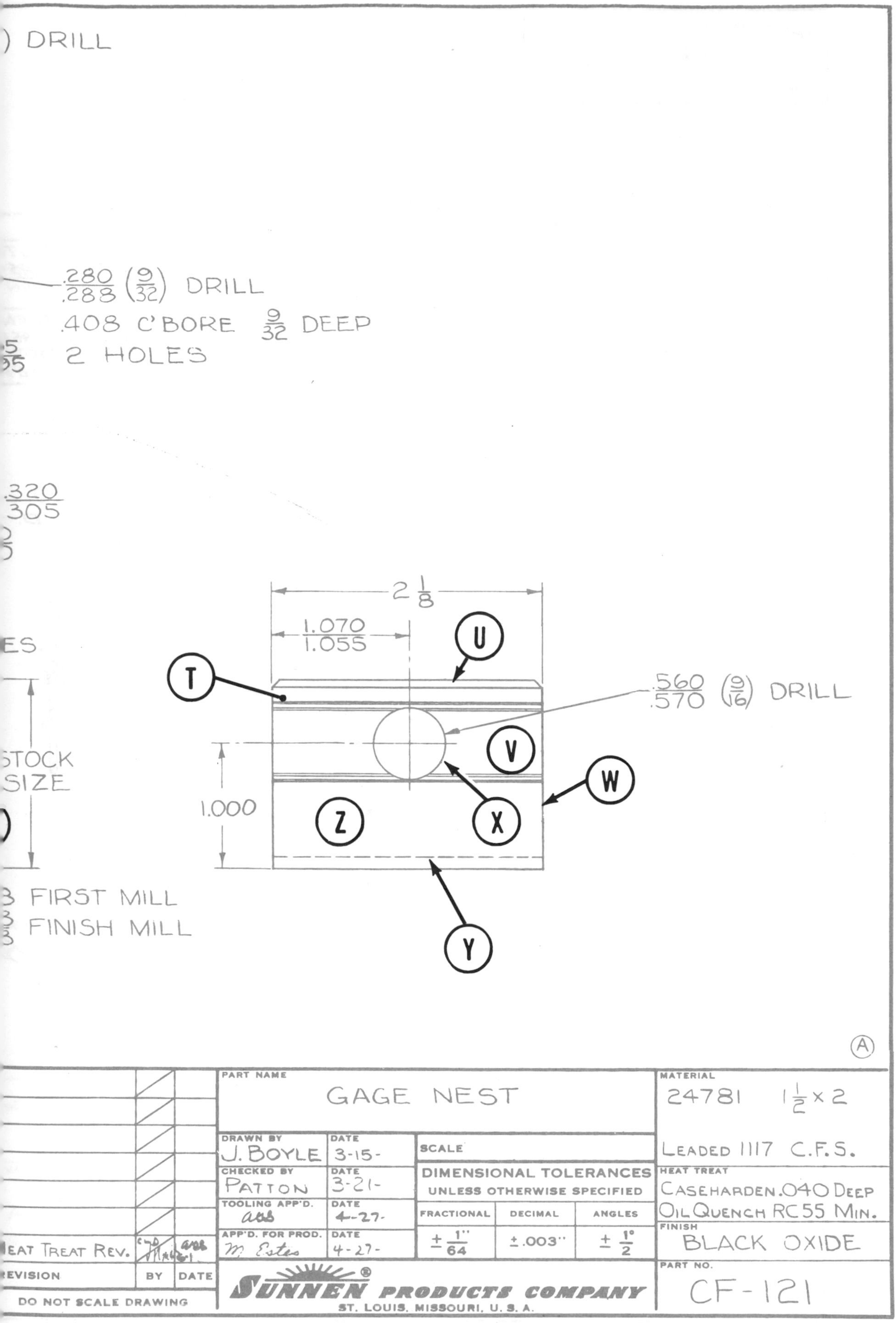

) DRILL

$\frac{.280}{.288}$ $(\frac{9}{32})$ DRILL
.408 C'BORE $\frac{9}{32}$ DEEP
5 / 35 2 HOLES

.320 / 305

ES

STOCK SIZE

3 FIRST MILL
3 FINISH MILL

2 $\frac{1}{8}$

$\frac{1.070}{1.055}$

$\frac{.560}{.570}$ $(\frac{9}{16})$ DRILL

1.000

T U V W X Y Z

Ⓐ

			PART NAME: GAGE NEST			MATERIAL: 24781 $1\frac{1}{2} \times 2$ LEADED 1117 C.F.S.
			DRAWN BY: J. BOYLE	DATE: 3-15-	SCALE	
			CHECKED BY: PATTON	DATE: 3-21-	DIMENSIONAL TOLERANCES UNLESS OTHERWISE SPECIFIED	HEAT TREAT: CASEHARDEN .040 DEEP OIL QUENCH RC55 MIN.
			TOOLING APP'D.: ABS	DATE: 4-27-	FRACTIONAL $\pm\frac{1''}{64}$ / DECIMAL ±.003" / ANGLES $\pm\frac{1°}{2}$	
HEAT TREAT REV.			APP'D. FOR PROD.: M. Estes	DATE: 4-27-		FINISH: BLACK OXIDE
REVISION	BY	DATE	SUNNEN PRODUCTS COMPANY ST. LOUIS, MISSOURI, U. S. A.			PART NO. CF-121
DO NOT SCALE DRAWING						

Blueprint Reading Activity A-8-BPR-2
DETAIL AND ASSEMBLY DRAWINGS

Refer to the blueprint in Fig. A-8-BPR-2 and answer the following questions.

1. What type of an assembly is represented? 1. ____________________

2. Give the name of the assembly. 2. ____________________

3. What is the number of the drawing? 3. ____________________

4. How many steps are involved in the adjusting of the brake? 4. ____________________

5. In what position is the brake handle when step "1" is performed? 5. ____________________

6. Circle the direction in which the knob is turned. 6. ↺ ↻

7. What position is the brake handle to be in during steps "2" and "3"? 7. ____________________

8. The lever on the brake assembly will be in what condition when step "3" is completed? 8. ____________________

9. In what position is the brake handle during step "4"? 9. ____________________

10. How is final adjustment made to the brake? 10. ____________________

11. When is the brake properly adjusted? 11. ____________________

Blueprint Reading Activity A-11-BPR-1
DIMENSIONS AND TOLERANCES

Refer to the blueprint in Fig. A-11-BPR-1 and answer the following questions.

1. What is the name of the part? 1. ______________________

2. What is the print number? 2. ______________________

3. What general tolerances are given? 3. ______________________

4. Name the type section shown. 4. ______________________

5. What is the scale of the original plan? 5. ______________________

6. What material is specified? 6. ______________________

7. Give the overall dimensions. 7. ______________________

8. Give the dimensions and/or specifications for the following features: 8. A. ______ F. ______ G. ______ H. ______

9. Give the line or surface in the rt. side view that represent the following in the front views: 9. A ______ B ______ C ______ J ______ L ______

10. What relationship must exist between features E and K? C and K? 10. E-K ______ C-K ______

11. Give the specification for C. 11. ______________________

12. What surface texture is required for surfaces E and J? A and B? 12. E-J ______ A-B ______

13. What feature is controlled by dimension at D? 13. ______________________

14. Give the heat treatment specification. 14. ______________________

15. What finish specification is given? 15. ______________________

16. How many of this part are required for the next assembly? 16. ______________________

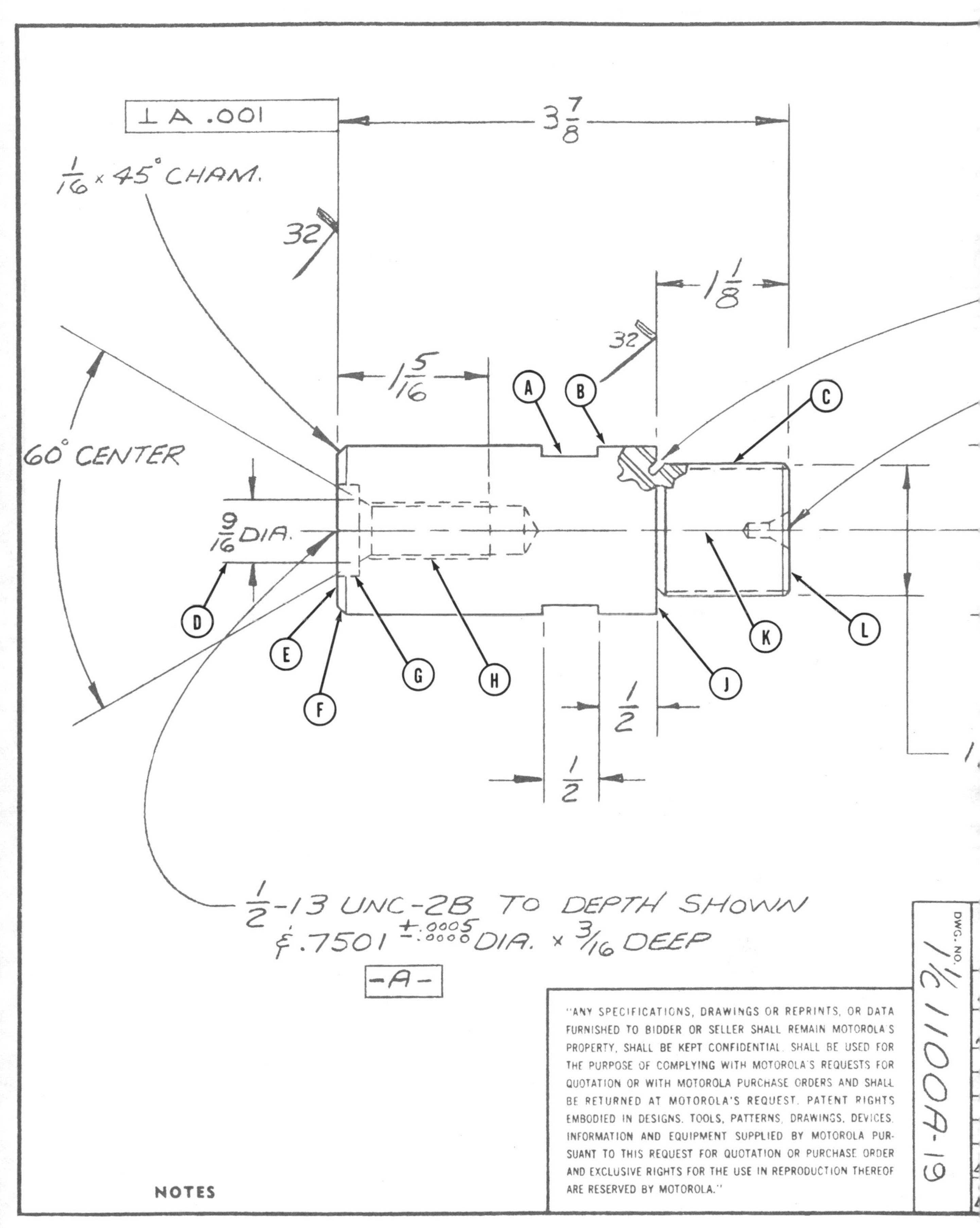

Fig. A-11-BPR-1. Industry blueprint. (Motorola, Inc.)

E.C.O.	ENG'R	LET.	CHANGE	BY	DATE
	L.F.R.	A	RELEASED	RICHARDSON	4-15-

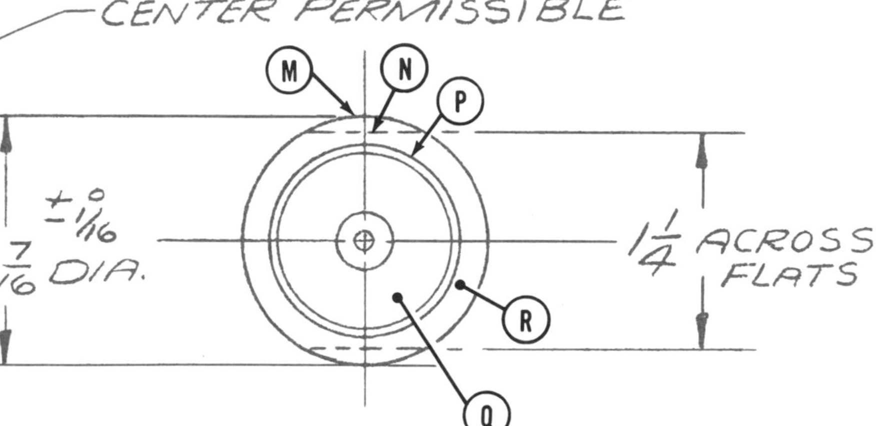

3/8-12 UNF-2A THREAD
P.D. 1.0691 / 1.0631

◎ A .002

MATERIAL SPEC.			ITEM	REQ'D	MATERIAL SPECIFICATION LIST
CARPENTER 610					

HEAT TREAT: HARDEN & DRAW R/C 58-60	
APPLIED FINISH: BLACK OXIDE PER MIL-C-13924B CLASS III	

	REQ'D	NEXT ASSEMBLY
	1	1½C1180A
DEVICE USED ON	REQ'D	NEXT ASSEMBLY

	DATE
DRAWN BY: KEN RICHARDSON	4-15-
CHECKED BY: KEN RICHARDSON	5-6-

UNLESS OTHERWISE SPECIFIED

1. 63 √ RMS ALL MACHINED SURFACES
2. FRAC. TOL. ± 1/64
3. DEC. TOL. ± .005
4. ANGULAR TOL. ± 3°
5. FEATURE CONTROL SYMBOLS PER MIL-STD-8 CURRENT REV.
6. BREAK ALL SHARP EDGES & CORNERS, REMOVE BURRS.
7. UNDERLINED DIMS NOT TO SCALE

TITLE: RAM ADAPTER FOR 2 STRIP 40 CAVITY MOLD

MOTOROLA INC.
Semiconductor Products Division
5005 EAST McDOWELL ROAD, PHOENIX, ARIZONA 85008

SCALE	SIZE	DWG. NO.
FULL	B	1½C1180A-19

Blueprint Reading Activity A-11-BPR-2
DIMENSIONS AND TOLERANCES

Refer to the blueprint in Fig. A-11-BPR-2 and answer the following questions. Give all answers to dimensions in millimeters.

1. What is the name of the part? 1. ______

2. Give the drawing number. 2. ______

3. What is the radius of the feature at E? 3. ______

4. Give the dimensions for the following: 4. F ______
G ______
H ______
L ______
M ______
N ______

5. Give the diameter and depth of the hole at J. Will this be through to the slot? By how many millimeters? 5. ______

6. What is the width of the slot at K? 6. ______

7. What is the diameter of the hole at P? 7. ______

8. Give the diameter and thickness of the boss after machining at P. 8. ______

9. Give the angle and tolerance which the pedal pad must make with the arm. 9. ______

10. What is the thickness of the part along the hole at E? 10. ______

Blueprint Reading Activity A-12-BPR-1
SECTIONAL VIEWS

Refer to the blueprint in Fig. A-12-BPR-1 and answer the following questions.

1. What is the name of the object? 1. ______

2. Give the number of the drawing. 2. ______

3. What material is used for the casting? 3. ______

4. What type of section is shown? 4. ______

5. Give the overall finished dimensions of the cast sheave. 5. ______

6. Give the inside diameter of the recess for the friction rings. 6. ______

7. What is the diameter, in inches, of the drill to be used for the rivet holes? 7. ______

8. How many rivet holes are to be drilled and what is their location? 8. ______

9. How many friction rings are required and where would their description be found? 9. ______

10. Give the dimension for the center hole and indicate what type of a dimension it is. 10. ______

11. How wide are the belt grooves at the 6 in. diameter? What angle are they to be? 11. ______

12. What is the center distance between belt grooves? 12. ______

13. Give the thickness and major diameter of the grooves in the center hole. 13. ______

14. What is the thickness of the sheave beyond the 3.275 diameter? 14. ______

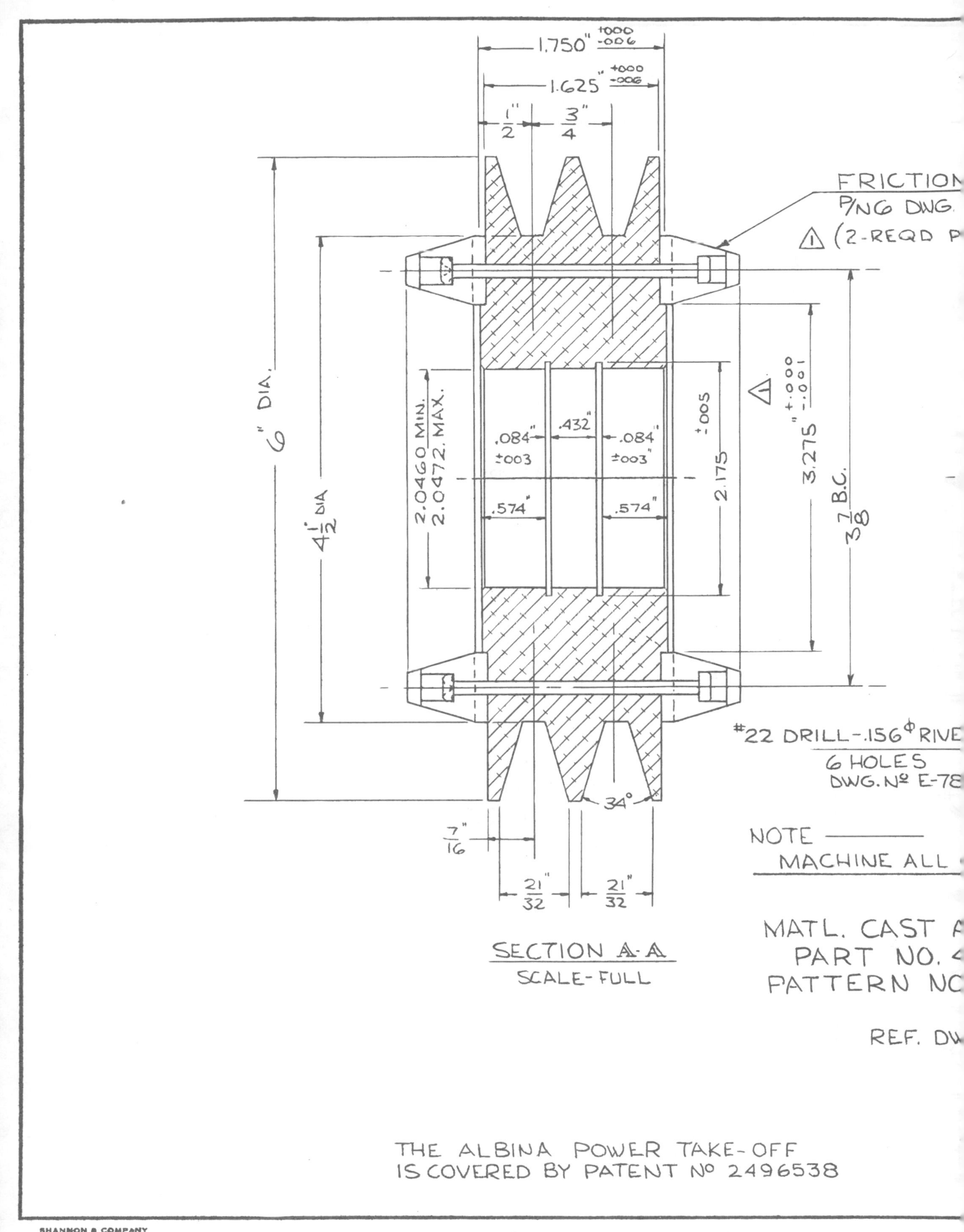

Fig. A-12-BPR-1. Industry blueprint. (Albina Engine and Machine Works)

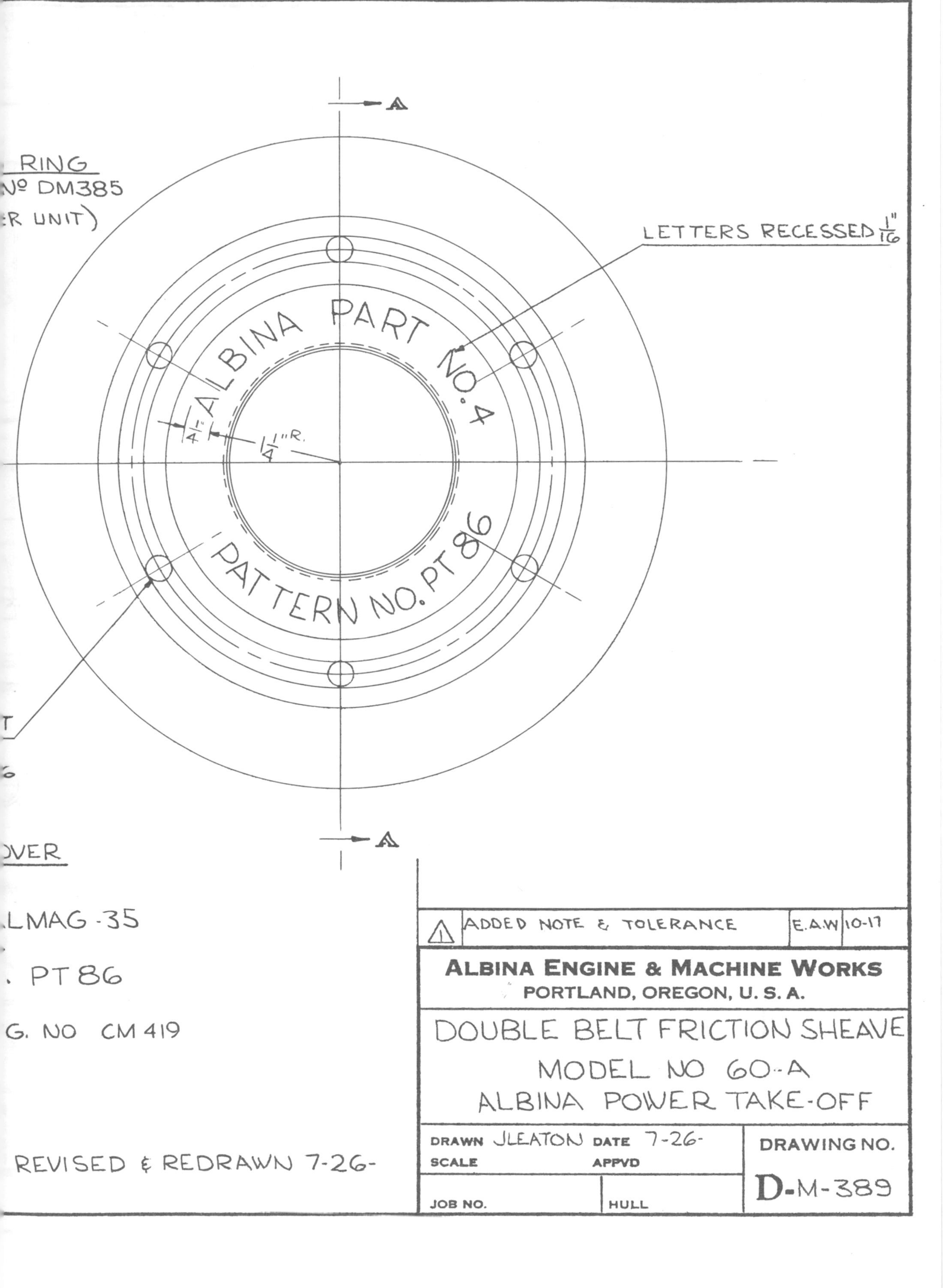

RING
№ DM385
R UNIT)
LETTERS RECESSED 1/16"
ALBINA PART NO. 4
PATTERN NO. PT 86
1/4"
1 1/4" R.
A
A
T
OVER
LMAG-35
. PT 86
G. NO CM 419
REVISED & REDRAWN 7-26-
ADDED NOTE & TOLERANCE E.A.W 10-17
ALBINA ENGINE & MACHINE WORKS
PORTLAND, OREGON, U. S. A.
DOUBLE BELT FRICTION SHEAVE
MODEL NO 60-A
ALBINA POWER TAKE-OFF
DRAWN J LEATON DATE 7-26-
SCALE APPVD
JOB NO. HULL
DRAWING NO.
D-M-389

Blueprint Reading Activity A-12-BPR-2
SECTIONAL VIEWS

Refer to the blueprint in Fig. A-12-BPR-2 and answer the following questions.

1. Give the name of the part and drawing number? 1. ____________

2. What material is specified? 2. ____________

3. What type sectional views are shown? 3. ____________

4. Give the overall size of the piston after machining. 4. ____________

5. What are the limit dimensions for the diameter just below the upper ring groove? 5. ____________

6. Give the diameter and width of the lower ring groove after machining. 6. ____________

7. What radius is specified for the inner diameter of the ring grooves? 7. ____________

8. What is the width and two diameters of the upper ring groove? 8. ____________

9. What is the measurement of the upper ring groove over .050 diameter gage? 9. ____________

10. Give the limit dimensions on the diameter and the depth of the reaming operation for the two holes in the ring grooves. 10. ____________

11. Give the size of the piston rod holes. 11. ____________

12. What is the center distance of the piston rod holes from the base? 12. ____________

Blueprint Reading Activity A-14-BPR-1
TITLE BLOCK

Refer to the blueprint in Fig. A-14-BPR-1 and answer the following questions.

1. What is the name and number of the blueprint? 1. ____________________
2. What is the scale of the original plan? 2. ____________________
3. How large was the drawing made? 3. ____________________
4. What material is specified? In what form? 4. ____________________
5. Give the heat treatment specification. 5. ____________________
6. What hardness is specified? 6. ____________________
7. This part is used on what model? 7. ____________________
8. Give the limit dimension for the hole. 8. ____________________
9. What is the thickness of the part through the hole? 9. ____________________
10. Give the dimension between the hole and the arc on the right side, as a limit dimension. 10. ____________________
11. This drawing has been released to production by how many individual approvals, exclusive of changes? 11. ____________________

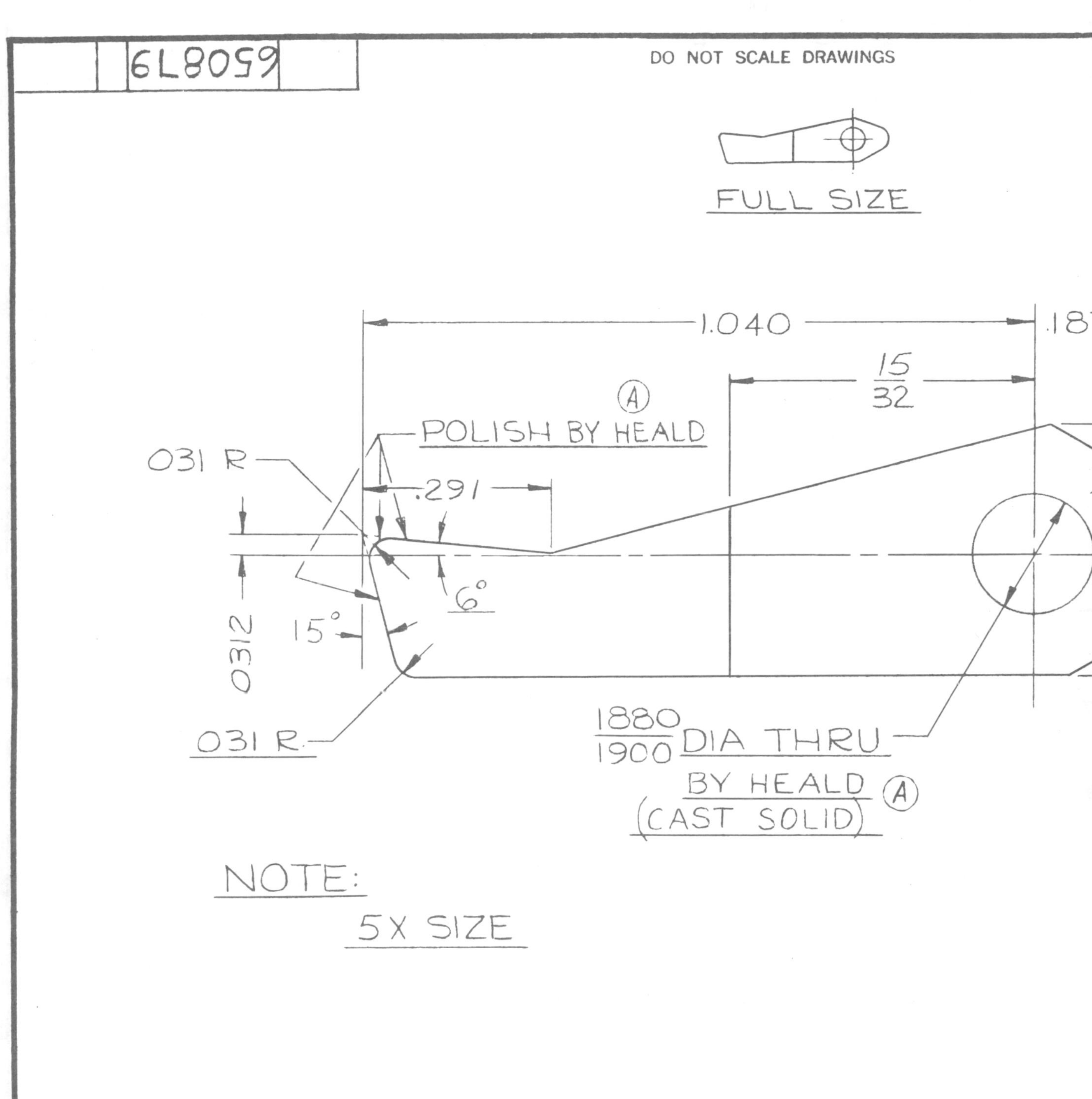

Fig. A-14-BPR-1. Industry blueprint. (Heald Machine Co.)

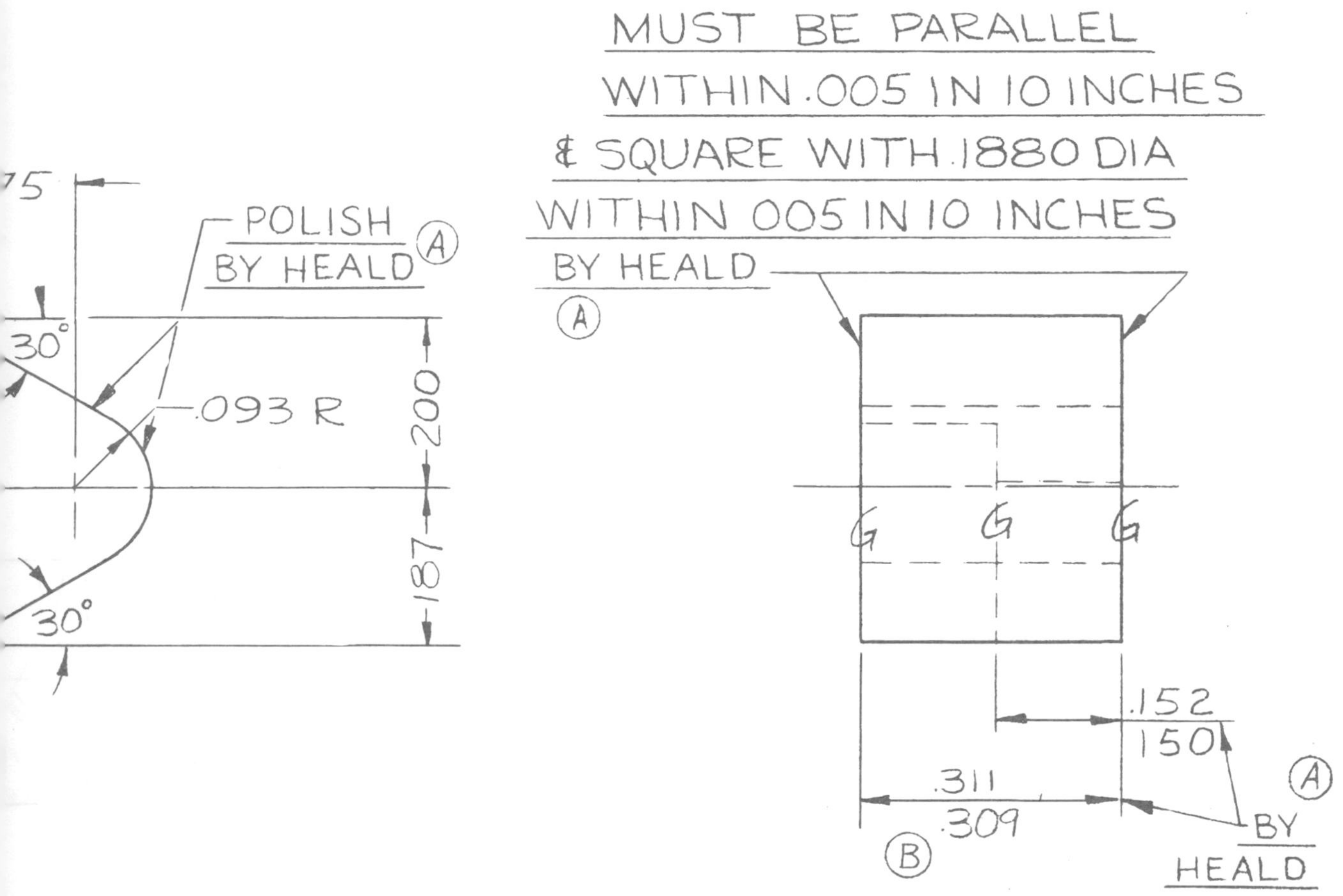

MUST BE PARALLEL WITHIN .005 IN 10 INCHES & SQUARE WITH .1880 DIA WITHIN .005 IN 10 INCHES BY HEALD (A)

POLISH BY HEALD (A)

.093 R

30° 30° .200 .187

G G G

.152 / .150

.311 / .309 (B)

(A) BY HEALD

ITEM	QTY.	PART NUMBER	DESCRIPTION	MFG. CODE	REP. PTS.

BILL OF MATERIALS

B

SCALE	MATERIAL	MATERIAL SIZE	HEAT TREAT	HARDNESS
5 X SIZE	(A) STEEL #8620	(A) PURCHASE PRECISION CASTING	200 & 500 1/32" DEEP BY HEALD (A)	60-63

USED ON	MODEL
ICF	90

DRAWN BY	DATE	FIRST MADE FOR	ASSY NO.	QTY.	PROPOSAL NO.
E BERNARD	12-1-	60-10391			

CHECKED BY: R. WESTON — DEPT.: GRIND ENG

STDS.: C Barrett — METH.: R SLACK

NAME: PAWL

UNLESS OTHERWISE NOTED TOLERANCES ARE:
DECIMALS: ± .005
FRACTIONS: ± 1/64
ANGLES: ± 1/2°

650879

CLASS A

SIMILAR TO

SUPERSEDED BY

SUPERSEDES

HEALD LIMITED

INTER. (NOT / IS) RETAINED

THE HEALD MACHINE COMPANY

ASSOCIATE OF THE CINCINNATI MILLING MACHINE COMPANY

Blueprint Reading Activity A-14-BPR-2
TITLE BLOCK

Refer to the blueprint in Fig. A-14-BPR-2 and answer the following questions.

1. Give name and number of the drawing. 1. ____________________
2. What is the scale of the original plan? 2. ____________________
3. Give the material specification. 3. ____________________
4. What heat treatment is specified? 4. ____________________
5. On what assembly is this part used? How many? 5. ____________________
6. Give the turned dimension of diameter D and G which are to be hardened. 6. ____________________
7. Give the location dimension of hole 1C in relation to the center lines. 7. ____________________
8. What kind of a dimension is this? 8. ____________________
9. Give the limit dimensions for diameter B after grinding. 9. ____________________
10. What is the depth of the bored hole on the right end? 10. ____________________

Blueprint Reading Activity A-15-BPR-1
LIST OF MATERIALS

Refer to the blueprint in Fig. A-15-BPR-1 and answer the following questions.

1. What is the name of the part or assembly? 1. ____________________

2. Give the drawing number. 2. ____________________

3. What is the scale of the original plan? 3. ____________________

4. What is the block called that lists the materials? 4. ____________________

5. How many different parts are listed? 5. ____________________

6. What is the name of Part No. -6? 6. ____________________

7. How many support flange parts are required? 7. ____________________

8. What material is used for Part No. -4? 8. ____________________

9. Give the specification for Part No. -7. 9. ____________________

10. What is the weight of Part No. -3? 10. ____________________

11. Give the overall dimensions for the length of the assembly. The diameter. 11. ____________________

12. What part has the greatest diameter? 12. ____________________

13. Give the dimension and the tolerance limits on the inside diameter of the: (a) Tube, (b) Band, and (c) Part No. -3. 13. (a) ____________________ (b) ____________________ (c) ____________________

14. What is the part number for the bellows? 14. ____________________

15. What is the indicated weight of the assembly? Is this weight actual or calculated? 15. ____________________

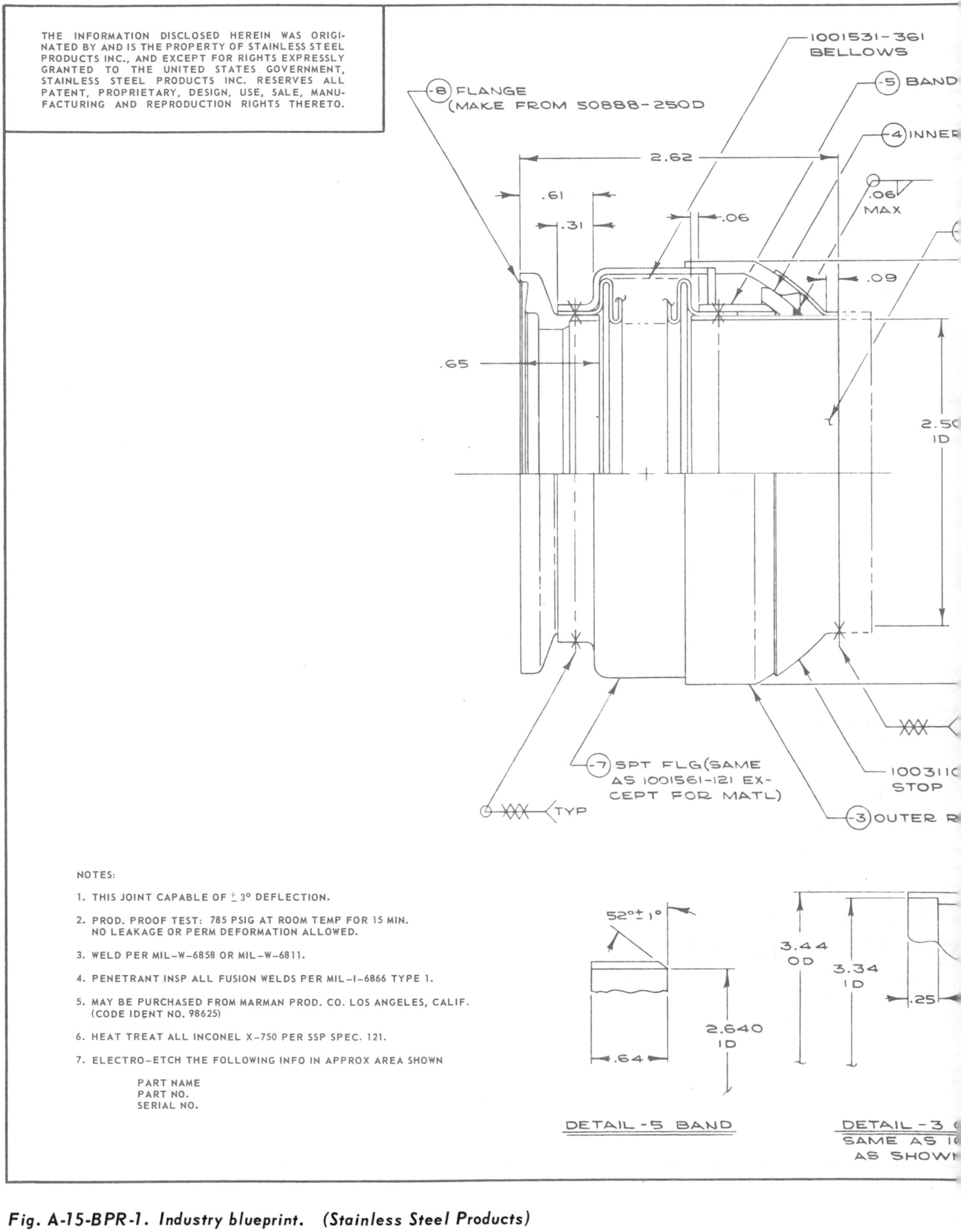
THE INFORMATION DISCLOSED HEREIN WAS ORIGINATED BY AND IS THE PROPERTY OF STAINLESS STEEL PRODUCTS INC., AND EXCEPT FOR RIGHTS EXPRESSLY GRANTED TO THE UNITED STATES GOVERNMENT, STAINLESS STEEL PRODUCTS INC. RESERVES ALL PATENT, PROPRIETARY, DESIGN, USE, SALE, MANUFACTURING AND REPRODUCTION RIGHTS THERETO.
1001531-361
BELLOWS
-8 FLANGE
(MAKE FROM 50888-250D
-5 BAND
-4 INNER
2.62
.61
.31
.06
.06
MAX
.09
.65
2.50
ID
-7 SPT FLG (SAME AS 1001561-121 EXCEPT FOR MATL)
TYP
-3 OUTER
STOP
NOTES:
1. THIS JOINT CAPABLE OF ± 3° DEFLECTION.
2. PROD. PROOF TEST: 785 PSIG AT ROOM TEMP FOR 15 MIN. NO LEAKAGE OR PERM DEFORMATION ALLOWED.
3. WELD PER MIL-W-6858 OR MIL-W-6811.
4. PENETRANT INSP ALL FUSION WELDS PER MIL-I-6866 TYPE 1.
5. MAY BE PURCHASED FROM MARMAN PROD. CO. LOS ANGELES, CALIF. (CODE IDENT NO. 98625)
6. HEAT TREAT ALL INCONEL X-750 PER SSP SPEC. 121.
7. ELECTRO-ETCH THE FOLLOWING INFO IN APPROX AREA SHOWN
PART NAME
PART NO.
SERIAL NO.
52° ± 1°
2.640
ID
.64
DETAIL -5 BAND
3.44
OD
3.34
ID
.25
DETAIL -3
SAME AS
AS SHOWN

Fig. A-15-BPR-1. Industry blueprint. (Stainless Steel Products)

REVISIONS				
SYM	ZONE	DESCRIPTION	DATE	APPROVED

RACE

TUBE

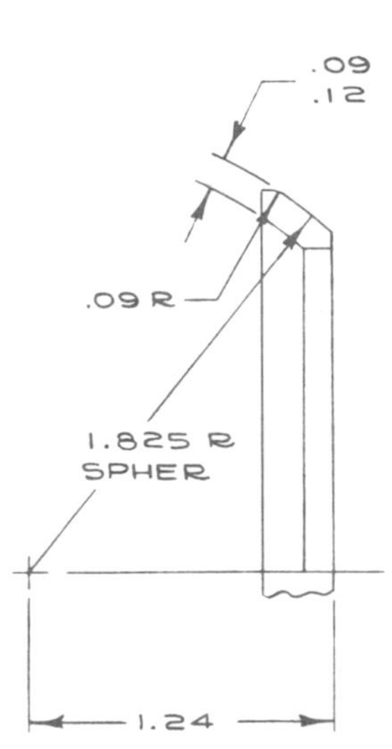

3.44
OD
(REF)

DETAIL -4 INNER RACE
SAME AS 1003110-10 EXCEPT
AS SHOWN

TRIM THRU CENTER OF
NUGGET AT DIM SHOWN

-13

ACE

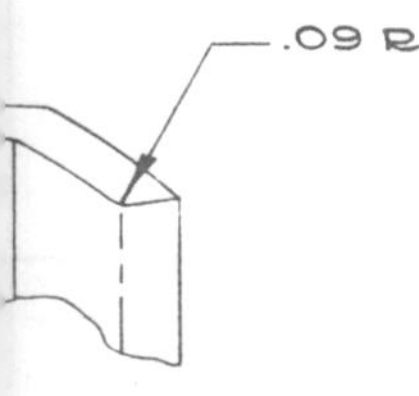

UTER RACE
03110-5 EXCEPT

	1	1003110-13		STOP	————	————	.046
	1	1001531-361		BELLOWS	————	————	.265
	1	-8		FLANGE	MAKE FROM 50888-250D	————	.22
	1	-7		SUPPORT FLANGE	.050 THK SH INCONEL X 750	MIL-N-7786	.191
	1	-6		TUBE	.032 THK SH CRES N 155	AMS 5532	.090
	1	-5		BAND	.060 THK SH CRES 321	MIL-S-6721(Ti)	.094
	1	-4		INNER RACE	.125 THK SH INCONEL X 750	MIL-N-7786	.115
	1	-3		OUTER RACE	.125 THK SH INCONEL X 750	MIL-N-7786	.334
	+	-1		ASSY	————	————	
QTY REQD	-1	PART OR IDENTIFYING NO.	ZONE	NOMENCLATURE OR DESCRIPTION	MATERIAL OR NOTE	SPECIFICATION	UNIT WT

LIST OF MATERIAL OR PARTS LIST

UNLESS OTHERWISE SPECIFIED DIMENSIONS ARE IN INCHES TOLERANCES ON DECIMALS .X ± .06 .XX ± .03 .XXX ± .010 ANGLES ± 0° 30′	DRAWN	SCHLIMGEN	12/9/	STAINLESS STEEL PRODUCTS INCORPORATED BURBANK, CALIFORNIA
	CHECK			
	ENGR			JOINT-UNIVERSAL 2.50 DIA - ASSY OF
	STRESS			
	WEIGHT			
125 RMS FINISH UNLESS OTHERWISE SPECIFIED	PROD			
	APPR	DJ Vuly	12/9/	
DO NOT SCALE THIS DRAWING	APPROVED			CODE IDENT NO. 98769 — SIZE D — 1004312
RELEASE DATE	APPROVED			SCALE 2/1 — WEIGHT 1.34 LB (CALC) — SHEET 1

Blueprint Reading Activity A-15-BPR-2
LIST OF MATERIALS

Refer to the blueprint in Fig. A-15-BPR-2 and answer the following questions.

1. What is the name of the drawing? 1. ____________________

2. Give the drawing number. 2. ____________________

3. What is the scale of the original plan? 3. ____________________

4. Give the name of the block that lists the materials. 4. ____________________

5. How many different parts are listed? 5. ____________________

6. Give the material, specification and size for the -5 Tube. 6. ____________________

7. What is the tensile strength of the above material? 7. ____________________

8. What is the unit weight of the -1 part? 8. ____________________

9. Give the dimension between rod end center holes for part -3. 9. ____________________

10. How many MS20470AD3 rivets are placed in each end of the tube? 10. ____________________

11. What is the model on which the -3 part is used? How many are required on the final assembly? 11. ____________________

Blueprint Reading Activity A-16-BPR-1
DRAWINGS NOTES

Refer to the blueprint in Fig. A-16-BPR-1 and answer the following questions.

1. What is the name of the part? 1. ____________________
2. Give the drawing number. 2. ____________________
3. What size drawing was made? 3. ____________________
4. What is the scale of the original plan? 4. ____________________
5. Give the material and specification for the casting. 5. ____________________
6. What tensile strength is required of the material? 6. ____________________
7. What part number is to be stamped on the part as required by note ⚠5 ? How is this to be done? 7. ____________________
8. How many general notes are included on the blueprint? Local notes? 8. ____________________
9. What linear tolerance is allowed on the casting? 9. ____________________
10. What military standard is to be used in inspecting the casting? 10. ____________________
11. Give the diameter of the spot face for the .37 DIA holes. 11. ____________________
12. How high above the surface are the numerals indicating the part numbers? 12. ____________________

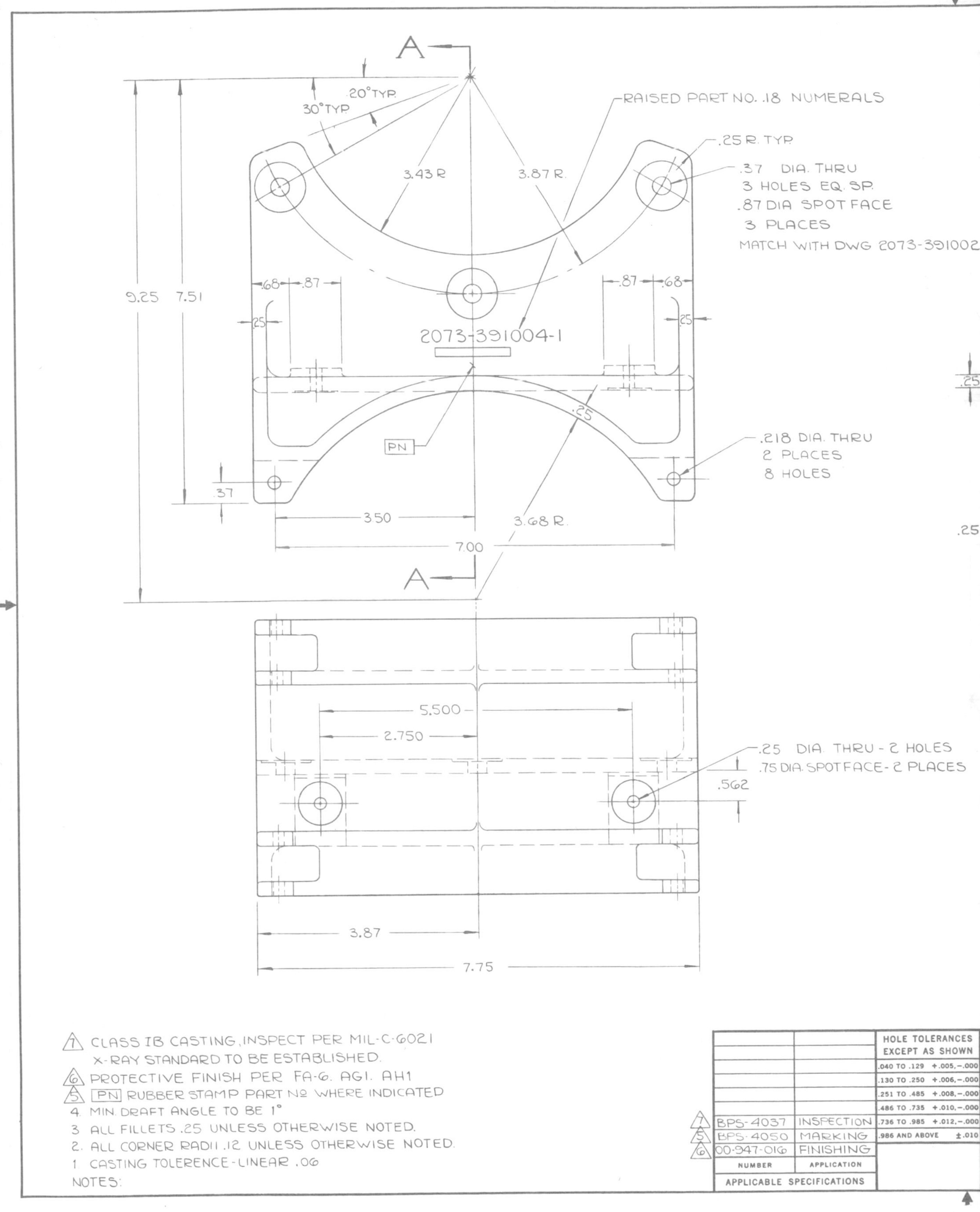

Fig. A-16-BPR-1. Industry blueprint. (Bell Aero Systems)

REVISIONS

SYM	DESCRIPTION	DATE	APPROVED

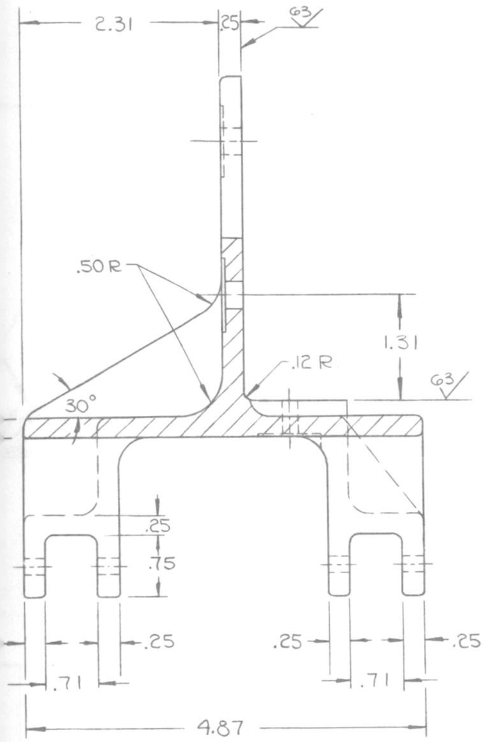

2.31 .25 63
.50 R
1.31
.12 R
63
30°
.25
.75
.25 .25 .25
.71 .71
4.87

SECTION A-A

2073-391004

QTY REQD	CODE IDENT	PART OR IDENTIFYING NO.	NOMENCLATURE OR DESCRIPTION	MATERIAL OR NOTE	SPECIFICATION	DIA	THK	W	LG	TS 1000 PSI	UNIT WT
1	06633	2073-391004-1C	CASTING	AL ALLOY 195T-6	QQ-A-601 CLASS 4M COND T6					32	
1	06633	2073-391004-1	BRACKET								2.50

LIST OF MATERIALS OR PARTS LIST

DASH NO.	NEXT ASSY	USED ON	NEXT ASSY	FINAL ASSY	EO NO.	SH OF SH
-1	2073-382003	SKMR-1	1	1		
	APPLICATION		QTY REQD		EO TO COMPLETE PRINT	

UNLESS OTHERWISE SPECIFIED DIMENSIONS ARE IN INCHES TOLERANCES ON .X ±.1 .XX ±.03 .XXX ±.010 ANGLES ±1/2°

UNLESS OTHERWISE SPECIFIED BREAK ALL SHARP EDGES APPROX .015 R OR CHAM

MACHINED SURFACES EXCEPT AS NOTED ✓

SUPERSEDES | SUPERSEDED BY

DESIGN deCORSE 9-14-
GROUP T.R. LAWSON 9/14/
APPD [signature] 10/5/
CHK J. Hunt 9/21/
STRESS F. C. Anderson 9/14/
WT C. S. Tilyou 9/18/
REL E. Kelly 10-5-
[signature] 10/4/

BELL AEROSYSTEMS COMPANY
DIVISION OF BELL AEROSPACE CORPORATION
AEROSPACE-ROCKETS DIVISION
POST OFFICE BOX 1 BUFFALO 5, NEW YORK

BRACKET-RESERVOIR
HYDRAULIC SYSTEM

20361

CODE IDENT NO. 06633 SIZE D 2073-391004

SCALE FULL SHEET

Blueprint Reading Activity A-16-BPR-2
DRAWINGS NOTES

Refer to the blueprint in Fig. A-16-BPR-2 and answer the following questions.

1. What is the name of the assembly? 1. ____________________

2. Give the drawing number. 2. ____________________

3. What is the scale of the original plan? 3. ____________________

4. What size drawing was made? 4. ____________________

5. How many separate parts are required for the assembly? 5. ____________________

6. Give the material and specification of the -3 tube. 6. ____________________

7. List the length of the assembly as a limit dimension. 7. ____________________

8. What kind of a line is used to illustrate the continuance of the bellows? 8. ____________________

9. What standards must be met with the production proof test? 9. ____________________

10. List the welding specification. 10. ____________________

11. How are the fusion welds to be inspected? 11. ____________________

12. What heat treatment is specified? 12. ____________________

Blueprint Reading Activity A-17-BPR-1
THE CHANGE SYSTEM

Refer to the blueprint in Fig. A-17-BPR-1 and answer the following questions.

1. Give the name of the part or assembly. 1. ____________________

2. What is the drawing number? 2. ____________________

3. Give the material shape, size and specification for the -3 part. 3. ____________________

4. How many of the -5 parts are used on assembly 265-530003? 4. ____________________

5. The original drawing was revised on how many different occasions? 5. ____________________

6. What was the change in A_2? 6. ____________________

7. What tolerance is given for the 3/8 inch diameter holes on the parts? 7. ____________________

8. What surface finish is specified for the machined surfaces? 8. ____________________

9. Why are the visible surfaces called out on the parts? 9. ____________________

10. How does finished -3 part differ from finished -5 part? 10. ____________________

11. May parts completed and on hand at the time of the change be reworked? What indication is there? 11. ____________________

12. How many "dash 3" parts are required for the next assembly number 265-530003? 12. ____________________

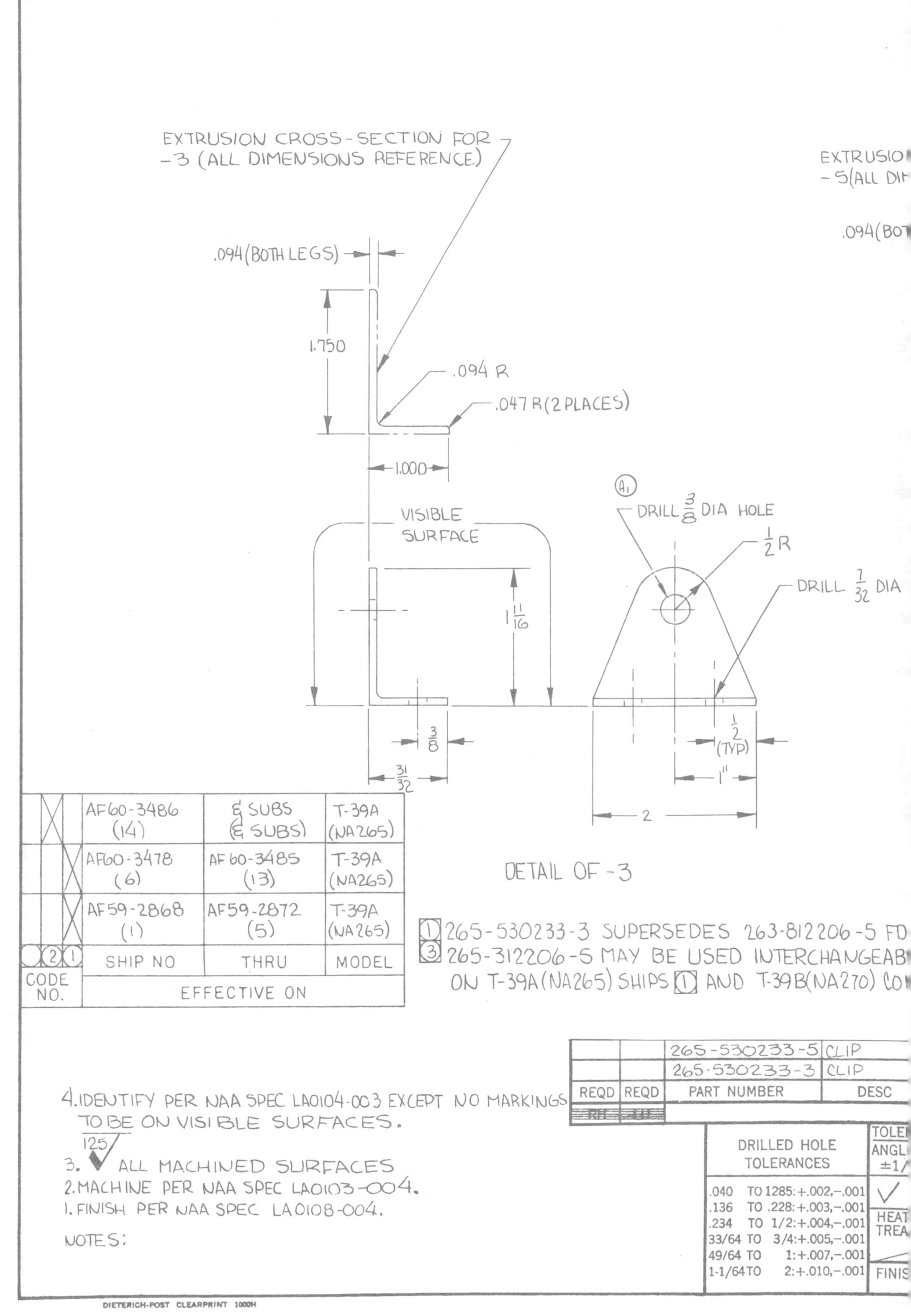

Fig. A-17-BPR-1. Industry blueprint. (North American Aviation)

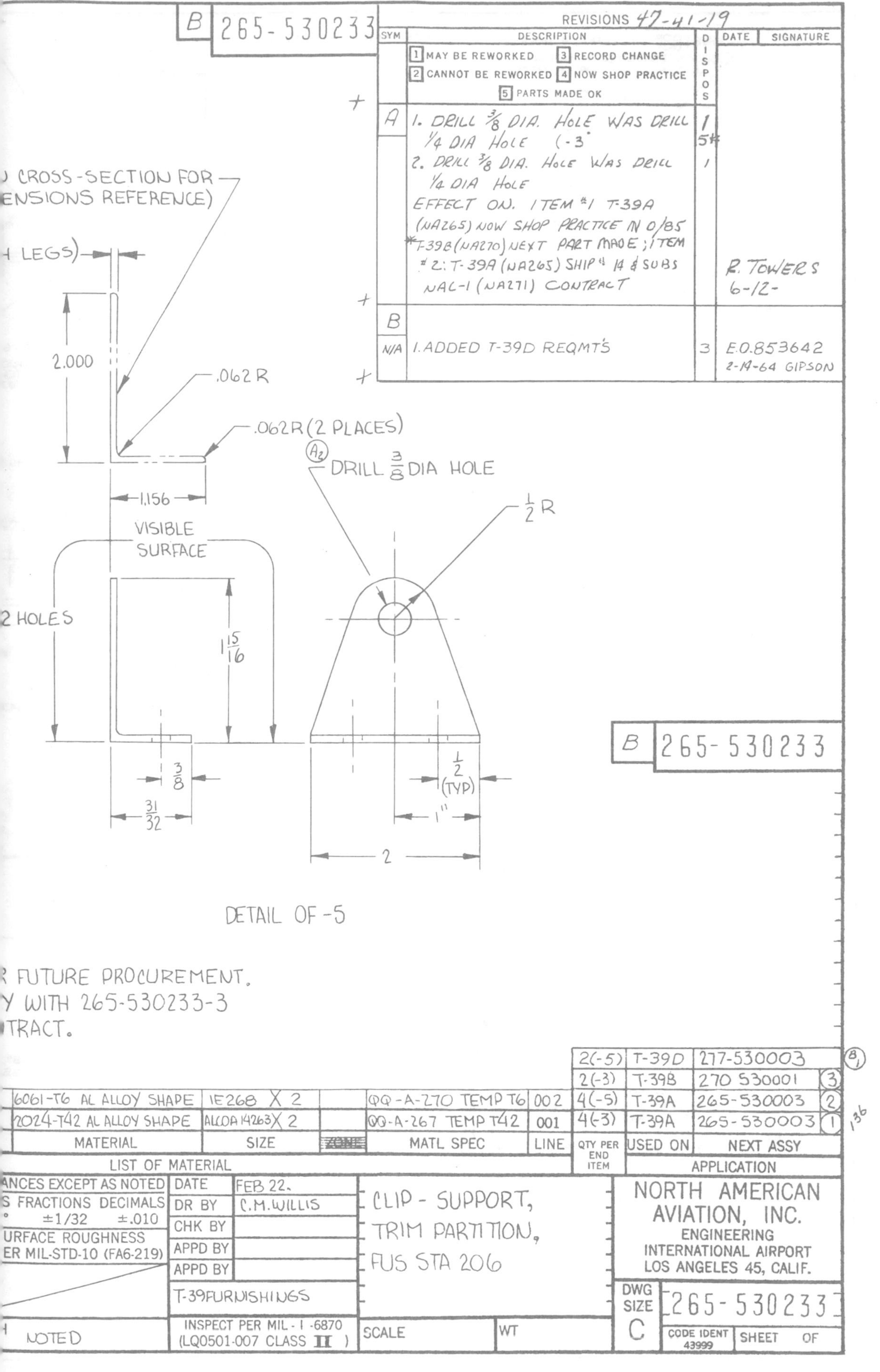

B 265-530233
REVISIONS 47-41-19
SYM
DESCRIPTION
1 MAY BE REWORKED
2 CANNOT BE REWORKED
3 RECORD CHANGE
4 NOW SHOP PRACTICE
5 PARTS MADE OK
DISPOS
DATE
SIGNATURE
A
1. DRILL 3/8 DIA. HOLE WAS DRILL 1/4 DIA HOLE (-3
2. DRILL 3/8 DIA. HOLE WAS DRILL 1/4 DIA HOLE
EFFECT ON. ITEM #1 T-39A (NA265) NOW SHOP PRACTICE IN 0/85 *T-39B (NA270) NEXT PART MADE; ITEM #2: T-39A (NA265) SHIP 14 & SUBS NAC-1 (NA271) CONTRACT
1
5
1
R. TOWERS
6-12-
B
N/A
1. ADDED T-39D REQMT'S
3
E.O. 853642
2-14-64 GIPSON
) CROSS-SECTION FOR
ENSIONS REFERENCE)
H LEGS)
2.000
.062 R
.062R (2 PLACES)
A2
DRILL 3/8 DIA HOLE
1/2 R
1.156
VISIBLE SURFACE
2 HOLES
1 15/16
3/8
31/32
1/2 (TYP)
1"
2
DETAIL OF -5
B 265-530233
R FUTURE PROCUREMENT.
Y WITH 265-530233-3
TRACT.
2(-5) T-39D 277-530003 B1
2(-3) T-39B 270 530001 3
6061-T6 AL ALLOY SHAPE 1E268 X 2 QQ-A-270 TEMP T6 002 4(-5) T-39A 265-530003 2
2024-T42 AL ALLOY SHAPE ALCOA 14263 X 2 QQ-A-267 TEMP T42 001 4(-3) T-39A 265-530003 1
136
MATERIAL
SIZE
ZONE
MATL SPEC
LINE
QTY PER END ITEM
USED ON
NEXT ASSY
LIST OF MATERIAL
APPLICATION
ANCES EXCEPT AS NOTED
S FRACTIONS DECIMALS
° ±1/32 ±.010
URFACE ROUGHNESS
ER MIL-STD-10 (FA6-219)
DATE FEB 22.
DR BY C.M. WILLIS
CHK BY
APPD BY
APPD BY
T-39 FURNISHINGS
CLIP - SUPPORT,
TRIM PARTITION,
FUS STA 206
NORTH AMERICAN
AVIATION, INC.
ENGINEERING
INTERNATIONAL AIRPORT
LOS ANGELES 45, CALIF.
DWG SIZE C
265-530233
NOTED
INSPECT PER MIL-I-6870
(LQ0501-007 CLASS II)
SCALE
WT
CODE IDENT 43999
SHEET
OF

Blueprint Reading Activity A-17-BPR-2
THE CHANGE SYSTEM

Refer to the blueprint in Fig. A-17-BPR-2 and answer the following questions.

1. Is this a detail or assembly drawing? 1. ______________________________
2. What is the name of the part? 2. ______________________________
3. Give the drawing number. 3. ______________________________
4. What is the scale of the original plan? 4. ______________________________
5. What material is specified? 5. ______________________________
6. What tolerance is specified for datum D? 6. ______________________________
7. Revision B made a change in a dimension. What was this change? 7. ______________________________
8. How did Revision D change the finish specification? 8. ______________________________
9. What was the radius dimension at H prior to revision? 9. ______________________________
10. What change was made in Revision L? 10. ______________________________
11. How did Revision N change the drawing? 11. ______________________________
12. What is the next assembly on which this part is used? 12. ______________________________

Blueprint Reading Activity A-18-BPR-1
THREADS

Refer to the blueprint in Fig. A-18-BPR-1 and answer the following questions.

1. What is the name and number of the part? 1. ______________________

2. What material is specified? 2. ______________________

3. Section AA is what type section? 3. ______________________

4. What type of section is shown in the left side view? 4. ______________________

5. Interpret the thread specification for A. 5. ______________________

6. Give the limit dimensions on the major diameter of the above thread. 6. ______________________

7. Give the size and tolerance for the drilled hole at B. 7. ______________________

8. Interpret the screw thread callout for B. 8. ______________________

9. What is the diameter and depth of the counterbore at B? 9. ______________________

10. Check the letter drill size chart in the back of this text for the correct tap drill at C. 10. ______________________

11. Interpret the thread specification at C. 11. ______________________

12. What is the diameter and tolerance at D. 12. ______________________

13. Give the diameter at E. 13. ______________________

14. What size drill is to be used at F? How is the hole finished and to what size? 14. ______________________

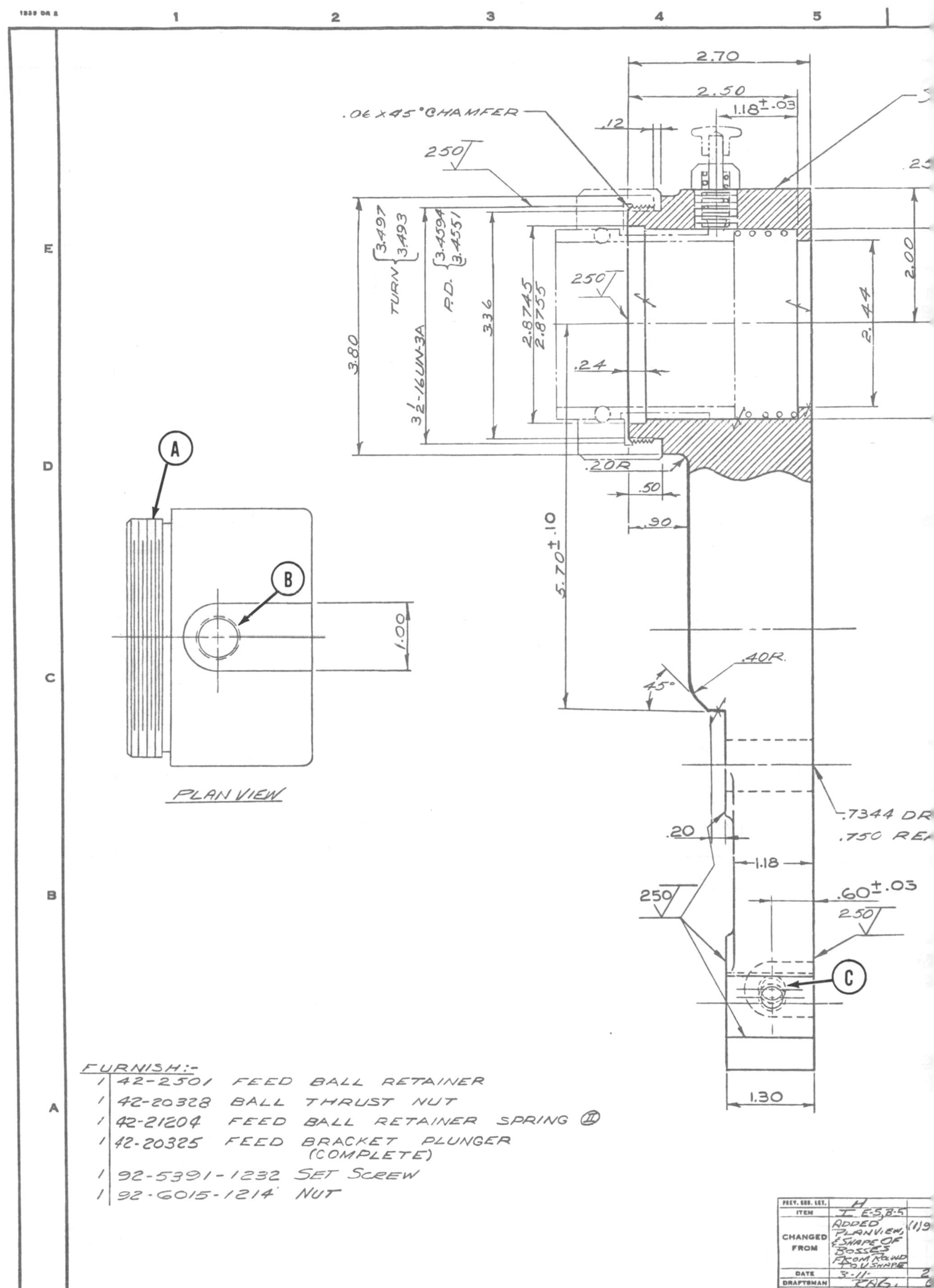

Fig. A-18-BPR-1. Industry blueprint. (Brown and Sharpe)

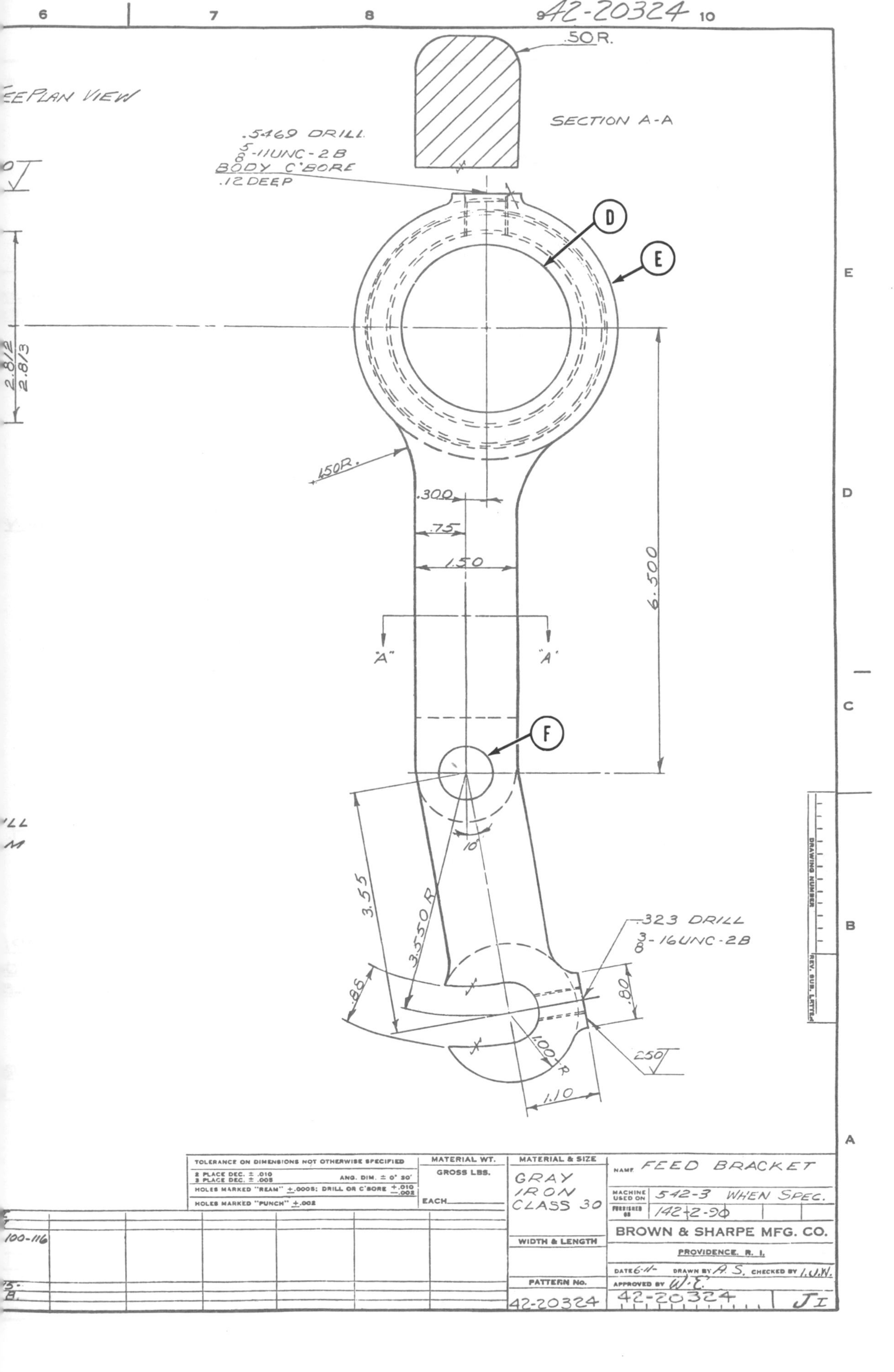
42-20324
.50R.
SECTION A-A
.5469 DRILL
5/8-11UNC-2B
BODY C'BORE
.12 DEEP
D
E
.150R.
.300
.75
1.50
6.500
"A"
"A"
F
10°
3.55
3.550R
.323 DRILL
3/8-16UNC-2B
.86
.80
.250
1.00R
1.10
TOLERANCE ON DIMENSIONS NOT OTHERWISE SPECIFIED
2 PLACE DEC. ± .010
3 PLACE DEC. ± .005
ANG. DIM. ± 0° 30'
HOLES MARKED "REAM" ± .0005; DRILL OR C'BORE +.010 −.002
HOLES MARKED "PUNCH" ± .002
MATERIAL WT.
GROSS LBS.
EACH
MATERIAL & SIZE
GRAY IRON CLASS 30
WIDTH & LENGTH
PATTERN No.
42-20324
NAME FEED BRACKET
MACHINE USED ON 542-3 WHEN SPEC.
FURNISHED 142+2-90
BROWN & SHARPE MFG. CO.
PROVIDENCE, R. I.
DATE 6-11- DRAWN BY A.S. CHECKED BY I.U.W.
APPROVED BY W.E.
42-20324
JI
DRAWING NUMBER
REV. SUB. LTTR.
E
D
C
B
A

Blueprint Reading Activity A-18-BPR-2
THREADS

Refer to the blueprint in Fig. A-18-BPR-2 and answer the following questions.

1. What is the name and number of the assembly? 1. ______________________ ______________________
2. What is the scale of the original plan? 2. ______________________
3. What material is specified? 3. ______________________
4. How many separate parts are required for the assembly? 4. ______________________
5. What change was made in the drawing by the last change order? 5. ______________________ ______________________
6. Interpret the thread specification at S. 6. ______________________ ______________________ ______________________ ______________________ ______________________
7. Interpret the thread specification at T. 7. ______________________ ______________________
8. Give the name and part number for U. 8. ______________________
9. Give the dimensions of the feature at V. 9. ______________________
10. Interpret the callout for the thread at W. 10. ______________________ ______________________
11. What size tap drill is to be used? How deep? 11. ______________________
12. What size hole is to be drilled at X? How deep? How is it to be finished? 12. ______________________ ______________________
13. Give the size of the counterbore at X. 13. ______________________
14. Interpret the thread required at Y. 14. ______________________ ______________________
15. Give the specifications for the feature at Z. 15. ______________________ ______________________

Blueprint Reading Activity A-19-BPR-1
CALLOUTS FOR MACHINE PROCESSES

Refer to the blueprint in Fig. A-19-BPR-1 and answer the following questions.

1. What is the name of the part? 1. ______________________

2. Give the print number. 2. ______________________

3. What is the material specification? 3. ______________________

4. Give the specification for the threaded part and interpret. 4. ______________________

____

____

____

5. What angle does the tapered surface make with the center line of the part? 5. ______________________

6. Calculate the length of the taper. 6. ______________________

7. What surface finish is to be given the taper? 7. ______________________

8. Give the diameter and tolerance of the .06 groove. 8. ______________________

9. State the diameter of the .056 groove as a limits dimension. 9. ______________________

10. What radius is to be held on the inside corners of the above groove? 10. ______________________

11. What concentricity is specified for diameter .3748? 11. ______________________

____

12. What was the last change made to the drawing? 12. ______________________

____

UNLESS OTHERWISE SPECIFIED					
DECIMALS		FRACTIONS	ANGLES	SURFACE ROUGHNESS	DIMENSIONS ARE IN INCHES
2 PLACE	3 PLACE				ALL DIAS. ON SAME AXIS CONC. WITHIN .006 T.I.R.
± .01	± .005	± 1/64	± 1/2°	63 √ A.A. MAX.	BREAK SHARP EDGES .01R. OR .01 × 45° MAX.
					ALL DIMENSIONS AND TOL. APPLY BEFORE FINISH

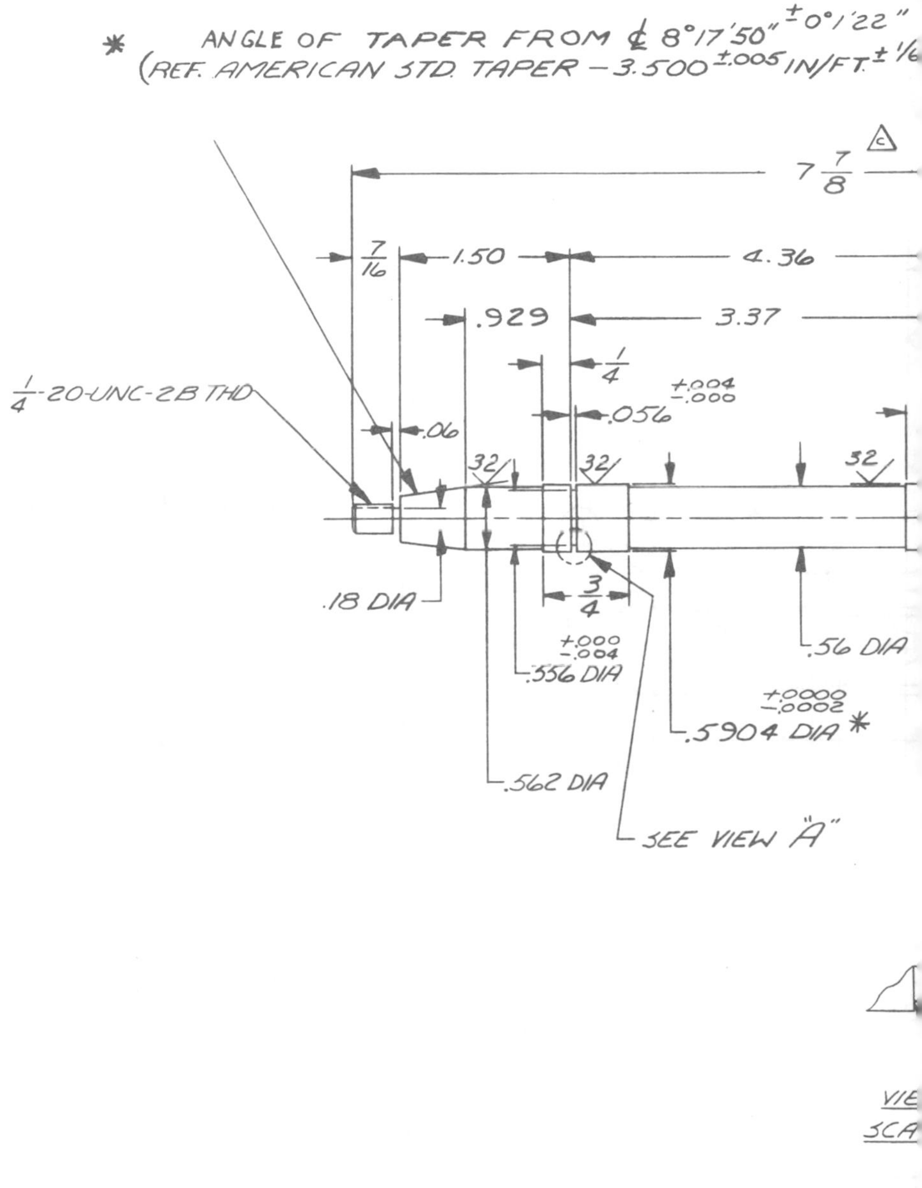

CHG.	DESCRIPTION	DATE	C.N. NO.
C	WAS 7 5/8	1-20-	12974
B	REVISED	12-17	12978
A	DR. #4502	9-6-	

Fig. A-19-BPR-1. Industry blueprint. (Perkin-Elmer)

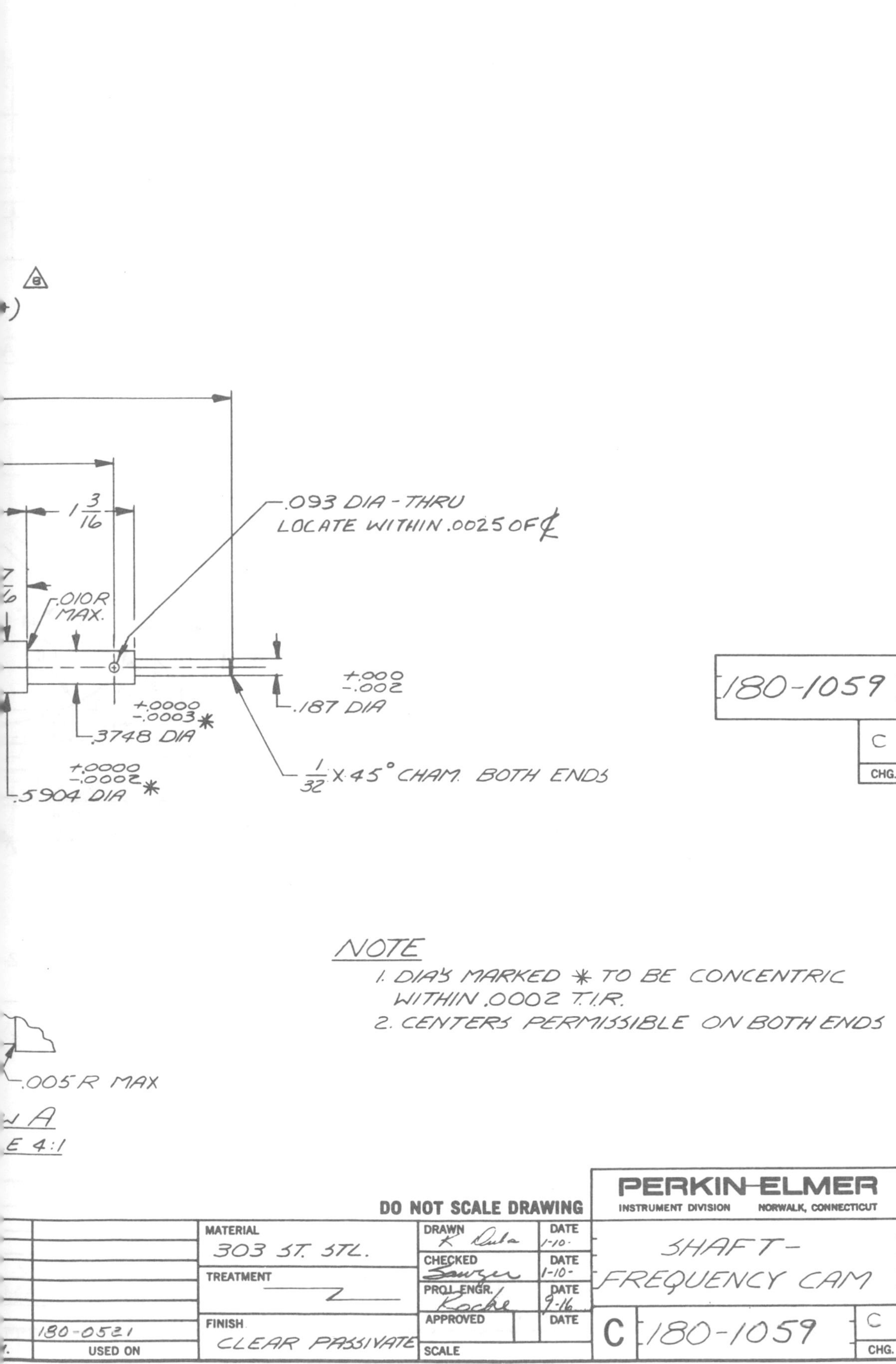

.093 DIA - THRU
LOCATE WITHIN .0025 OF ℄
$1\frac{3}{16}$
.010R MAX.
+.000 -.002
.187 DIA
+.0000 -.0003 *
.3748 DIA
+.0000 -.0002 *
.5904 DIA
$\frac{1}{32}$ X 45° CHAM. BOTH ENDS
180-1059
C
CHG.
NOTE
1. DIA'S MARKED * TO BE CONCENTRIC WITHIN .0002 T.I.R.
2. CENTERS PERMISSIBLE ON BOTH ENDS
.005 R MAX
DO NOT SCALE DRAWING
PERKIN-ELMER
INSTRUMENT DIVISION
NORWALK, CONNECTICUT
MATERIAL
303 ST. STL.
TREATMENT
FINISH
CLEAR PASSIVATE
DRAWN
DATE
1-10-
CHECKED
DATE
1-10-
PROJ. ENGR.
DATE
9-16
APPROVED
DATE
SCALE
SHAFT - FREQUENCY CAM
C
180-1059
C
CHG.
180-0521
USED ON
FORM 1-0118-01

Blueprint Reading Activity A-19-BPR-2
CALLOUTS FOR MACHINE PROCESSES

Refer to the blueprint in Fig. A-19-BPR-2 and answer the following questions.

1. What type of a drawing is this? 1. ______

2. What is the name of the object? 2. ______

3. Give the drawing number. 3. ______

4. What is the scale of the original plan? 4. ______

5. What is the material specification? 5. ______

6. Give the overall finished dimensions of the piece. 6. ______

7. Changes have been made to the original drawing on how many different occasions? 7. ______

8. How many features were affected by the changes in A? 8. A ______

9. What is the callout for D? 9. D ______

10. Give the callout for E. 10. E ______

11. How is the surface texture specified for G? 11. ______

12. What is the callout for H? 12. H ______

13. Give the machine process specification for hole J. 13. J ______

14. What is the relationship specified in the local note for K and L? 14. ______

15. How does top part of feature K differ from bottom part? 15. K ______

16. What surface finish is specified for hole L? 16. L ______

17. Interpret the chamfer callout for hole L. 17. L ______

18. Give the callout for M. 18. M ______

Blueprint Reading Activity A-20-BPR-1
TOLERANCES OF POSITION AND FORM

Refer to the blueprint in Fig. A-20-BPR-1 and answer the following questions.

1. Give name and number of the print. 1. ____________________

2. Has the drawing been revised? 2. ____________________

3. What material is specified? 3. ____________________

4. Interpret the complete machine process callout and feature control symbol at C. 4. ____________________

5. Give the surface texture for D and interpret the feature control symbol controlling this diameter. 5. ____________________

6. Give the specification for the hole at E and interpret the feature control symbol. 6. ____________________

7. Interpret the feature control symbol controlling the relationship between F and H. 7. ____________________

8. Give the dimension at G. 8. ____________________

9. What kind of a dimension is J? 9. ____________________

10. Interpret the complete specification and feature control symbol at K. 10. ____________________

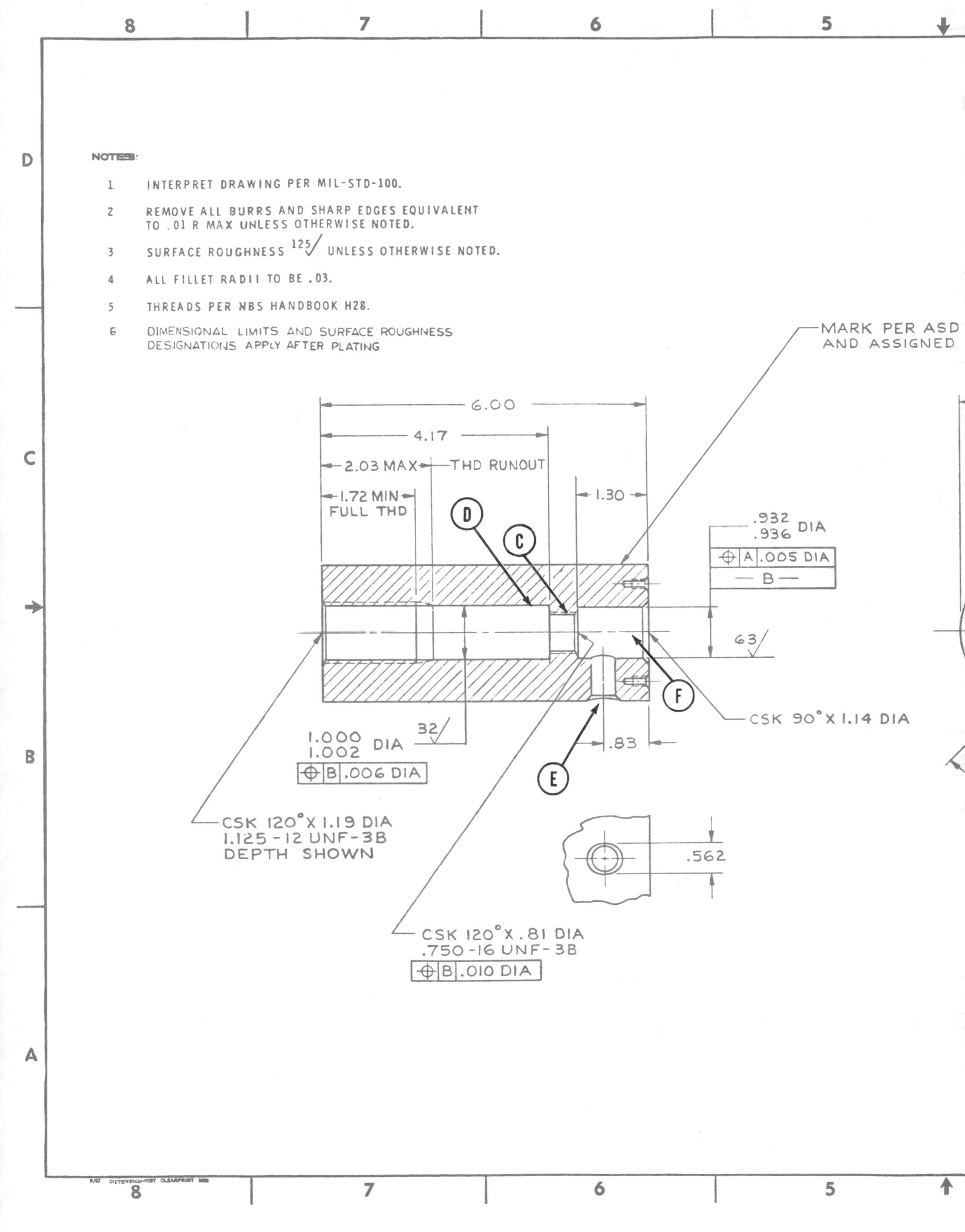

Fig. A-20-BPR-1. Industry blueprint. (Aerojet – General)

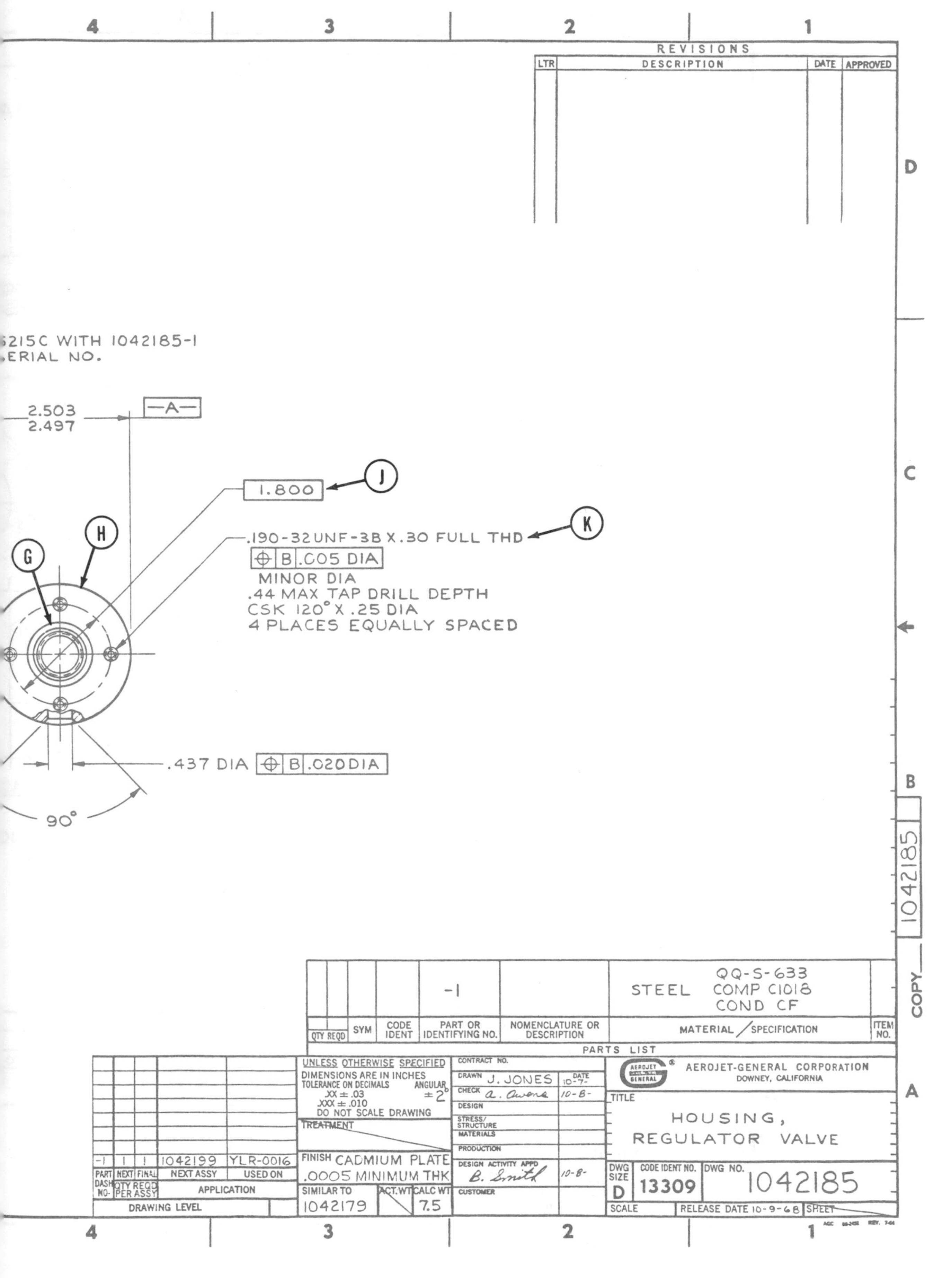

4
3
2
1
REVISIONS
LTR
DESCRIPTION
DATE
APPROVED
D
215C WITH 1042185-1
ERIAL NO.
2.503
2.497
—A—
1.800
J
C
H
K
G
.190-32UNF-3B X .30 FULL THD
⊕ B .005 DIA
MINOR DIA
.44 MAX TAP DRILL DEPTH
CSK 120° X .25 DIA
4 PLACES EQUALLY SPACED
.437 DIA ⊕ B .020 DIA
90°
B
1042185
COPY
-1
STEEL
QQ-S-633
COMP C1018
COND CF
QTY REQD
SYM
CODE IDENT
PART OR IDENTIFYING NO.
NOMENCLATURE OR DESCRIPTION
MATERIAL / SPECIFICATION
ITEM NO.
PARTS LIST
UNLESS OTHERWISE SPECIFIED
DIMENSIONS ARE IN INCHES
TOLERANCE ON DECIMALS
.XX ± .03
.XXX ± .010
ANGULAR ± 2°
DO NOT SCALE DRAWING
TREATMENT
FINISH CADMIUM PLATE
.0005 MINIMUM THK
SIMILAR TO
1042179
ACT.WT
CALC WT
7.5
CONTRACT NO.
DRAWN J. JONES
DATE
10-7-
CHECK A. Owens
10-8-
DESIGN
STRESS/STRUCTURE
MATERIALS
PRODUCTION
DESIGN ACTIVITY APPD
B. Smith
10-8-
CUSTOMER
AEROJET GENERAL
AEROJET-GENERAL CORPORATION
DOWNEY, CALIFORNIA
A
TITLE
HOUSING,
REGULATOR VALVE
DWG SIZE
D
CODE IDENT NO.
13309
DWG NO.
1042185
SCALE
RELEASE DATE 10-9-68
SHEET
-1
1
1
1042199
YLR-0016
PART DASH NO.
NEXT
FINAL
QTY REQD PER ASSY
NEXT ASSY
USED ON
APPLICATION
DRAWING LEVEL

Blueprint Reading Activity A-20-BPR-2
TOLERANCES OF POSITION AND FORM

Refer to the blueprint in Fig. A-20-BPR-2 and answer the following questions.

1. What is the name of the part and print number? 1. ______________________

2. Has the print been revised? 2. ______________________

3. Give the heat treatment specification. 3. ______________________

4. This part may be made from what forging part number? 4. ______________________

5. What surface texture is specified for Datums A, B and C? 5. A. __________ B. __________ C. __________

6. Interpret the relationship between Datum B and Datums A and C. 6. ______________________

7. Interpret the feature control symbol controlling the two 15/32 holes. 7. ______________________

8. Interpret the meaning of the 3.000 BASIC dimension. 8. ______________________

9. Interpret the feature control symbol at Datum C. 9. ______________________

10. Interpret the callout for the threaded hole shown in Section B-B. 10. ______________________

Blueprint Reading Activity A-21-BPR-1
SPUR GEAR

Refer to the blueprint in Fig. A-21-BPR-1 and answer the following questions.

1. Give the name of the part and the part number. 1. ____________________

2. What is the material specification? 2. ____________________

3. What surface texture is specified? 3. ____________________

4. How many teeth are specified in each side of the gear? 4. ____________________

5. What is the pitch diameter? 5. ____________________

6. If you were to measure the tooth with the gear-tooth caliper, what reading would you expect to get for:
   A. Tooth thickness? 6. A. ____________________
   B. Height of tooth face? B. ____________________

7. Give the tolerance limits for the keyway dimensions and its relationship to the center line. 7. ____________________

8. What does the 1/8 inch offset above the center line indicate? 8. ____________________

9. What relationship must exist between the pitch diameter and faces of the teeth with the bore? 9. ____________________

10. Give the width of the tooth face as a limit dimension. 10. ____________________

11. What machine processes are to be performed on the teeth after they have been cut? 11. ____________________

12. Give the total width of the gear as a limit dimension. 12. ____________________

13. Give the addendum circle dimension. 13. ____________________

14. What are the dimensions of the eccentric part in the gear? 14. ____________________

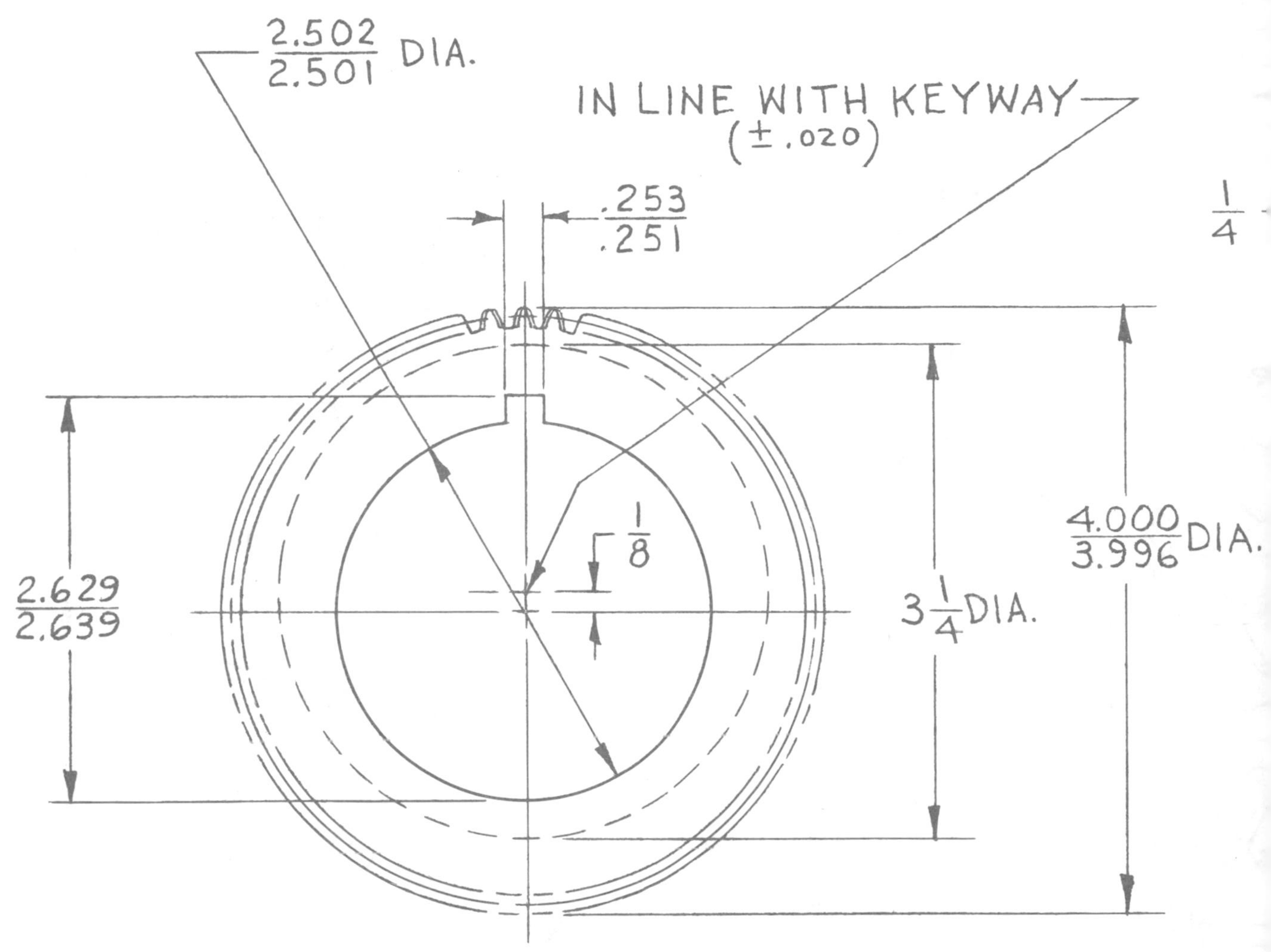

INVOLUTE SPUR GEAR DATA	
NO. OF TEETH	54
DIAMETRAL PITCH	14
CIRCULAR PITCH	.2244
PRESSURE ANGLE	20
PITCH DIAMETER	3.8571
CHORDAL TOOTH THICK.	.1105-.1097
ADDENDUM	.0714
WHOLE DEPTH	.1541
PIN DIAMETER	.120
PIN MEASUREMENT	4.0148-4.0128
REMARKS:	

PITCH DIAMETE
CONCENTRIC W/IN .
SQUARE WITH BOR

UNLESS OTHE
TOLERANCES

V THREADS:

ANGLES: ± 1

Fig. A-21-BPR-1. Industry blueprint. (South Bend Lathe)

PART NO. PT8570DD1

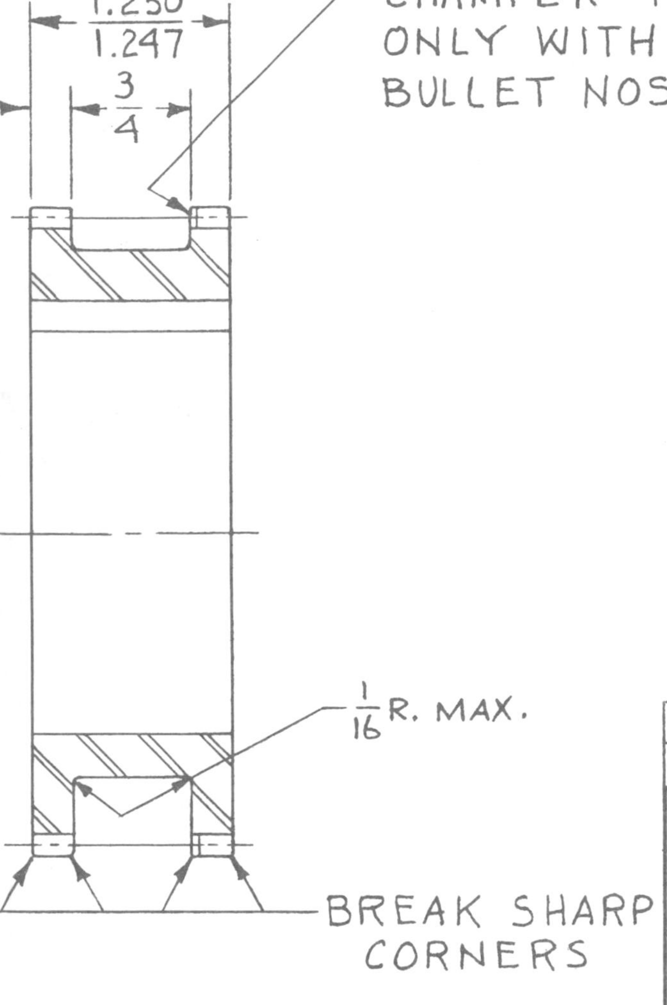

R OF TEETH & BORE TO BE
002" T.I.R., FAÇES TO BE
E W/IN .002" T.I.R.

RWISE SPECIFIED:
MICROINCH FINISH ON ALL MACHINED SURFACES 125 MAX.
FRACTIONAL DIMENSIONS ± .010". ALL OTHER DIMENSIONS TO LIMITS GIVEN.
OUNTERSINK TAPPED HOLES TO MAJOR DIAMETER 120° INCLUDED
NGLE. CHAMFER MALE THREADED PARTS TO MINOR DIAMETER
20° INCLUDED ANGLE. THREAD LENGTH OR DEPTH DIMENSION
NDICATES FULL THREADS.
2 DEGREE. COUNTERSINKS: 82°. REMOVE ALL BURRS.

SYM.	BY	CHANGE	DATE
		EXP. 839-048	

SOUTH BEND LATHE — ONE OF THE Amsted INDUSTRIES
SOUTH BEND, IND. U.S.A.

PART OR UNIT: GEAR

SUPERSEDED NUMBER: — DATE:

MAT'L. C-1213-C.D.S.

D'WN. BY SIGNATURE: John P. Nicodemus — DATE: 1-14-

C'KD. BY SIGNATURE: — DATE:

SCALE: — SHT. NO. 1 OF 1 — CODE NO.:

PART NO. PT8570DD1 — B

QTY.

ing)

Blueprint Reading Activity A-21-BPR-2
BEVEL GEAR

Refer to the blueprint in Fig. A-21-BPR-2 and answer the following questions.

1. What is the name of the part and the print number? 1. ______________________

2. Has the print been revised to reflect the change order? Where on print? 2. ______________________

3. What material is specified? 3. ______________________

4. What heat treatment is called out? 4. ______________________

5. Give the finish specification. 5. ______________________

6. How many teeth are to be cut in the gear? 6. ______________________

7. What type bevel gear is to be made? 7. ______________________

8. What is the value for the pressure angle? 8. ______________________

9. Give the value of the face angle. Pitch angle. Root angle. 9. Face __________ Pitch __________ Root __________

10. What tolerance is permitted on these angles? 10. ______________________

11. Give the size of the pitch diameter and the allowed tolerance. 11. ______________________

12. What relationship is to exist between surface C and the bore? 12. ______________________

13. Give the diameter and tolerance of the chamfer at hole D. 13. ______________________

14. What is the specification for the chamfer at E? 14. ______________________

Blueprint Reading Activity A-21-BPR-3
SPUR AND WORM

Refer to the blueprint in Fig. A-21-BPR-3 and answer the following questions.

1. What is the name and number of the part? 1. ______________________
____
____
____

2. Give the specified material and size. 2. ______________________
____

3. What heat treatment is specified? 3. ______________________
____

4. How did Revision B3 change the part? 4. ______________________
____

5. For the spur gear, give the following data: 5. No. of teeth ______________
Pitch dia. ______________
Addendum ______________
Max. Circular tooth thickness ______________
Dia. of measuring wires ______________
Part No. of mating gear ______________
Width of tooth face ______________

6. Give the following information for the worm: 6. No. of threads ______________
Lead angle ______________
Lead ______________
Pitch dia. ______________
Pitch dia. runout T.I.R. ______________
Part No. of mating gear ______________
Length of worm ______________

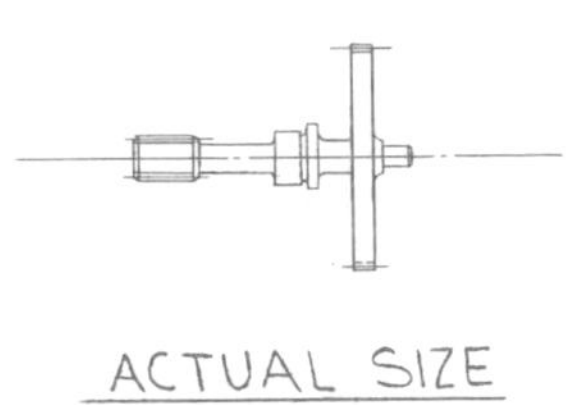

ACTUAL SIZE

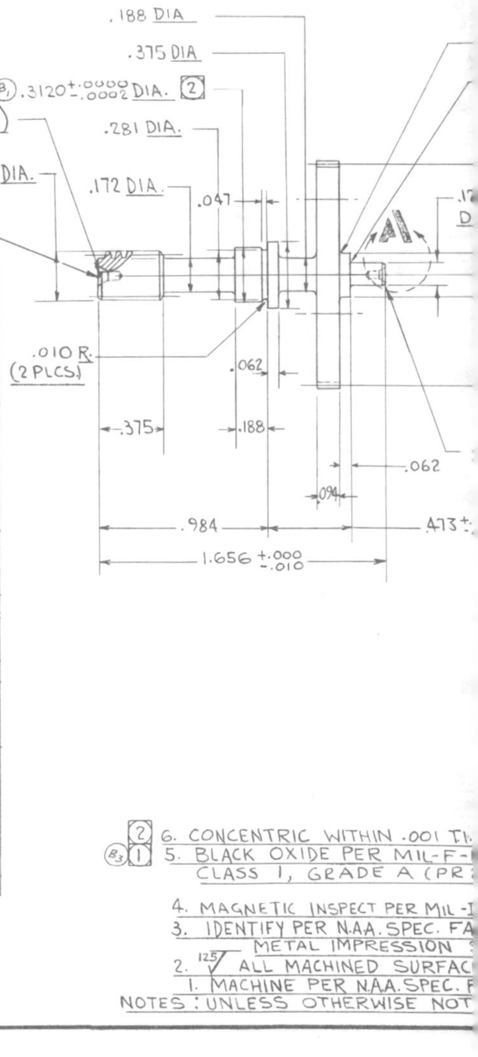

OPTIONAL CENTER HOLE SIZE I OR SMALLER PER NAA SPEC FA6-34

CHAMFER 45°×.030 (2 PLCS)

.188 DIA

.375 DIA

(B1) .3120 +.0000 −.0002 DIA. [2]

.281 DIA.

[2] .259 +.000 −.002 DIA.

.172 DIA.

.047

.010 R. (2 PLCS)

.062

.375

.188

.062

.094

.984

.473 ±

1.656 +.000 −.010

FINE PITCH SPUR GEAR DATA

INVOLUTE TOOTH FORM PER ASA B6.7			
PRESSURE ANGLE			20°
DIAMETRAL PITCH			64
NUMBER OF TEETH			80
PITCH DIAMETER (THEORETICAL)			1.25000
INVOLUTE BASE CIRCLE DIAMETER			1.17462
ADDENDUM (STANDARD)			.0156
WORKING DEPTH OF TOOTH			.0312
WHOLE DEPTH OF TOOTH		MIN	.0364
CIRCULAR PITCH (THEORETICAL)			.04909
CIRCULAR TOOTH THICKNESS	AT PITCH DIA	THEOR	.0245
	MACHINE TO	MAX	.0237
		MIN	.0230
CLASS PER ASA B6.11	COMM.		4B
DIAMETER OF MEASURING WIRES			.02700
MEASUREMENT OVER WIRES		THEOR	1.2881
		MAX	1.2861
		MIN	1.2843
TOTAL COMPOSITE ERROR		MAX	.0015
TOOTH TO TOOTH COMPOSITE ERROR		MAX (~~MIN~~)	.0007
WORKING PROFILE SURFACE FINISH (TEETH TO BE FREE OF BURRS)			32 √
MATING GEAR	PART NUMBER		223-54784
	NUMBER OF TEETH		16
	CENTER DISTANCE		.7510
	TOTAL SPECIFIED BACKLASH	MAX	.0030
		MIN	.0015

WORM DATA

FINE PITCH WORM PER ASA B6.9			
PRESSURE ANGLE (NORMAL)			20°
AXIAL PITCH			.0500
NUMBER OF THREADS			1
DIRECTION OF THREADS			R.H
LEAD ANGLE			4.0°
LEAD			.0500
PITCH DIAMETER (THEORETICAL)			.22760
ADDENDUM			.0159
WORKING DEPTH OF THREAD			.0318
WHOLE DEPTH OF THREAD		MIN	.0369
NORMAL PITCH (CIRCULAR)		THEO.	.04988
THICKNESS OF THREAD (NORMAL)	ON PITCH DIA.	THEO.	.0249
	MACHINE TO	MAX.	.0241
		MIN.	.023A.
CLASS PER ASA 36.11	(COMM.)		4B
DIA. OF MEASURING WIRES			.02667
MEASUREMENT OVER WIRES		THEO.	.2637
		MAX.	.2615
		MIN.	.2596
WORKING PROFILE SURFACE FINISH (TEETH TO BE FREE OF BURRS)			32 √
PITCH DIA. RUNOUT T.I.R.			.0008
MATING WORM GEAR	PART NUMBER		223-547109
	NUMBER OF TEETH		32
	CENTER DISTANCE		.3684
	TOTAL SPECIFIED BACKLASH	MAX.	.0030
		MIN.	.0015

[2] 6. CONCENTRIC WITHIN .001 T.

(B3) [1] 5. BLACK OXIDE PER MIL-F- CLASS 1, GRADE A (PR

4. MAGNETIC INSPECT PER MIL-I

3. IDENTIFY PER N.A.A. SPEC. FA METAL IMPRESSION

2. 125√ ALL MACHINED SURFAC

1. MACHINE PER N.A.A. SPEC. F

NOTES: UNLESS OTHERWISE NOT

Fig. A-21-BPR-3. Industry blueprint. (North American Aviation)

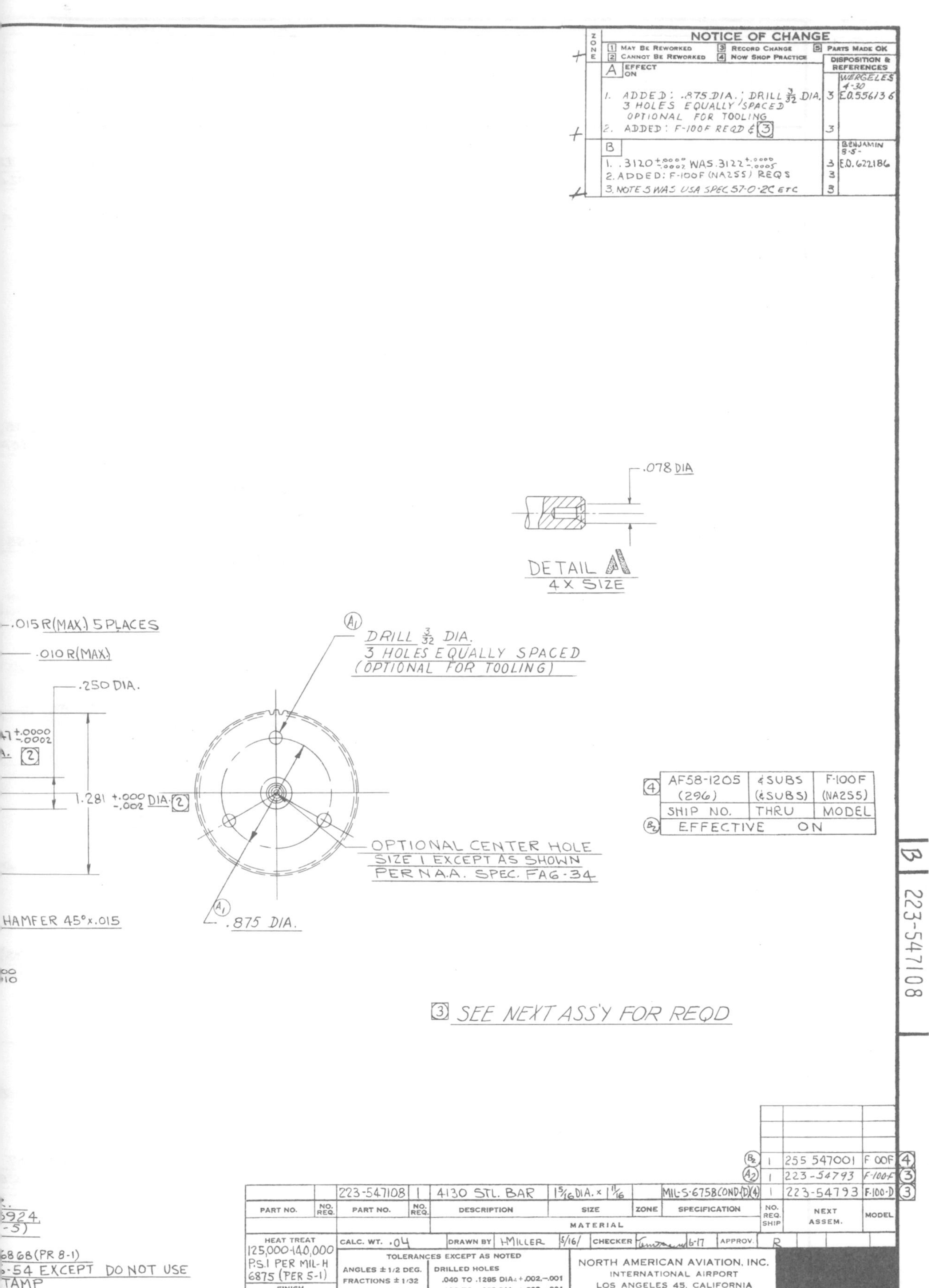

NOTICE OF CHANGE
1 MAY BE REWORKED
2 CANNOT BE REWORKED
3 RECORD CHANGE
4 NOW SHOP PRACTICE
5 PARTS MADE OK
DISPOSITION & REFERENCES
A EFFECT ON
1. ADDED: .875 DIA.; DRILL 3/32 DIA. 3 HOLES EQUALLY SPACED OPTIONAL FOR TOOLING
2. ADDED: F-100F REQD & 3
WERGELES 4-30 E.O. 556136
B
1. .3120 +.0000 -.0002 WAS .3122 +.0000 -.0005
2. ADDED: F-100F (NA255) REQS
3. NOTE 5 WAS USA SPEC 57-0-2C ETC
BENJAMIN 8-5- E.O. 622186
.078 DIA
DETAIL A
4 X SIZE
.015 R(MAX.) 5 PLACES
.010 R(MAX.)
.250 DIA.
1.281 +.000 -.002 DIA. 2
DRILL 3/32 DIA.
3 HOLES EQUALLY SPACED
(OPTIONAL FOR TOOLING)
OPTIONAL CENTER HOLE
SIZE 1 EXCEPT AS SHOWN
PER N.A.A. SPEC. FA6-34
.875 DIA.
HAMFER 45°x.015
AF58-1205 (296) | & SUBS (& SUBS) | F-100F (NA255)
SHIP NO. | THRU | MODEL
EFFECTIVE ON
3 SEE NEXT ASS'Y FOR REQD
B 223-547108
1 255 547001 F-100F 4
1 223-54793 F-100F 3
1 223-54793 F-100-D 3
223-547108 1 4130 STL. BAR 1 5/16 DIA. x 1 11/16 MIL-S-6758 COND D 4
PART NO. NO. REQ. PART NO. NO. REQ. DESCRIPTION SIZE ZONE SPECIFICATION NO. REQ. SHIP NEXT ASSEM. MODEL
MATERIAL
HEAT TREAT 125,000-140,000 P.S.I PER MIL-H 6875 (PER 5-1)
FINISH 1
SCALE:
DWG. SIZE: D
CALC. WT. .04
DRAWN BY H.MILLER 5/16/
CHECKER 6-17
APPROV.
TOLERANCES EXCEPT AS NOTED
ANGLES ± 1/2 DEG.
FRACTIONS ± 1/32
DECIMALS ± .010
SURFACE ROUGHNESS PER NAT'L AIRCRAFT STANDARD 30
DRILLED HOLES
.040 TO .1285 DIA. +.002, -.001
.136 TO .228 DIA. +.003, -.001
.234 TO 1/2 DIA. +.004, -.001
33/64 TO 3/4 DIA. +.005, -.001
49/64 TO 1" DIA. +.007, -.001
1-1/64 TO 2" DIA. +.010, -.001
NORTH AMERICAN AVIATION, INC.
INTERNATIONAL AIRPORT
LOS ANGELES 45, CALIFORNIA
SHAFT-ELEC PITCH CORRECTION
ROTARY-ACTUATOR-OUTPUT WORM
223-547108
132
6868 (PR 8-1)
-54 EXCEPT DO NOT USE
TAMP
6-125

38Th (2)

Blueprint Reading Activity A-21-BPR-4
SPLINES

Refer to the blueprint in Fig. A-21-BPR-4 and answer the following questions.

1. What is the name of the part and the drawing number? 1. ____________________
____

2. Has the print been revised? 2. ____________________

3. What material is specified? 3. ____________________
____

4. Give the heat treatment specification. 4. ____________________
____

5. What surface texture is specified? 5. ____________________

6. What type spline is to be machined? 6. ____________________

7. How many teeth are in the spline? 7. ____________________

8. Give the pressure angle. 8. ____________________

9. Give the pitch diameter. 9. ____________________

10. What relationship must exist between the splines and their axis? Ground diameters? 10. Axis: ____________________
Ground dia: ____________________
____

11. What relationship must exist between the hole shown in Section B-B and the shaft center line? 11. ____________________
____
____

12. Interpret the thread specification. 12. 9/16 ____________________
18 ____________________
UNF ____________________
2A ____________________
P.D. ____________________

Blueprint Reading Activity A-22-BPR-1
NUMERICAL CONTROL

Refer to the blueprint in Fig. A-22-BPR-1, and to pages 280, 281, 282, 283, 284 and 285, and answer the following questions.

1. What is the name of the part and the drawing number? 1. ______________________
2. Give the material specification. 2. ______________________
3. How is the part to be finished? 3. ______________________
4. How many tools are required in the machining of the part on the N/C machine? What is tool No. 13? 4. ______________________
5. How many parts are set up at one time? 5. ______________________
6. Where is machine zero for X and Y axes? 6. ______________________
7. Where is machine zero for the Z axis? 7. ______________________
8. Give letters indicating features to be machined with 1/4 end mill. 8. ______________________
9. What is the Z axis setting for this operation? 9. ______________________
10. Indicate by letters the part to be machined with the one inch end mill. 10. ______________________
11. What two operations are performed at B as indicated on the programming manuscript? 11. ______________________
12. What three operations are performed at No. 13 as indicated on programming manuscript? Which letter represents this feature on the blueprint? 12. ______________________
13. How many operations are performed at H and J? List the tools used. 13. ______________________
14. What is the Z axis setting and the R data for the counterbore operation? 14. ______________________
15. Is hole A programmed on this tape? 15. ______________________

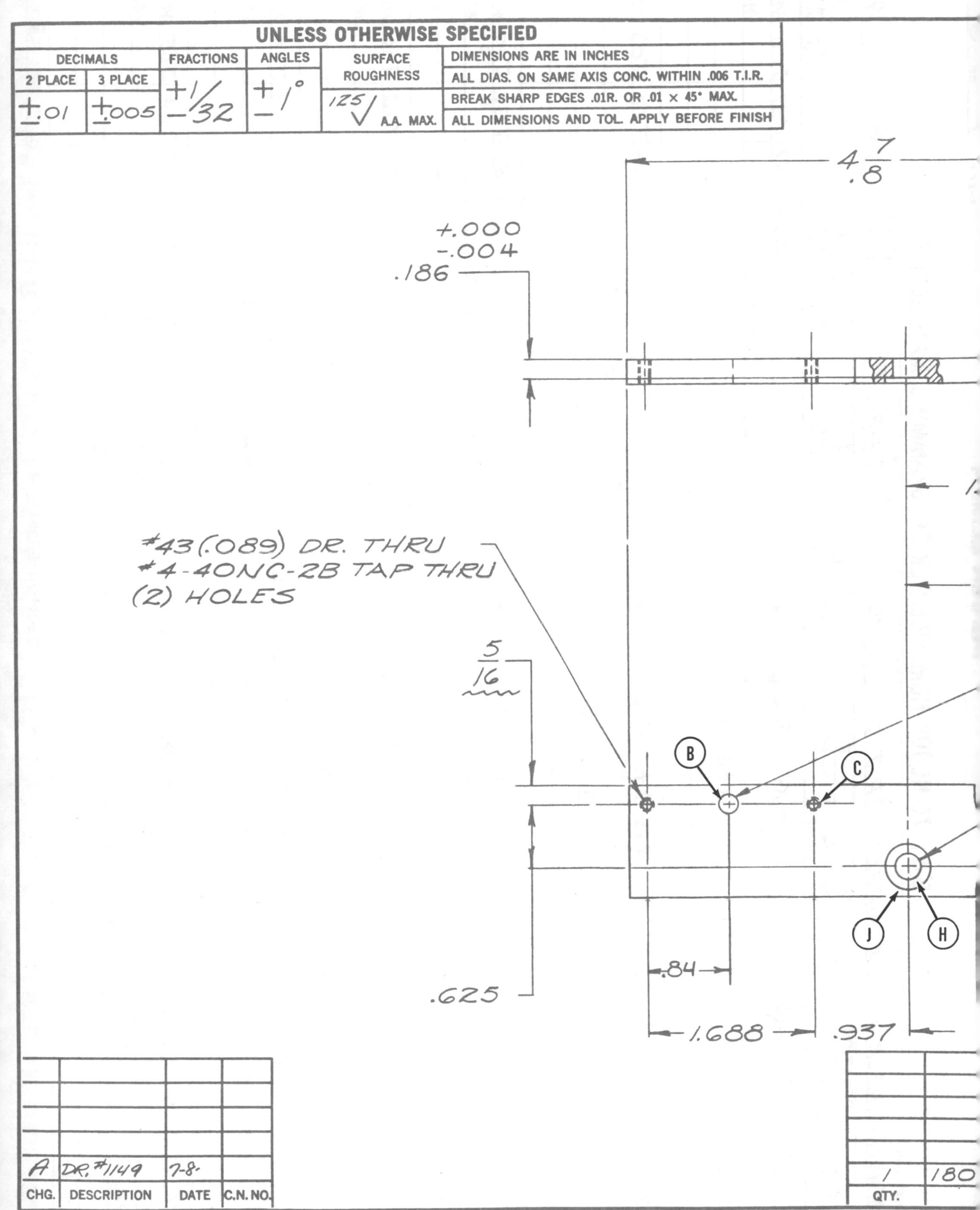

Fig. A-22-BPR-1. Industry blueprint. (Perkin – Elmer)

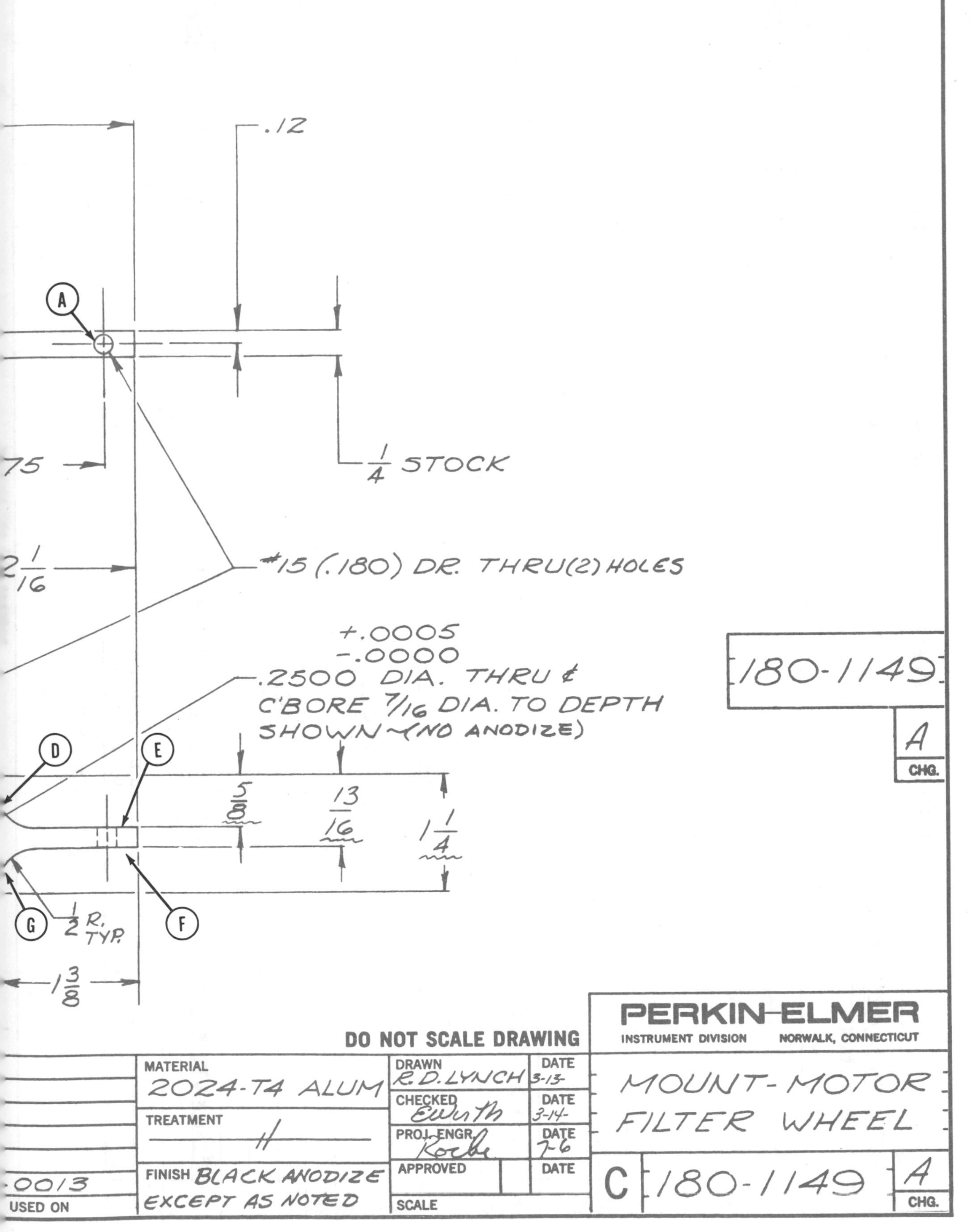

.12
A
75
1/4 STOCK
2 1/16
#15 (.180) DR. THRU (2) HOLES
+.0005
-.0000
.2500 DIA. THRU ¢
C'BORE 7/16 DIA. TO DEPTH
SHOWN ~(NO ANODIZE)
D
E
5/8
13/16
1 1/4
G
1/2 R. TYP.
F
1 3/8
180-1149
A
CHG.
DO NOT SCALE DRAWING
PERKIN-ELMER
INSTRUMENT DIVISION
NORWALK, CONNECTICUT
MATERIAL
2024-T4 ALUM
TREATMENT
FINISH BLACK ANODIZE
EXCEPT AS NOTED
-0013
USED ON
DRAWN
R.D. LYNCH
DATE
3-13-
CHECKED
DATE
3-14-
PROJ. ENGR.
DATE
7-6
APPROVED
DATE
SCALE
MOUNT-MOTOR
FILTER WHEEL
C
180-1149
A
CHG.
SHEET 6 OF 3
PERKIN-ELMER
1-0477-00

G & L 70-NC-10V NUMERICENTER PROGRAMMING MANUSCRIPT

PART NAME:	BY:	DATE:	TAPE NO:	PART NO:	REV.
MOUNT-MOTOR FILTER WHEEL	P.J.T.	9-9-	TAPE #1	180-1149	A

OPERATION			N/O	G	X	Y	F	Z	R	E	S	T	J	M
LOC. NO.	DESCRIPTION	CODE	SEQ. NO.	PREP. CODE	X – AXIS DATA	Y – AXIS DATA	MILL FEED	Z – AXIS DATA	R DATA	SPDL. FEED	SPDL. SPEED	TOOL NO.	TOOL COMP.	MISC. FUNC.
	UNLOAD PCS			g70	X+06700	Y+08000							J00	M02
001		-01												
	P.TISANO	-02												
PATTERN TO TAPE														
X=0.0, Y=0.0, N=5*														
X=6.0, Y=0.0*														
X=12.0, Y=0.0*														
X=18.0, Y=0.0*														
X=24.0, Y=0.0*														

1-0477-00

PERKIN-ELMER

G & L 70-NC-10V NUMERICENTER SETUP SHEET — 32 x 60 Table__________24 x 48 Travel

PART NAME	PART NO.	REV.	OP. NO.	SET-UP BY	DATE	TOOL/FIXTURE NO.
MOUNT-MOTOR FILTER WHEEL	180-1149	A	TAPE #1	P.-J.-T.	9-6-	T-000-2013

T-000-2007-64

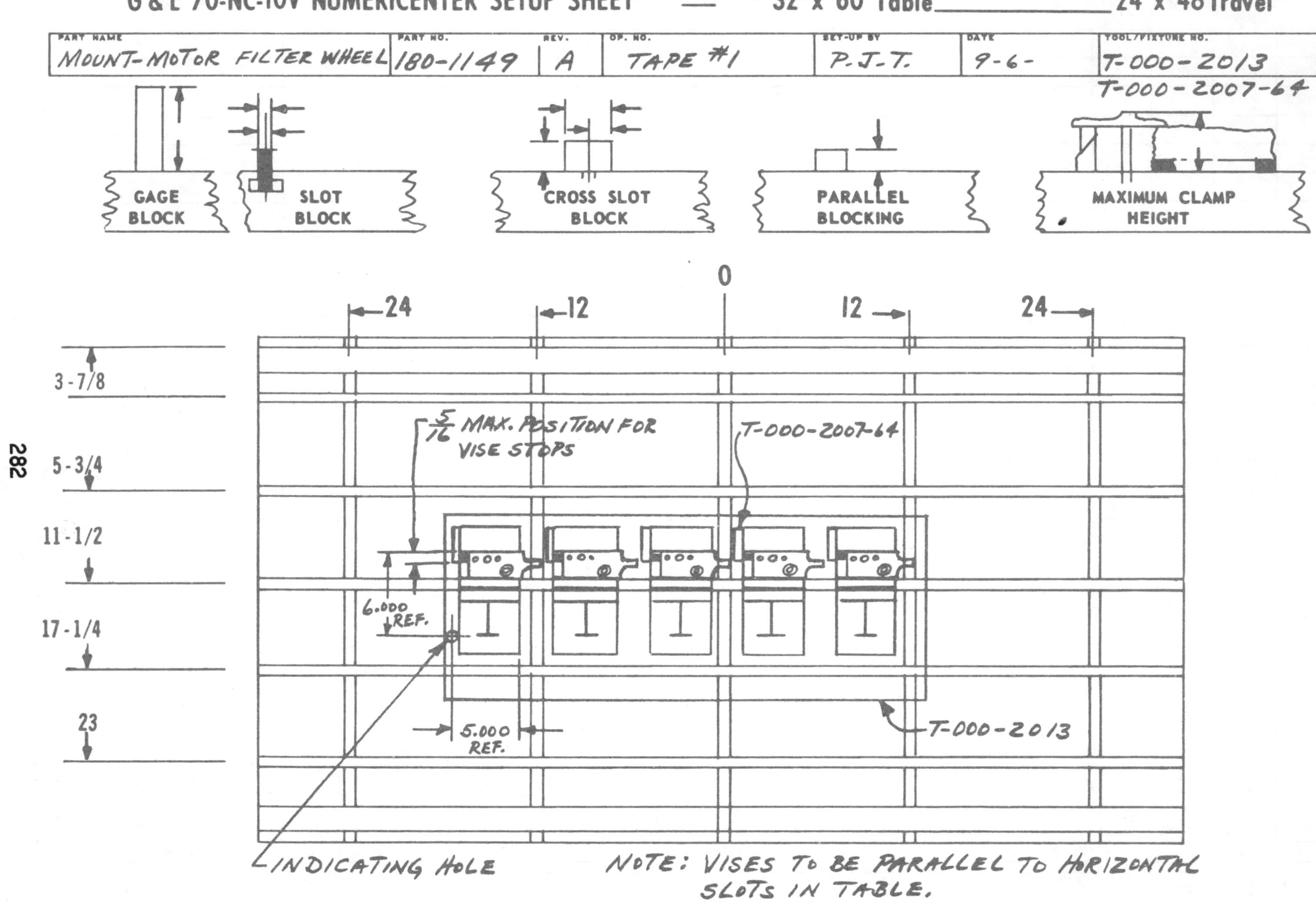

PERKIN-ELMER

1-0478-00

G & L 70-NC-10V NUMERICENTER SETUP & TOOLING SHEET

PART NAME	OP. NO.	W.C.	TAPE NO.	MACHINE	PART NO.	REV.
MOUNT-MOTOR FILTER WHEEL			#1	G&L	180-1149	A

MACHINE NUMBER	PROGRAM BY	DATE	CHECK BY	DATE	(THRU) ENGR. CHANGE NO.	DATE
	P. J. T.	9-6-				

SET UP AND ORIENTATION INSTRUCTIONS	T	RIGHT MATRIX	J	TOOL LENGTH	T	LEFT MATRIX	J	TOOL LENGTH
	TOOL POS.		TOOL COMP.		TOOL POS.		TOOL COMP.	
POSITION PIECES AS SHOWN	t10	1/4 END MILL	J01		t30	1" END MILL	J02	
ON VISES.	t11	#2 CENTER DR.			t31	#43(.089) DR.		
	t12	#4-40 TAP	J03		t32	#15(.180) DR.		
ZERO AT LOWER L.H. CORNER	t13	15/64 DR.			t33	7/16 DIA. C'BORE	J04	
OF VISES USING INDICATING	t14	.2500 REAMER	J05		t34			
HOLE.	t15				t35			
	t16				t36			
LONGEST TOOL TO CLEAR THE	t17				t37			
HIGHEST POINT OF THE SETUP.	t18				t38			
	t19				t39			
ALL TOOL ADJUST TOOLS TO	t20				t40			
BE PROGRAMMED 2" ABOVE	t21				t41			
VISE JAW STEP (.125 X.084)	t22				t42			
USING A 2.000 GAGE BLOCK.	t23				t43			
	t24				t44			
	t25				t45			
	t26				t46			
	t27				t47			
	t28				t48			
	t29				t49			

1-0476-00

PERKIN-ELMER

Interpretations of Column Headings and Codes for Giddings and Lewis 70-NC-10V Numericenter

(Reference to blueprints on pages 279 and 287.)

Setup and Tooling Sheet

T-Tool Pos. - - The position of the tool in the matrix. This number will be programmed into the N/C tape and the tool fed into the machine at the time required.

Matrix - - A holding device for the tools which will be fed automatically. There are two of these, a right and a left.

J-Tool Comp. - - A compensation for tool depth that can be manually set and the machine automatically compensates for this setting.

Programming Manuscript

Loc. No. - - Used for marking location of the particular operation or function on the shop copy of the blueprint as an assist to the machine operator. The two Perkin-Elmer blueprints used in this section have not been marked. These numbers also serve somewhat as sequence numbers.

Code - - Used by the company for any special code. The -01 and -02 indicate end of program sequence. The -5 indicates a manual indexing of a rotary fixture.

N/0 Seq. No. - - Sequence of operations.

G-Prep. Code - - This is a standard preparation code for this machine indicating certain machine functions as:

g00 - come to precise position and stop program.
g70 - position according to accuracy.
g72 - precise machining.
g80 - tells machine how far to retract above work.
g81 - automatic drilling with Hydro-sense, tool rapid traverses to workpiece and then slowly feeds into work, then returns to the R setting.
g82 - drill with dwell (provides for a number of revolutions of tool at bottom setting to produce a smooth finish).
g83 - automatic tapping with feed.
g85 - automatic reaming.
g87 - automatic boring cycle without retracting (tool feeds out instead of rapid traverse).
g89 - automatic boring cycle with dwell.

X- and Y- Axis Data - - Settings for X and Y dimensions from machine zero.

F-Mill Feed - - Coded in terms of feed rate.

Z- Axis Data - - Settings for Z dimensions from machine zero.

R- Data - - Rapid traverse distance from machine Z- zero. The R data motion is an approach or retract motion at rapid traverse rate.

E-Spdl. Feed - - Coded rate of spindle feed.

S-Spdl. Speed - - Coded rate of spindle speed.

T-Tool No. - - Number of tool position in matrix where tool is held.

J-Tool Comp. - - See above.

Misc. Func. - - A number of functions are possible, following are examples:

M02 - end of program.
M03 - right hand spindle turn.
M05 - spindle off.
M08 - coolant on.

Pattern To Tape - - This indicates that the location of the 5 vises are 6 inches apart which must be programmed into the tape.

G & L 70-NC-10V NUMERICENTER PROGRAMMING MANUSCRIPT

PART NAME:	BY:	DATE:	TAPE NO:	PART NO:	REV.
MOTOR MOUNT - GRATING DR.	P.J.T	8-23-	TAPE #1	180-1167	C

OPERATION			N/O	G	X	Y	F	Z	R	E	S	T	J	M
LOC. NO.	DESCRIPTION	CODE	SEQ. NO.	PREP. CODE	X - AXIS DATA	Y - AXIS DATA	MILL FEED	Z - AXIS DATA	R DATA	SPDL. FEED	SPDL. SPEED	TOOL NO.	TOOL COMP.	MISC. FUNC.
7					X+01078									
8						Y+01047								
9					X+03172									
				g80										
9	#6-32 TAP			g70	X+03172	Y+01047				e63	S67	T32	J03	M03
9				g83				Z04250	R03280					M08
8					X+01078									
7						Y-01047								
6					X+03172									
				g80										
	UNLOAD PC.			g70	X+03000	Y+04000							J00	M02
001		-01												
	P. TISANO	-02												

1-0477-00

PERKIN-ELMER

SHEET 3 OF 3

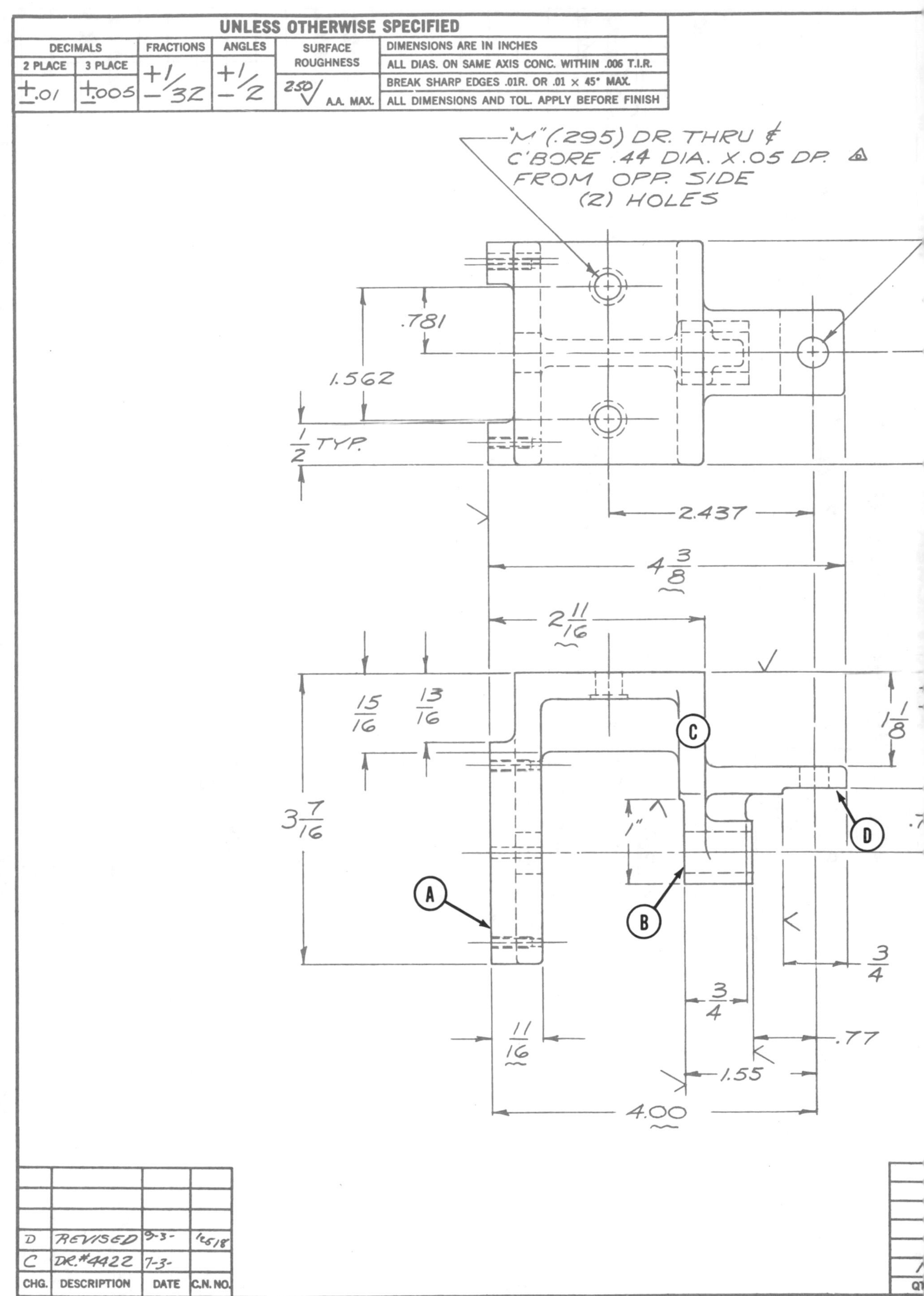

Fig. A-22-BPR-2. Industry blueprint. (Perkin – Elmer)

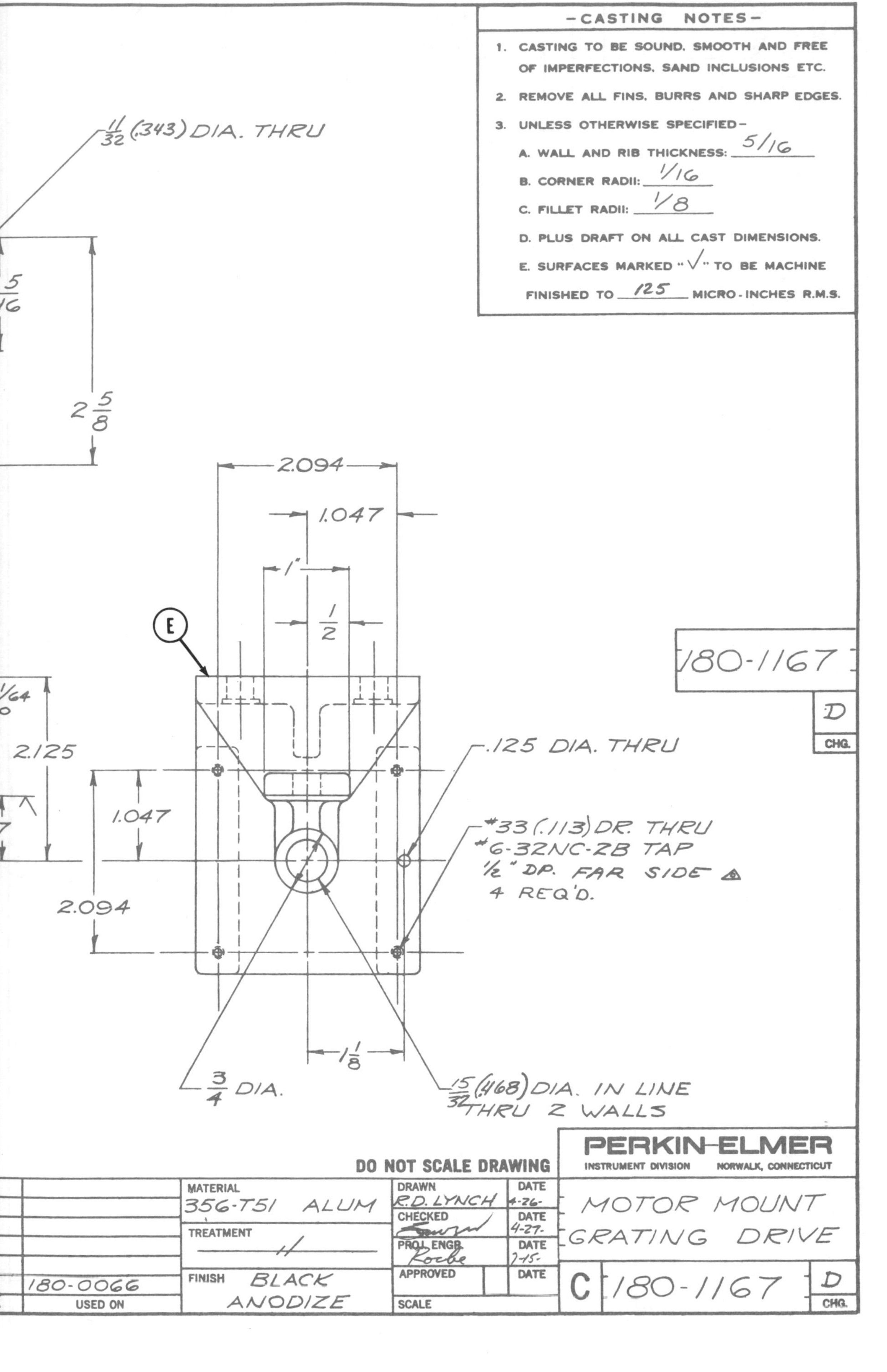

-CASTING NOTES-
1. CASTING TO BE SOUND, SMOOTH AND FREE OF IMPERFECTIONS, SAND INCLUSIONS ETC.
2. REMOVE ALL FINS, BURRS AND SHARP EDGES.
3. UNLESS OTHERWISE SPECIFIED-
A. WALL AND RIB THICKNESS: 5/16
B. CORNER RADII: 1/16
C. FILLET RADII: 1/8
D. PLUS DRAFT ON ALL CAST DIMENSIONS.
E. SURFACES MARKED "√" TO BE MACHINE FINISHED TO 125 MICRO-INCHES R.M.S.
11/32 (.343) DIA. THRU
5/16
2 5/8
2.094
1.047
1"
1/2
E
1/64
2.125
1.047
2.094
.125 DIA. THRU
#33 (.113) DR. THRU
#6-32NC-2B TAP
1/2" DP. FAR SIDE
4 REQ'D.
1 1/8
3/4 DIA.
15/32 (.468) DIA. IN LINE THRU 2 WALLS
180-1167
D
CHG.
DO NOT SCALE DRAWING
PERKIN-ELMER
INSTRUMENT DIVISION
NORWALK, CONNECTICUT
MATERIAL
356-T51 ALUM
TREATMENT
FINISH BLACK ANODIZE
DRAWN R.D. LYNCH
DATE 4-26-
CHECKED
DATE 4-27-
PROJ. ENGR. Roebe
DATE 7-15-
APPROVED
DATE
SCALE
MOTOR MOUNT
GRATING DRIVE
C 180-1167
D
CHG.
180-0066
USED ON

Blueprint Reading Activity A-22-BPR-2
NUMERICAL CONTROL

Refer to the blueprint in Fig. A-22-BPR-2, and to pages 284, 285, 288, 289, 290 and 291, and answer the following questions.

1. What is the name of the part and print number? 1. ______________________

2. What material is specified? 2. ______________________

3. Where is machine zero for X and Y axes? 3. ______________________

4. Where is machine zero for the Z axis? 4. ______________________

5. Interpret what is being done at location No. 1 with the 1" Ski-Kut (a spiral-type end mill). 5. ______________________

6. What does the R Data of R02800 mean for the operation at location No. 1? 6. ______________________

7. What operation is being performed with the Ski-Kut at location No. 2? 7. ______________________

8. Two additional operations are performed at No. 2. What are they? 8. ______________________

9. What does the g00 indicate? 9. ______________________

10. Describe the position of the part at station "B." 10. ______________________

11. What operations are performed at location No. 3? 11. ______________________

12. Describe the position of the part at station "C." 12. ______________________

13. What operations are performed at locations Nos. 6, 7, 8, and 9? 13. ______________________

14. What does the g81 and the R00000 indicate for location No. 10 with the 1/8 drilling operation? 14. ______________________

15. The M08 code in the miscellaneous function column of the program manuscript programs what function? 15. ______________________

G & L 70-NC-10V NUMERICENTER SETUP SHEET — 32 x 60 Table ______ 24 x 48 Travel

PART NAME	PART NO.	REV.	OP. NO.	SET-UP BY	DATE	TOOL/FIXTURE NO.
MOTOR MOUNT-GRATING DR.	180-1167	C	TAPE #1	P.J.T.	8-23-	T-180-1167-1 T-180-1095-2 T-180-1095-3

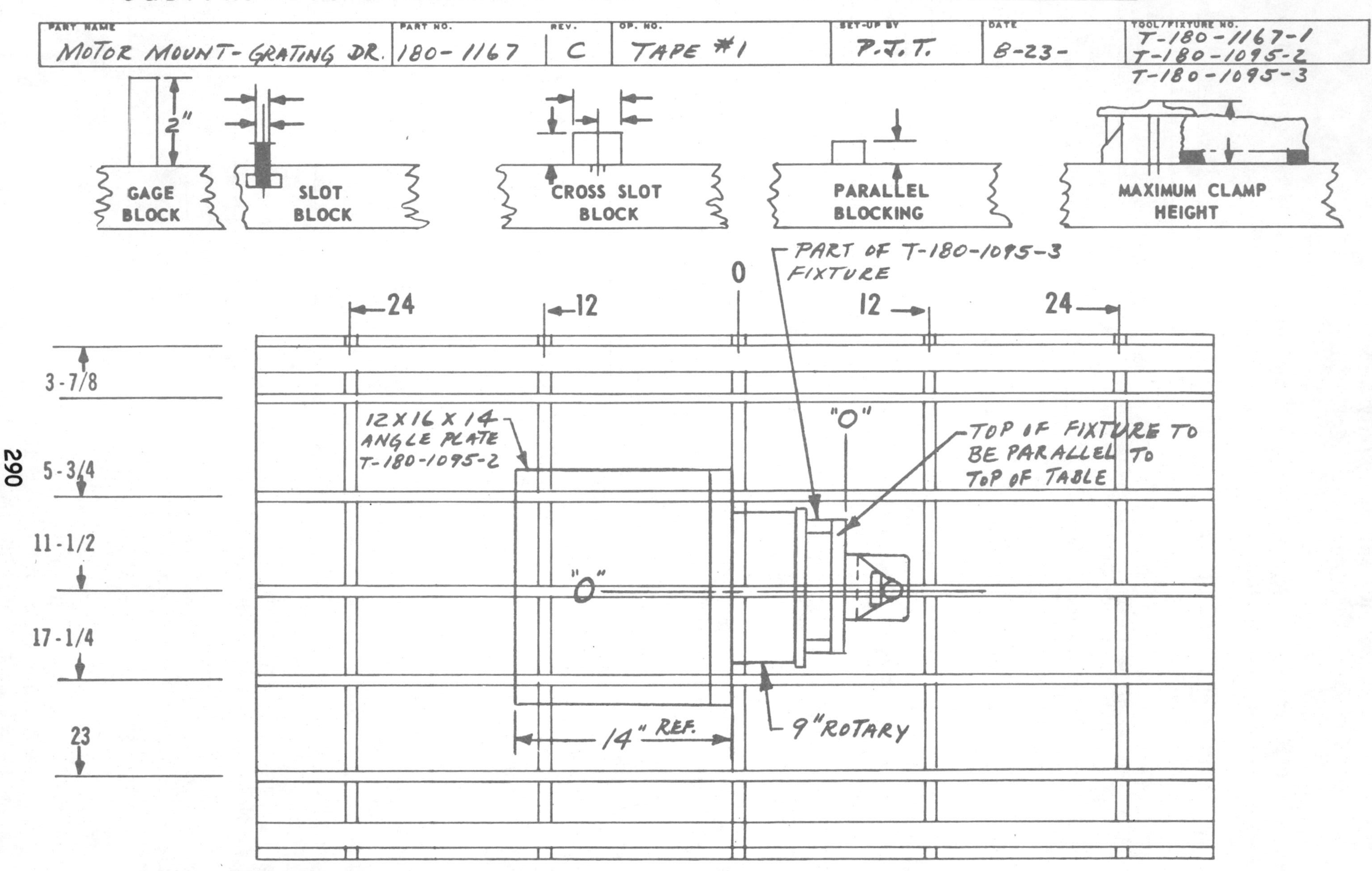

NOTES: 1. FACE OF ROTARY TO BE PARALLEL WITH CROSS SLOTS.
2. START MACHINING WITH STATION "A" AS SHOWN.

1-0478-00

PERKIN-ELMER

G & L 70-NC-10V NUMERICENTER SETUP & TOOLING SHEET

PART NAME	OP. NO.	W.C.	TAPE NO.	MACHINE	PART NO.	REV.
MOTOR MOUNT- GRATING DR.			#1	G&L	180-1167	C

MACHINE NUMBER	PROGRAM BY	DATE	CHECK BY	DATE	(THRU) ENGR. CHANGE NO.	DATE
	P. J. T.	8-23-..				

SET UP AND ORIENTATION INSTRUCTIONS	T	RIGHT MATRIX	J	TOOL	T	LEFT MATRIX	J	TOOL
	TOOL POS.		TOOL COMP.	LENGTH	TOOL POS.		TOOL COMP.	LENGTH
POSITION ANGLE PLATE, 9" ROTARY,	t10	1" SKI-KUT	J01 J04 J05		t30	#2 CENTER DR.		
FIXTURE & WORKPIECE AS SHOWN.	t11	15/32 DR.	J02		t31	1/8 DR.		
	t12	#33(.113) DR.			t32	#6-32 TAP	J03	
ZERO "X" & "Y" AS SHOWN.	t13				t33			
	t14				t34			
LONGEST TOOL TO CLEAR THE	t15				t35			
HIGHEST POINT OF THE SETUP.	t16				t36			
	t17				t37			
ALL TOOL ADJUST TOOLS TO BE	t18				t38			
PROGRAMMED 2" ABOVE FLAT	t19				t39			
SURFACE (A STATION) OF FIXTURE	t20				t40			
USING A 2.000 GAGE BLOCK.	t21				t41			
	t22				t42			
	t23				t43			
	t24				t44			
	t25				t45			
	t26				t46			
	t27				t47			
	t28				t48			
	t29				t49			

PERKIN-ELMER

Blueprint Reading Activity A-23-BPR-1
PRECISION SHEET METAL

Refer to the blueprint in Fig. A-23-BPR-1 and answer the following questions.

1. Give the name of the piece and the drawing number. 1. ______________________
2. How many separate parts are needed to make one bracket? 2. ______________________
3. What material is specified? 3. ______________________
4. Give the hole size at B and the positional tolerance. 4. ______________________
5. What kind of a dimension is at C and what is its meaning? 5. ______________________
6. Interpret the welding to be done at A. 6. ______________________
7. Give the welding rod specification. 7. ______________________
8. Using the formula given in Unit 23, calculate the length of the support piece in the flat. 8. ______________________
9. What is the next assembly on which this bracket is used? 9. ______________________
10. The bracket is used on what model number? 10. ______________________

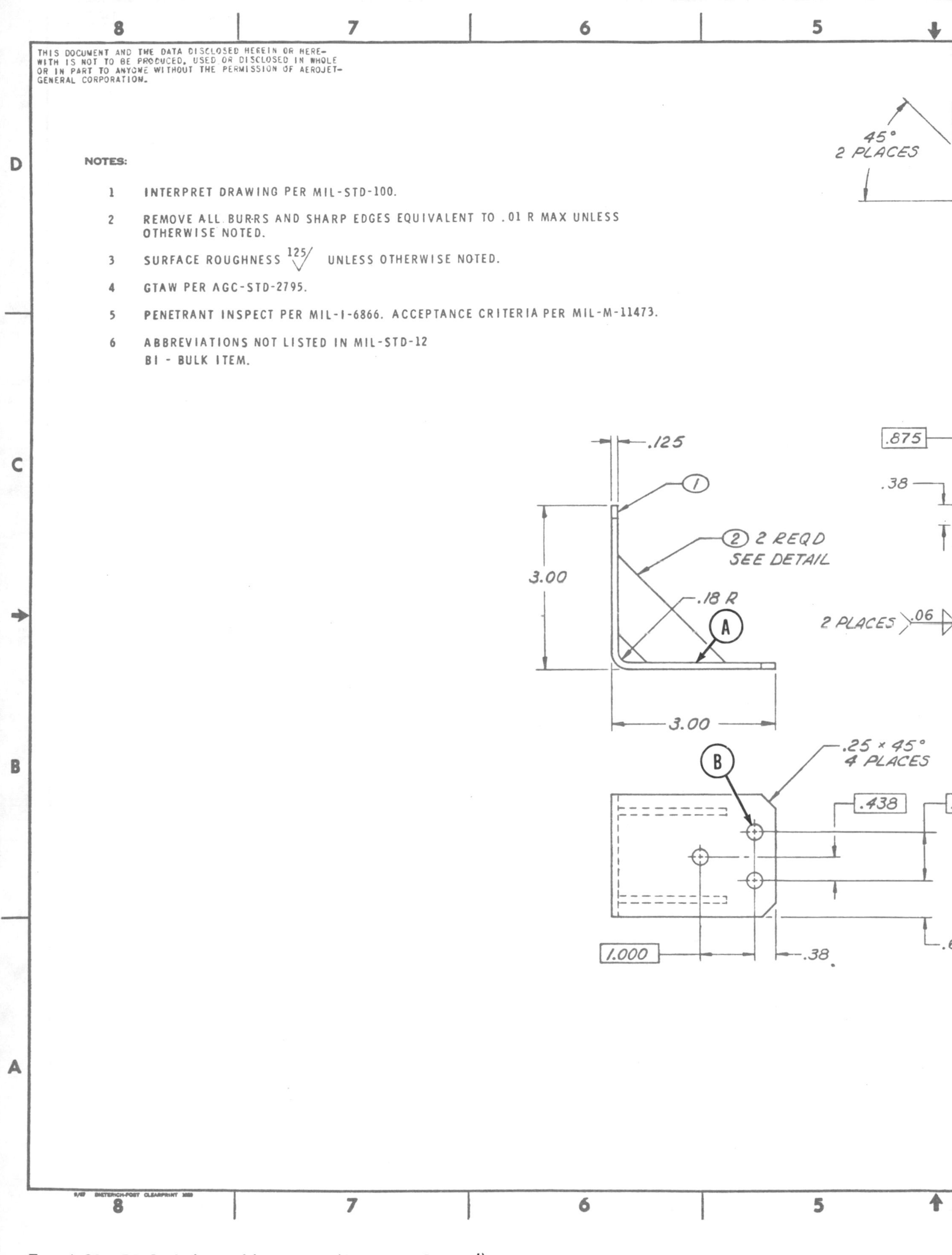

Fig. A-23-BPR-1. Industry blueprint. (Aerojet – General)

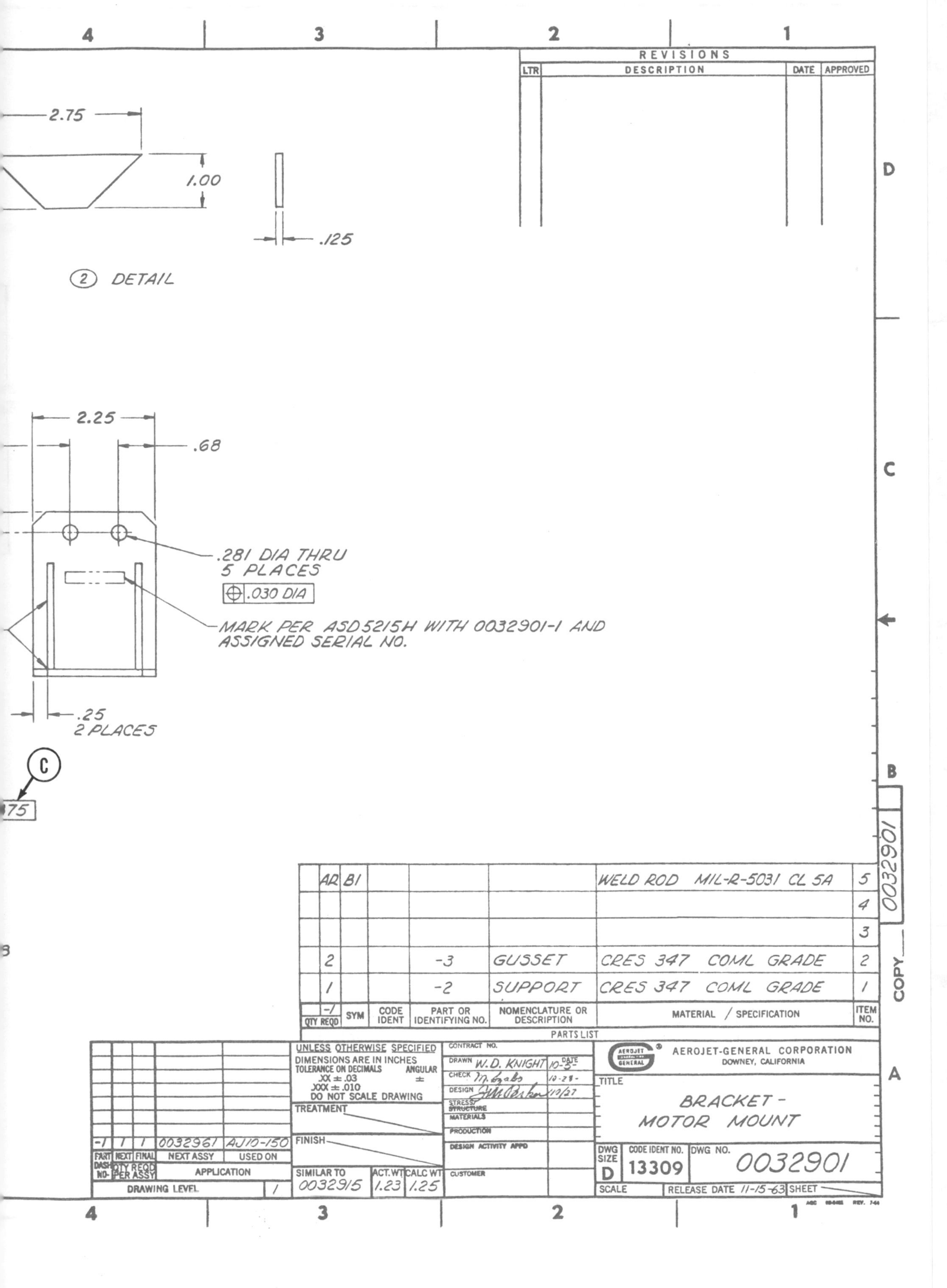

QTY REQD -1	SYM	CODE IDENT	PART OR IDENTIFYING NO.	NOMENCLATURE OR DESCRIPTION	MATERIAL / SPECIFICATION	ITEM NO.
AR	B1				WELD ROD MIL-R-5031 CL 5A	5
						4
						3
2			-3	GUSSET	CRES 347 COML GRADE	2
1			-2	SUPPORT	CRES 347 COML GRADE	1

PARTS LIST

PART DASH NO.	NEXT QTY REQD PER ASSY	FINAL	NEXT ASSY	USED ON
-1	1	1	0032961	AJ10-150

APPLICATION — DRAWING LEVEL 1

UNLESS OTHERWISE SPECIFIED
DIMENSIONS ARE IN INCHES
TOLERANCE ON DECIMALS .XX ± .03, .XXX ± .010; ANGULAR ±
DO NOT SCALE DRAWING

TREATMENT
FINISH
SIMILAR TO 0032915 — ACT. WT 1.23 — CALC WT 1.25

CONTRACT NO.
DRAWN W. D. KNIGHT 10-3-
CHECK 10-28-
DESIGN 10/27
STRESS/STRUCTURE
MATERIALS
PRODUCTION
DESIGN ACTIVITY APPD
CUSTOMER

AEROJET-GENERAL CORPORATION
DOWNEY, CALIFORNIA

TITLE: BRACKET - MOTOR MOUNT

DWG SIZE D — CODE IDENT NO. 13309 — DWG NO. 0032901

SCALE — RELEASE DATE 11-15-63 — SHEET

Blueprint Reading Activity A-23-BPR-2
PRECISION SHEET METAL

Refer to the blueprint in Fig. A-23-BPR-2 and answer the following questions.

1. Give the title and number of the drawing. 1. ______

2. What material is specified for the Mask (item 2)? What size? 2. ______

3. How are items 3 and 4 fastened to item 2? 3. ______

4. Where are items 3 and 4 located with respect to datum C? 4. ______

5. Interpret the feature control symbols at X. 5. ______

6. Give the dimension and tolerance for datum A. 6. ______

7. What is the distance between centers of the two items 4 and how must they be located relative to datum B? 7. ______

8. Interpret the feature control symbol at Y. 8. ______

9. What are the hole sizes and tolerances at Z? 9. ______

10. Is an overspray of finish material permitted on the rear of item 2? 10. ______

11. Give the finished length of items 5 and 6. 11. ______

12. What is to be done to the edges of the parts? 12. ______

Blueprint Reading Activity A-24-BPR-1
WELDING

Refer to the blueprint in Fig. A-24-BPR-1 and answer the following questions.

1. What is the name and number of the assembly? 1. ______________________ ______________________

2. What was the last change made to the drawing? 2. ______________________ ______________________

3. What is the standard by which the welding is to be done? 3. ______________________

4. Indicate the type of welds required at A, B and D. 4. A ______________________ B ______________________ D ______________________

5. Give the part number, description, material and size of the items at C and F. 5. C ______________________ F ______________________

6. What type of weld is required at E, G and H. 6. E ______________________ G ______________________ H ______________________

7. Interpret the type of welds required at J, K, L and M. 7. J ______________________ K ______________________ L ______________________ M ______________________

8. Interpret the type of welds required at N, P, R and T. 8. N ______________________ P ______________________ R ______________________ T ______________________

9. Give the part number, description, material and size of the item at S. 9. S ______________________ ______________________

10. Interpret the type of welds required at U and V. 10. U ______________________ V ______________________

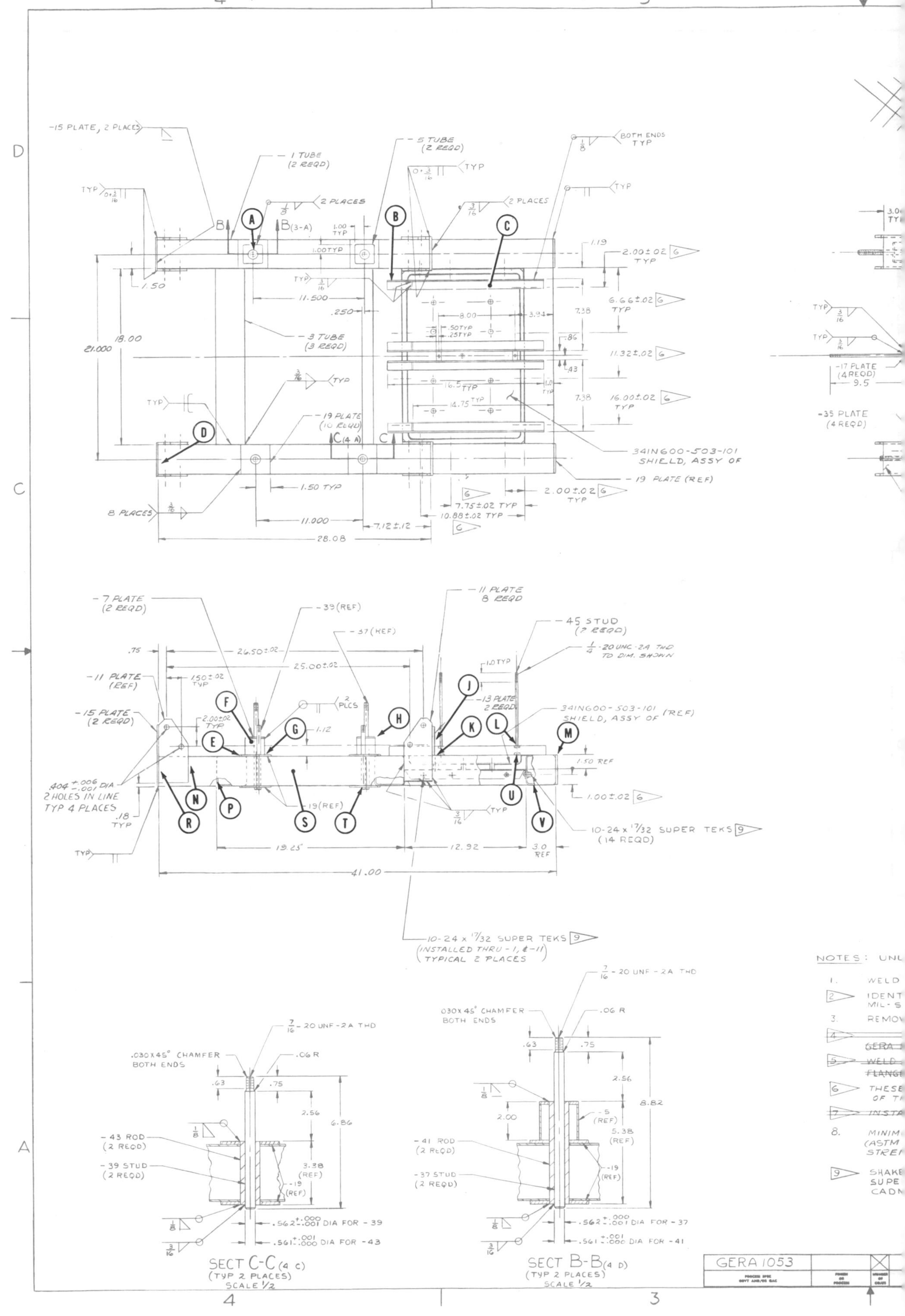

Fig. A-24-BPR-1. Industry blueprint. (Goodyear Aerospace)

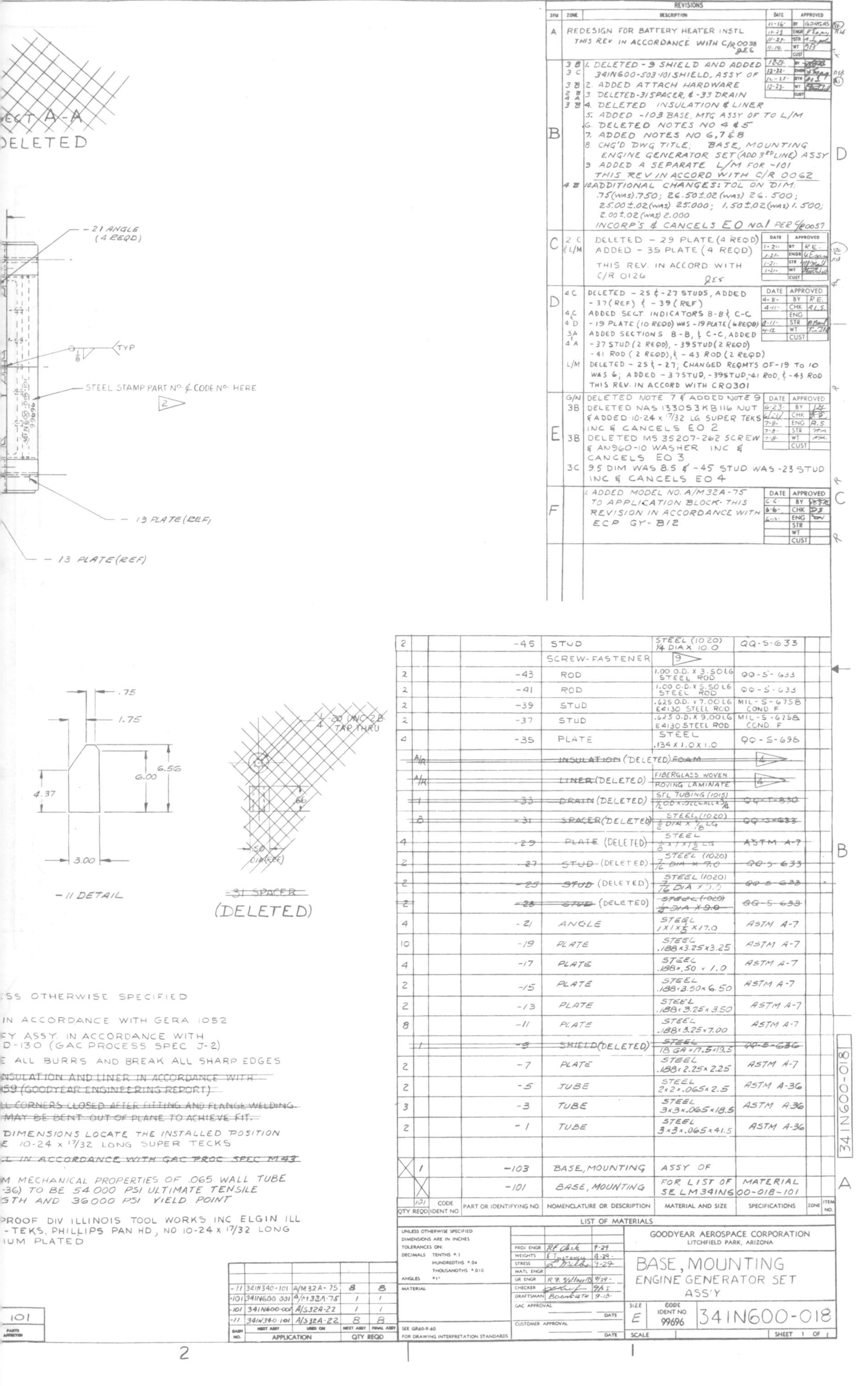
REVISIONS
A REDESIGN FOR BATTERY HEATER INSTL THIS REV IN ACCORDANCE WITH C/R 0038
B 1. DELETED -9 SHIELD AND ADDED 341N600-503-101 SHIELD, ASSY OF
2. ADDED ATTACH HARDWARE
3. DELETED -31 SPACER, & -33 DRAIN
4. DELETED INSULATION & LINER
5. ADDED -103 BASE, MTG ASSY OF TO L/M
6. DELETED NOTES NO 4 & 5
7. ADDED NOTES NO 6, 7 & 8
8. CHG'D DWG TITLE; BASE, MOUNTING ENGINE GENERATOR SET (ADD 3RD LINE) ASSY
9. ADDED A SEPARATE L/M FOR -101
THIS REV IN ACCORD WITH C/R 0062
10. ADDITIONAL CHANGES: TOL ON DIM: .75 (WAS) .750; 26.50±.02 (WAS) 26.500; 25.00±.02 (WAS) 25.000; 1.50±.02 (WAS) 1.500; 2.00±.02 (WAS) 2.000
INCORP'S & CANCELS EO No1 PER C/R 0057
C DELETED -29 PLATE (4 REQD) ADDED -35 PLATE (4 REQD) THIS REV. IN ACCORD WITH C/R 0126
D DELETED -25 & -27 STUDS, ADDED -37 (REF) & -39 (REF)
ADDED SECT. INDICATORS B-B & C-C
-19 PLATE (10 REQD) WAS -19 PLATE (6 REQD)
ADDED SECTIONS B-B, & C-C, ADDED -37 STUD (2 REQD), -39 STUD (2 REQD) -41 ROD (2 REQD), & -43 ROD (2 REQD)
DELETED -25 & -27; CHANGED REQMTS OF -19 TO 10 WAS 6; ADDED -37 STUD, -39 STUD, -41 ROD, & -43 ROD
THIS REV. IN ACCORD WITH CR0301
E DELETED NOTE 7 & ADDED NOTE 9
DELETED NAS 1330S3KB116 NUT & ADDED 10-24 x 17/32 LG SUPER TEKS INC & CANCELS EO 2
DELETED MS 35207-262 SCREW & AN960-10 WASHER INC & CANCELS EO 3
9.5 DIM WAS 8.5 & -45 STUD WAS -23 STUD INC & CANCELS EO 4
F 1. ADDED MODEL NO. A/M32A-75 TO APPLICATION BLOCK. THIS REVISION IN ACCORDANCE WITH ECP GY-B12
SECT A-A DELETED
-21 ANGLE (4 REQD)
TYP
STEEL STAMP PART No. & CODE No. HERE
-13 PLATE (REF)
-13 PLATE (REF)
-11 DETAIL
-31 SPACER (DELETED)
LIST OF MATERIALS
2 | -45 | STUD | STEEL (1020) 1/4 DIA X 10.0 | QQ-S-633
SCREW-FASTENER | 9
2 | -43 | ROD | 1.00 O.D. X 3.50 LG STEEL ROD | QQ-S-633
2 | -41 | ROD | 1.00 O.D. X 5.50 LG STEEL ROD | QQ-S-633
2 | -39 | STUD | .625 O.D. X 7.00 LG E4130 STEEL ROD | MIL-S-6758 COND F
2 | -37 | STUD | .625 O.D. X 9.00 LG E4130 STEEL ROD | MIL-S-6758 COND F
4 | -35 | PLATE | STEEL .134 X 1.0 X 1.0 | QQ-S-698
INSULATION (DELETED) FOAM
LINER (DELETED) FIBERGLASS WOVEN ROVING LAMINATE
-33 DRAIN (DELETED)
-31 SPACER (DELETED)
-29 PLATE (DELETED)
-27 STUD (DELETED)
-25 STUD (DELETED)
-23 STUD (DELETED)
4 | -21 | ANGLE | STEEL 1 X 1 X 1/8 X 17.0 | ASTM A-7
10 | -19 | PLATE | STEEL .188 X 3.25 X 3.25 | ASTM A-7
4 | -17 | PLATE | STEEL .188 X .50 X 1.0 | ASTM A-7
2 | -15 | PLATE | STEEL .188 X 3.50 X 6.50 | ASTM A-7
2 | -13 | PLATE | STEEL .188 X 3.25 X 3.50 | ASTM A-7
8 | -11 | PLATE | STEEL .188 X 3.25 X 7.00 | ASTM A-7
-9 SHIELD (DELETED)
2 | -7 | PLATE | STEEL .188 X 2.25 X 2.25 | ASTM A-7
2 | -5 | TUBE | STEEL 2 X 2 X .065 X 2.5 | ASTM A-36
3 | -3 | TUBE | STEEL 3 X 3 X .065 X 18.5 | ASTM A-36
2 | -1 | TUBE | STEEL 3 X 3 X .065 X 41.5 | ASTM A-36
1 | -103 | BASE, MOUNTING | ASSY OF
-101 | BASE, MOUNTING | FOR LIST OF MATERIAL SE LM341N600-018-101
QTY REQD | CODE IDENT NO | PART OR IDENTIFYING NO. | NOMENCLATURE OR DESCRIPTION | MATERIAL AND SIZE | SPECIFICATIONS | ZONE | ITEM NO.
...SS OTHERWISE SPECIFIED
...IN ACCORDANCE WITH GERA 1052
...FY ASSY. IN ACCORDANCE WITH ...D-130 (GAC PROCESS SPEC J-2)
...E ALL BURRS AND BREAK ALL SHARP EDGES
...DIMENSIONS LOCATE THE INSTALLED POSITION
...E 10-24 x 17/32 LONG SUPER TECKS
...M MECHANICAL PROPERTIES OF .065 WALL TUBE ...-36) TO BE 54 000 PSI ULTIMATE TENSILE ...GTH AND 36000 PSI YIELD POINT
...PROOF DIV ILLINOIS TOOL WORKS INC ELGIN ILL. ...-TEKS, PHILLIPS PAN HD, NO 10-24 x 17/32 LONG ...IUM PLATED
-11 | 341N340-101 | A/M32A-75 | 8 | 8
-101 | 341N600-001 | A/M32A-75 | 1 | 1
-101 | 341N600-001 | A/S32A-22 | 1 | 1
-11 | 341N340-101 | A/S32A-22 | 8 | 8
APPLICATION | QTY REQD
GOODYEAR AEROSPACE CORPORATION LITCHFIELD PARK, ARIZONA
BASE, MOUNTING ENGINE GENERATOR SET ASS'Y
SIZE E | CODE IDENT NO 99696 | 341N600-018
SCALE | SHEET 1 OF 1
341N600-018

Blueprint Reading Activity A-24-BPR-2
WELDING

Refer to the blueprint in Fig. A-24-BPR-2 and answer the following questions.

1. Give the name and number of the assembly.

1. ______________________________

2. Interpret the welding symbols indicated on the blueprint by the following:

2. P ______________________________

R ______________________________

S ______________________________

T ______________________________

U ______________________________

V ______________________________

W ______________________________

X ______________________________

Y ______________________________

3. Interpret the form tolerance feature control symbols at Z.

3. ______________________________

Blueprint Reading Activity A-25-BPR-1
INSTRUMENTATION AND CONTROL DIAGRAMS - PNEUMATIC

Refer to the blueprint in Fig. A-25-BPR-1 and to the Sequence of Operations and Bill of Materials on page 302, and answer the following questions.

1. What is the name of the blueprint? 1. ____________________

2. Give the drawing number. 2. ____________________

3. Describe briefly the function of the machine operation. 3. ____________________

4. What actuates the automatic cycle? 4. ____________________

5. When the transfer cylinders are in their down-position, what starts the drill motors 1, 2, 3, and 4? 5. ____________________

6. While the drilling cycle is in process, what happens to valves 7 and 3? 6. ____________________

7. What activates valves 8? 7. ____________________

8. When valves 8 are shifted to in-line position with 3, what takes place? 8. ____________________

(Questions continued on page 303)

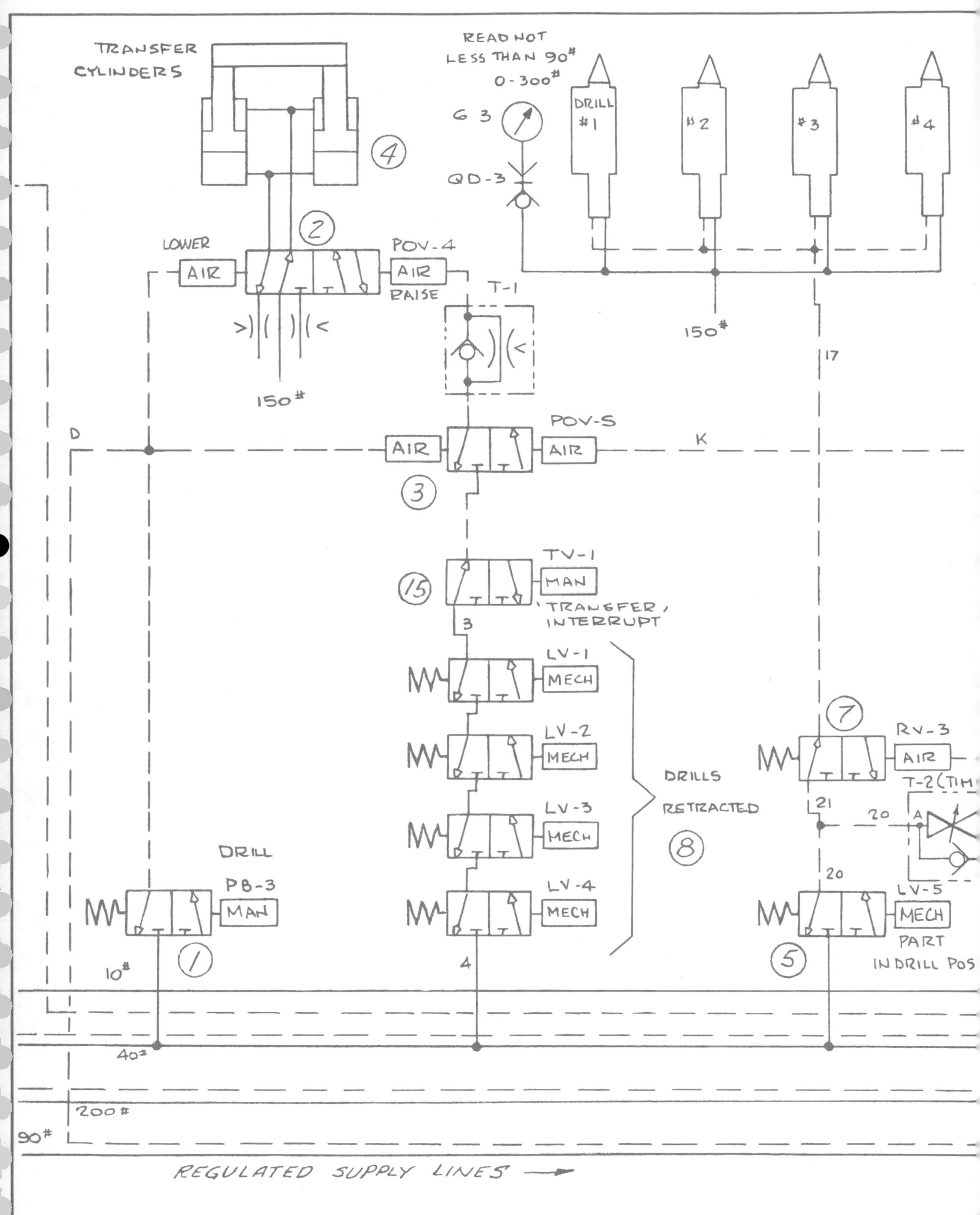

Fig. A-25-BPR-1. Industry blueprint. (Unidynamics)

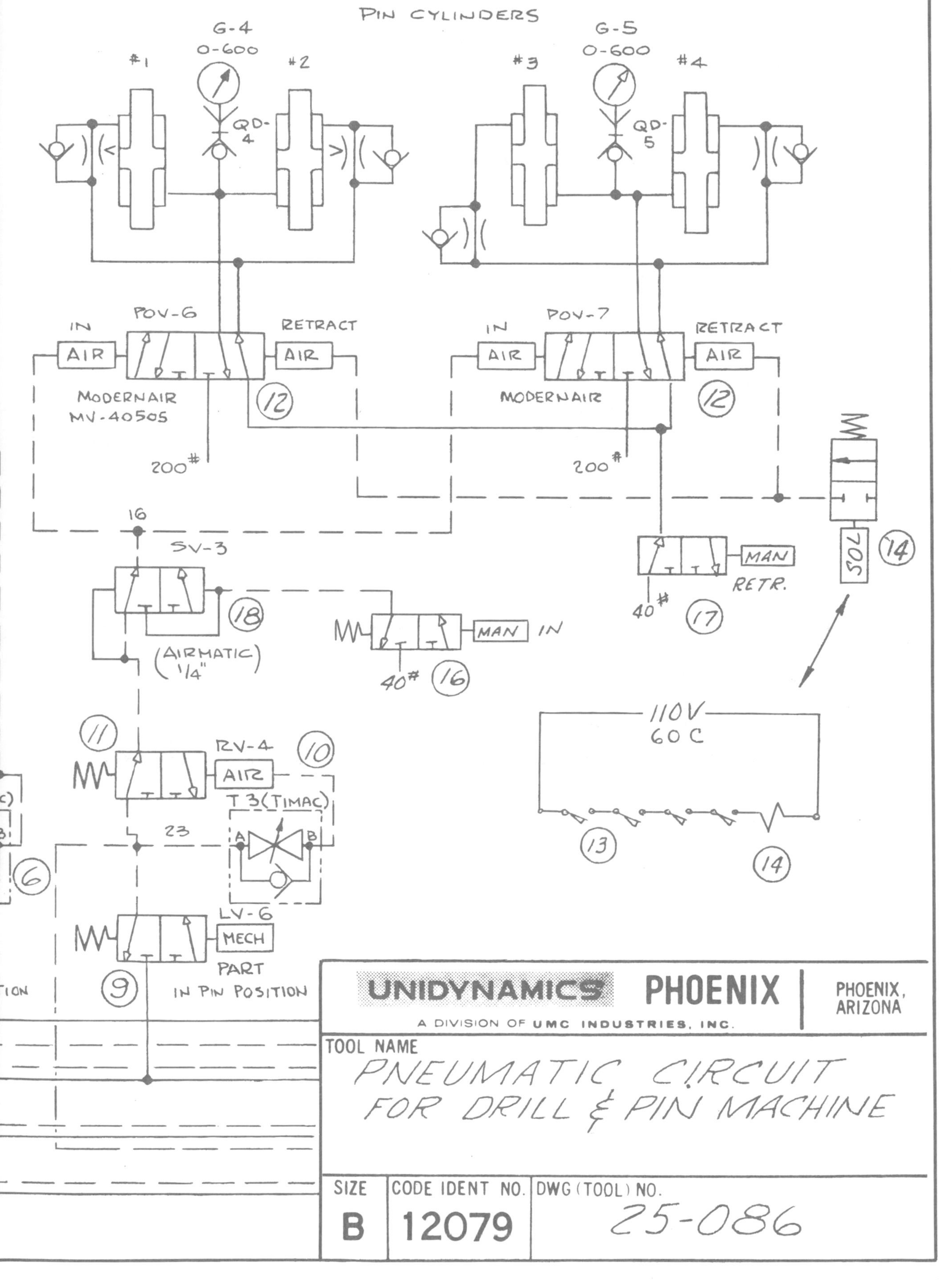

PIN CYLINDERS
G-4
0-600
#1
#2
QD-4
G-5
0-600
#3
#4
QD-5
POV-6
IN
AIR
RETRACT
AIR
MODERNAIR
MV-40505
12
200#
POV-7
IN
AIR
RETRACT
AIR
MODERNAIR
12
200#
16
SV-3
18
(AIRMATIC 1/4")
MAN
IN
40#
16
MAN
RETR.
40#
17
SOL
14
110V
60 C
13
14
11
RV-4
AIR
10
T 3 (TIMAC)
A
B
23
6
LV-6
MECH
PART
IN PIN POSITION
9
TION
UNIDYNAMICS PHOENIX
PHOENIX, ARIZONA
A DIVISION OF UMC INDUSTRIES, INC.
TOOL NAME
PNEUMATIC CIRCUIT FOR DRILL & PIN MACHINE
SIZE
B
CODE IDENT NO.
12079
DWG (TOOL) NO.
25-086

9. What happens with the return of cylinders 4 to the "up" position?

9. ____________________

10. What causes the pin cylinders to retract?

10. ____________________

11. When are valves 15 thru 18 used?

11. ____________________

Blueprint Reading Activity A-25-BPR-2
INSTRUMENTATION AND CONTROL DIAGRAMS - HYDRAULICS

Refer to the blueprint in Fig. A-25-BPR-2 and to the Sequence of Operations on page 306, and answer the following questions.

1. Give the name of the graphic diagram and drawing number. 1. ____________________

2. State briefly the purpose of the fluid power circuit. 2. ____________________

3. What actuates the clamping of the workpiece in the fixture? 3. ____________________

4. When push button valve 2 is depressed alone, why doesn't this actuate valve 5? 4. ____________________

5. Explain what happens when cylinder 6 is extended and either valve 2 or 3 are released? 5. ____________________

6. What causes the workpiece to rotate 180 deg. for second dimpling operation? 6. ____________________

7. What kind of a valve is valve 10? 7. ____________________

8. Is cylinder 6 extended or retracted by high pressure fluid? 8. ____________________

9. What is the normal position (as shown) for cylinder 8? 9. ____________________

10. Does rotary actuator 11 operate on high or low pressure fluid? 10. ____________________

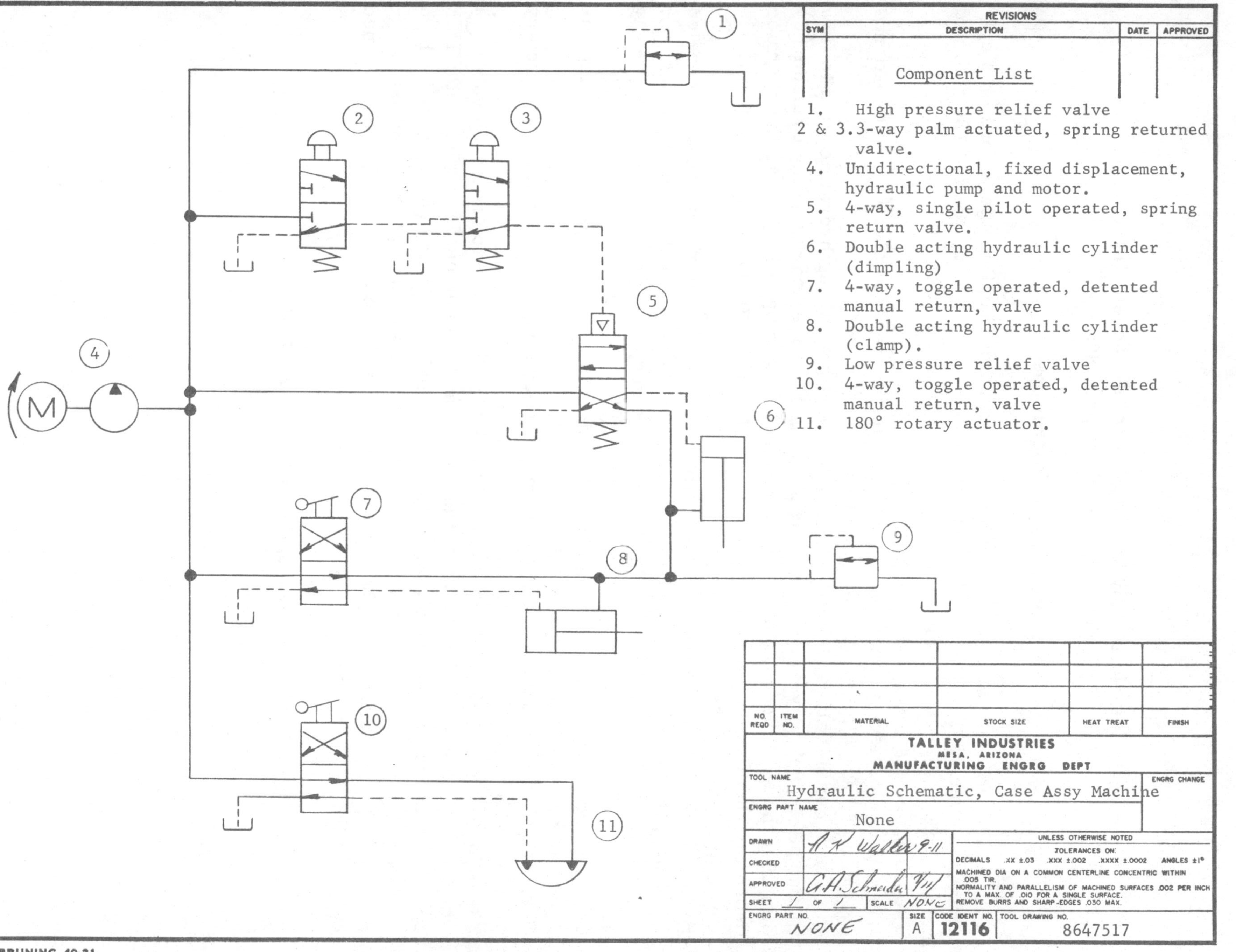

Fig. A-25-BPR-2. Industry blueprint. (Talley Industries)

Case Assembly Machine Hydraulic Schematic

(Refer to blueprint on page 305.)

Sequence of Operations

DESCRIPTION: A fluid power system to perform two dimpling operations on a cylinderical assembly. The part is loaded into a fixture, dimpled, rotated 180 deg. and dimpled again. This is accomplished by the following sequence:

1. The workpiece is loaded into the fixture.
2. When the workpiece is in position, the operator depresses valve (7) which extends cylinder (8) and clamps the piece in position.
3. Both push button valves (2) and (3) are depressed permitting hydraulic fluid to flow to the pilot of 4-way valve (5), thus shifting the valve and extending cylinder (6) dimpling the workpiece. Note: This step cannot be accomplished unless the operator depresses both palm buttons. This safety feature is to insure that the operator's hands are clear of the mechanism.
4. By releasing either of the push button valves (2) and (3) the fluid will be exhausted from the pilot of 4-way valve (5), thus shifting the valve and retracting cylinder (6).
5. Once the dimple cylinder (6) has been retracted, the operator depresses 4-way valve (10) which rotates the rotary actuator (11) 180 deg.
6. When the workpiece has been rotated, operations 3 and 4 are repeated, making another dimple 180 deg. from the first.
7. Upon completion of operation 6, the operator releases valve 10 which returns rotary actuator (11) to its original position.
8. When the part has returned to its original position, the operator releases valve (7) which unclamps the workpiece from the fixture and completes the cycle. Note: This circuit is a dual pressure circuit. The two cylinders (6) and (8) operate on high pressure fluid on their extending sequences. This is accomplished by pressure relief valve (1). Retraction of these cylinders is done on low pressure, accomplished by pressure relief valve (7).

Sequence of Operations for Fig. A-25-BPR-2.

Blueprint Reading Activity A-25-BPR-3
INSTRUMENTATION AND CONTROL DIAGRAMS - FLUIDICS

Refer to the blueprint in Fig. A-25-BPR-3 and to the Sequence of Operations on page 310, and answer the following questions.

1. What is the name of the circuit diagram? The drawing number? 1. ______

2. Briefly, state the purpose of the control circuit. 2. ______

3. What is the purpose of the preference circuit? 3. ______

4. What device in the preference circuit sets up the initial preference signal? 4. ______

5. How is this preference signal removed? 5. ______

6. State the purpose of the interlock / indexing circuit. 6. ______

7. What three conditions must exist before the index table will index? 7. ______

8. What is the device that actuates the interface valve that actuates the indexing cylinder? Is the control signal coming from the 0_1 or 0_2 port? 8. ______

9. What actuates the press station circuit? 9. ______

10. State the purpose of the timer. 10. ______

(Questions continued on page 308)

11. What causes the staking cylinder to retract?

11. ____________________

12. Supposing the staking station completes its operation before the pressing station and is ready for the port, will the table index? Why?

12. ____________________

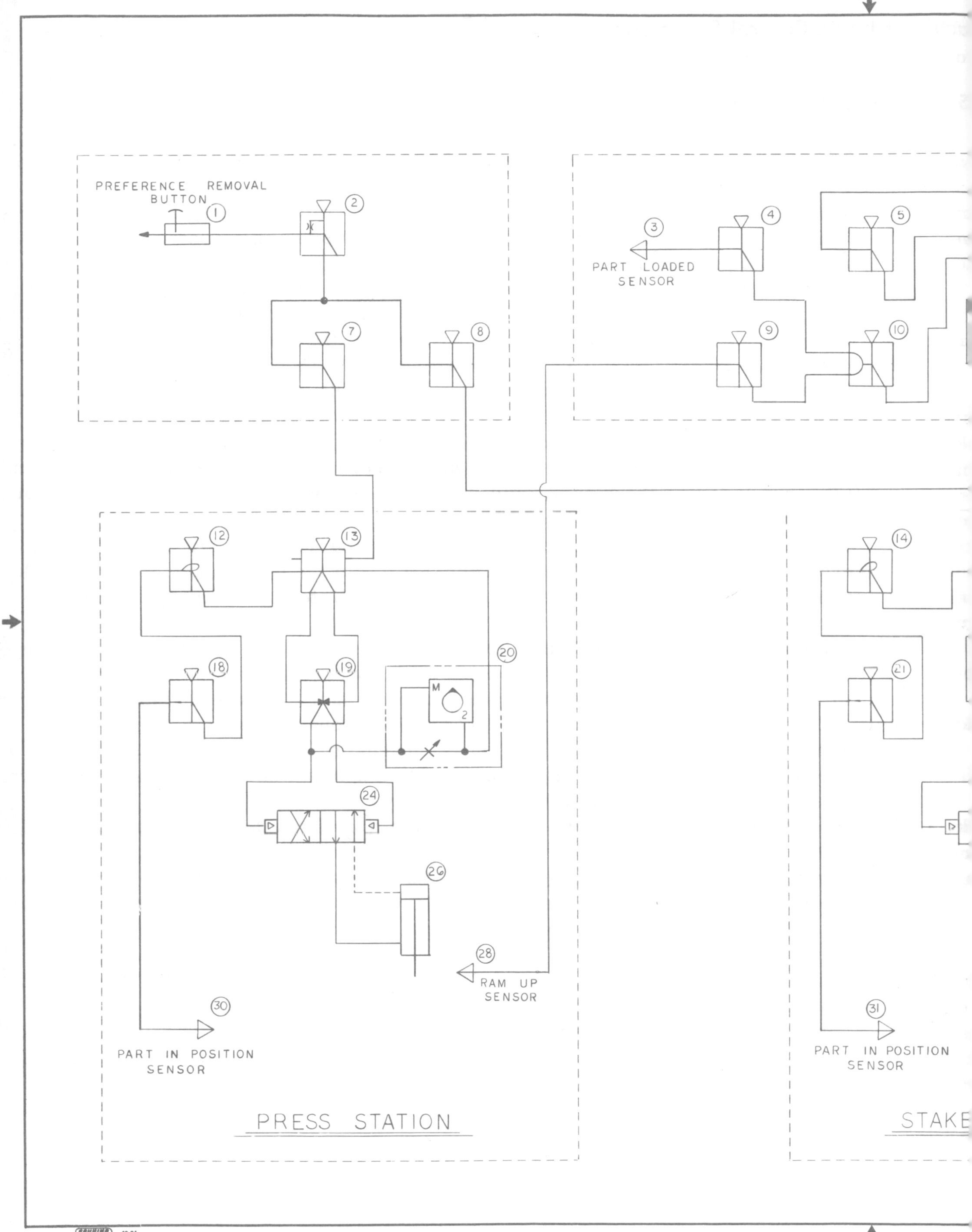

Fig. A-25-BPR-3. Industry blueprint. (Talley Industries)

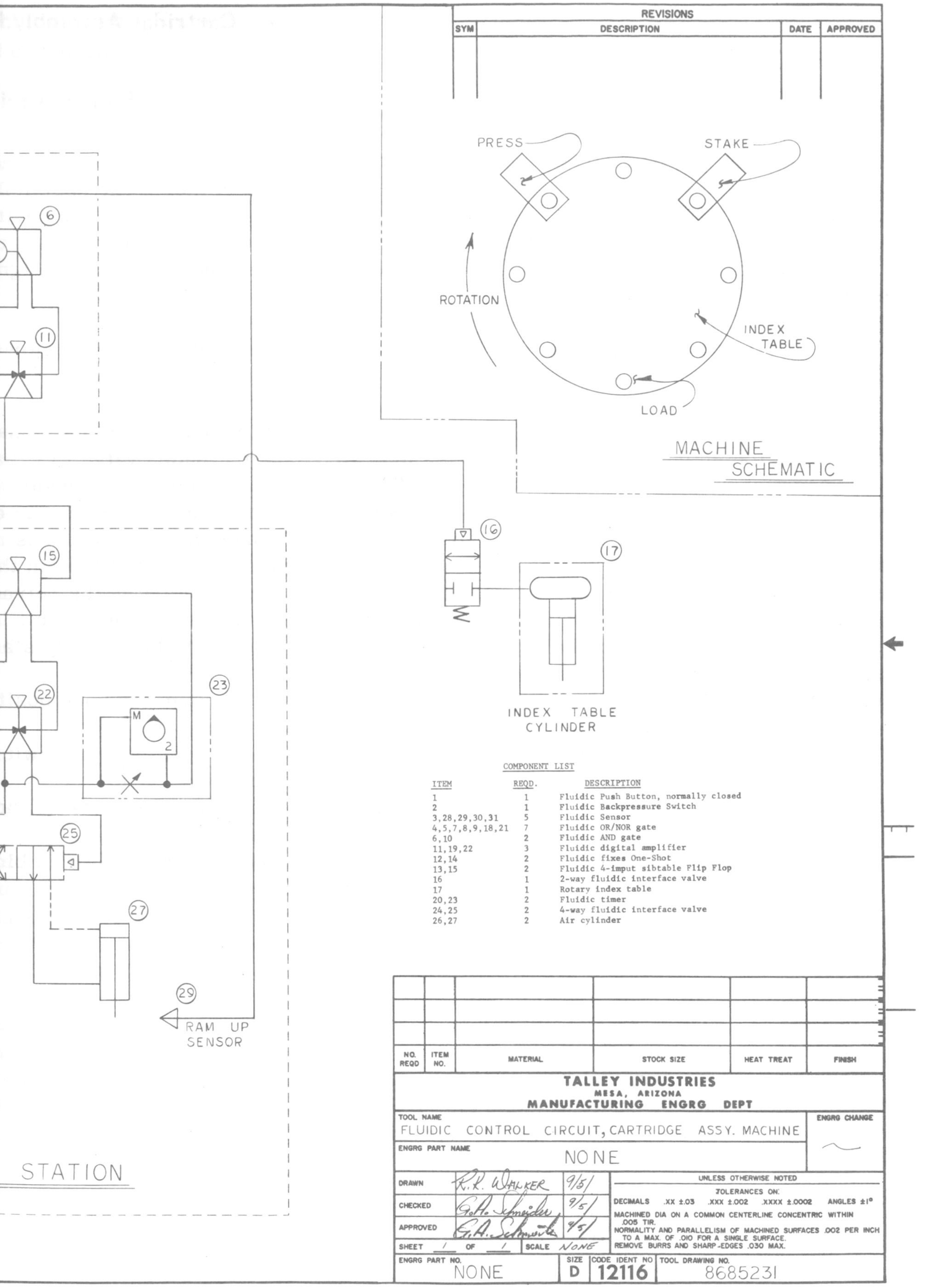

COMPONENT LIST

ITEM	REQD.	DESCRIPTION
1	1	Fluidic Push Button, normally closed
2	1	Fluidic Backpressure Switch
3,28,29,30,31	5	Fluidic Sensor
4,5,7,8,9,18,21	7	Fluidic OR/NOR gate
6,10	2	Fluidic AND gate
11,19,22	3	Fluidic digital amplifier
12,14	2	Fluidic fixes One-Shot
13,15	2	Fluidic 4-imput sibtable Flip Flop
16	1	2-way fluidic interface valve
17	1	Rotary index table
20,23	2	Fluidic timer
24,25	2	4-way fluidic interface valve
26,27	2	Air cylinder

Evaluation Activities

Blueprint Reading Activity
PART 2 - EVALUATION BLUEPRINT

Refer to the blueprint on page 313 and answer the following questions:

1. What is the name of the object? 1. ______________________
____

2. Give the number of the blueprint. 2. ______________________

3. Is this a detail or an assembly drawing? 3. ______________________

4. What three main views are shown? 4. ______________________

5. What type of sectional views are shown? 5. ______________________

6. Give the material specification. 6. ______________________

7. What is the overall size of the part? 7. ______________________

8-14. Identify the types of lines at:

8. A ______________________

9. C ______________________

10. D ______________________

11. E ______________________

12. F ______________________

13. G ______________________

14. M ______________________

(Questions continued on page 312)

15. What is the thickness of the surface at B? 15.__________

16. Give the dimension for H. 16.__________

17. What is the diameter of the hole at J? 17.__________

18. What is the thickness of the material at K? 18.__________

19. Give the height, width and length of the pads at L. 19.__________

20. The threaded hole at N is to be drilled thru for what size hole? 20.__________

CHANGE NO.	LETTER	CHANGE	DATE	CHKD BY	CHANGE NO.	LETTER	CHANGE	DATE	CHKD BY

CORE .204 ±.003 DIA. MINUS .006 TAPER — .30 DEEP
DRILL (#7) .201 $^{+.004}_{-.002}$ DIA. — THRU.
TAP 1/4-20 UNC-2B TH'D — THRU.
P.D. .2175 — .2223
2 HOLES IN TRUE POSITION WITHIN .024 DIA.

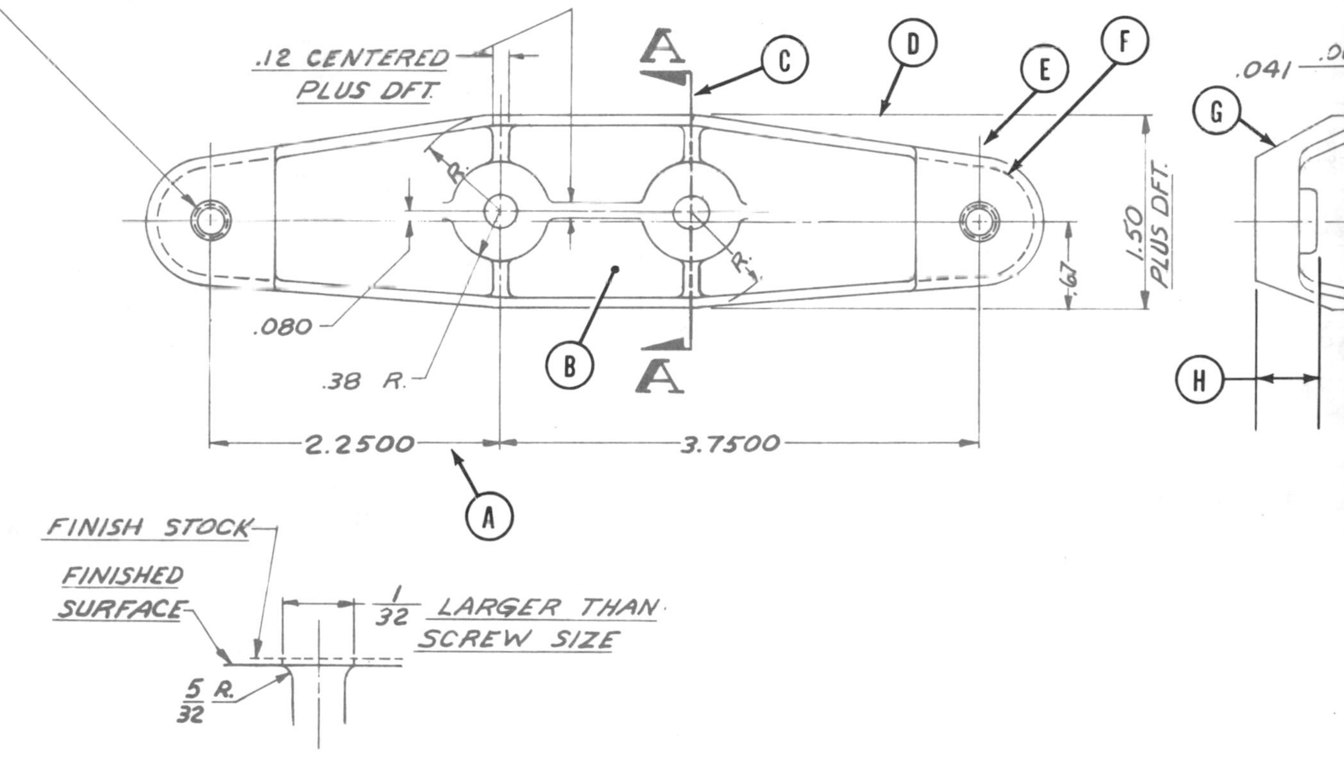

CORED HOLES THAT REQUIRE TAPPING MUST CONFORM WITH ABOVE SKETCH.
(EXCEPT AS OTHERWISE NOTED)

NOTE:
ALL SMALL RADII & FILLETS .06,
WALLS .078 ±.015, DRAFT 2°
UNLESS OTHERWISE SPECIFIED.

ENG. SPEC. #1

Evaluation Blueprint 2. (Marine Engineering)

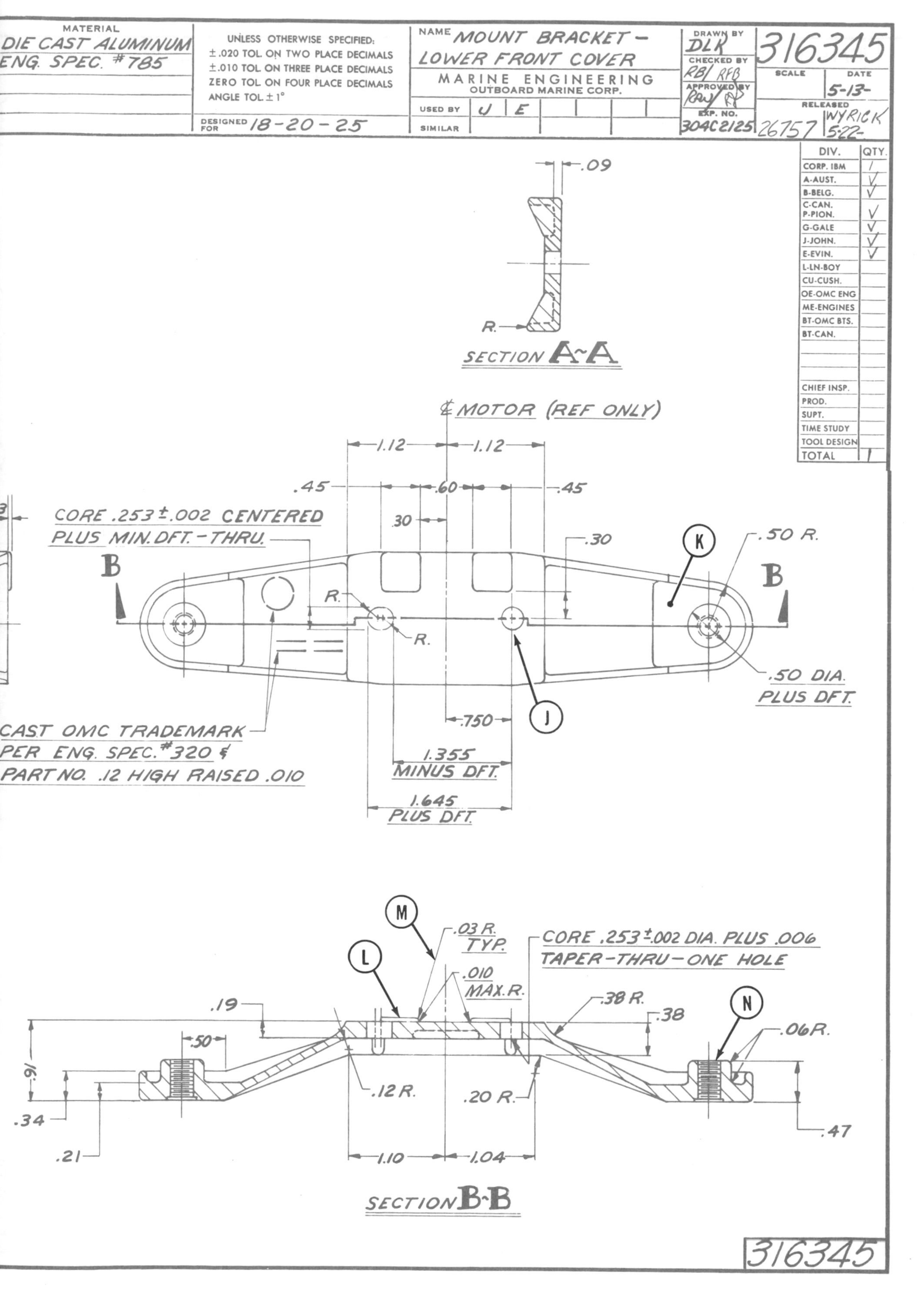

MATERIAL
DIE CAST ALUMINUM
ENG. SPEC. #785
UNLESS OTHERWISE SPECIFIED:
±.020 TOL. ON TWO PLACE DECIMALS
±.010 TOL. ON THREE PLACE DECIMALS
ZERO TOL. ON FOUR PLACE DECIMALS
ANGLE TOL. ± 1°
DESIGNED FOR 18-20-25
NAME MOUNT BRACKET - LOWER FRONT COVER
MARINE ENGINEERING
OUTBOARD MARINE CORP.
USED BY J E
SIMILAR
DRAWN BY DLK
CHECKED BY RB/ RFB
APPROVED BY
EXP. NO. 304C2125
316345
SCALE
DATE 5-13-
RELEASED WYRICK 5-22-
26757
DIV.
QTY.
CORP. IBM
A-AUST.
B-BELG.
C-CAN. P-PION.
G-GALE
J-JOHN.
E-EVIN.
L-LN-BOY
CU-CUSH.
OE-OMC ENG
ME-ENGINES
BT-OMC BTS.
BT-CAN.
CHIEF INSP.
PROD.
SUPT.
TIME STUDY
TOOL DESIGN
TOTAL
.09
R.
SECTION A-A
℄ MOTOR (REF ONLY)
1.12
1.12
.45
.60
.45
CORE .253±.002 CENTERED PLUS MIN. DFT. - THRU.
.30
.30
K
.50 R.
B
B
R.
R.
.50 DIA. PLUS DFT.
J
.750
CAST OMC TRADEMARK PER ENG. SPEC. #320 & PART NO. .12 HIGH RAISED .010
1.355 MINUS DFT.
1.645 PLUS DFT.
M
.03 R. TYP.
CORE .253±.002 DIA. PLUS .006 TAPER-THRU-ONE HOLE
L
.010 MAX. R.
.19
.38 R.
.38
N
.06 R.
.50
.91
.12 R.
.20 R.
.34
.47
.21
1.10
1.04
SECTION B-B
316345

Evaluation Activities

Blueprint Reading Activity
PART 3 - EVALUATION BLUEPRINT

Refer to the blueprint on page 314 and answer the following questions:

1. Is this an assembly or detail drawing? 1. ______________________

2. What is the name and number of the drawing? 2. ______________________

3. How many sheets are in the set? 3. ______________________

4. What size drawing was made? 4. ______________________

5. What is the scale of the original plan? 5. ______________________

6. What material is specified? 6. ______________________

7. The assembly will be used on what model? 7. ______________________

8. What is the meaning of A next to the $\frac{1.9997}{2.0000}$ DIA in Section A-A? 8. ______________________

9. What type of section is Section B-B? 9. ______________________

10. Give the part number and name for Z. 10. ______________________

11. Give the dimension for W. What does the D mean next to the dimension of this feature? 11. ______________________

12. What is the dimension of feature X? 12. ______________________

13. Give the dimensions for feature Y. 13. ______________________

14. How many general notes are included on the blueprint? 14. ______________________

PART 7
APPENDIX

Glossary

ABRASIVE: A material that cuts material that is softer than itself, such as emery, aluminum oxide and diamonds. It may be used in loose form, mounted on cloth, paper or bonded on a wheel.

ABSOLUTE SYSTEM: A system of numerically controlled machining that measures all coordinates from a fixed point of origin or zero point. Also known as point-to-point N/C machining.

ACCUMULATOR: A container in which fluid is stored under pressure as a source of fluid power.

ACTUATOR: A device for converting hydraulic energy into mechanical energy. A motor or cylinder.

ADDENDUM: The radial distance between the pitch circle and the top of the tooth.

ALCLAD: An aluminum alloy core with a thin coating of pure aluminum to prevent corrosion of the core metal.

ALLOWANCE: The intentional difference in the dimensions of mating parts to provide for different classes of fits.

ALLOY: A mixture of two or more metals fused or melted together to form a new metal.

ANNEAL: To soften metals by heating to remove internal stresses caused by rolling and forging.

ANODIZE: The process of protecting aluminum by oxidizing in an acid bath using a d-c current.

ARBOR: A shaft or spindle for holding cutting tools.

ASSEMBLY DRAWING: A drawing showing the working relationship of the various parts of a machine or structure as they fit together.

BACKLASH: The play (lost motion) between moving parts, such as threaded shaft and nut or the teeth of meshing gears.

BASIC DIMENSION: A theoretically exact value used to describe the size, shape or location of a feature.

BASIC SIZE: That size from which the limits of size are derived by the application of allowances and tolerances.

BEND ALLOWANCE: The amount of sheet metal required to make a bend over a specific radius.

BLANCHARDIZE: An operation which removes large amounts of stock through rotary grinding. Normally, it is a first operation for preparing castings for finish operations.

BLANKING: A stamping operation in which a press uses a die to cut blanks from flat sheets or strips of metal.

BORING: Enlarging a hole to a specified dimension by use of a boring bar. May be done on a lathe, jig bore, boring machine or mill.

BOSS: A small local thickening of the body of a casting or forging to allow more thickness for a bearing area or to support threads.

BRAZE: To join two close fitting metal parts with heat and a filler material of zinc and copper alloy.

BROACH: A tool for removing metal by pulling or pushing it across the work. The most common use is producing irregular hole shapes such as squares, hexagons, ovals or splines.

BURNISH: To smooth or polish metal by rolling or sliding tool over surface under pressure.

BURR: The ragged edge or ridge left on metal after a cutting operation.

BUSHING: A metal lining which acts as a bearing between rotating parts such as a shaft and pulley. Also used on jigs to guide cutting tool.

CALLOUT: A note on the blueprint giving a dimension, specification or a machine process.

CAM: A rotating or sliding device used to convert rotary motion into intermittent or reciprocating motion.

CARBURIZE: The heating of low-carbon steel for a period of time to a temperature below its melting point in carbonaceous solids, liquids or gases, then cooling slowly in preparation for heat treating.

CASE HARDENING: The process of hardening ferrous alloy so that the surface layer or case is made much harder than the interior core.

CASTING: An object made by pouring molten metal in a mold.

CHECK VALVE: A valve which permits flow of fluid in one direction only.

CHOKE: A restriction, the length of which is large with respect to its cross-sectional dimension.

CIRCUIT: The complete path of flow in a hydraulic system including the flow-generating device.

CIRCUIT DIAGRAM: A line drawing using graphic symbols or pictorial views to show the complete path of flow in a hydraulic system.

CIRCULAR PITCH: The length of the arc along the pitch circle between the center of one gear tooth to the center of the next.

CLOSED LOOP: A system in which the output of one or more elements is compared to some other signal to provide an actuating signal to control the output of the loop.

COMMAND SIGNAL (or input signal): An external signal to which the servo must respond.

COMPONENT: A single unit or part.

CONCENTRIC: Having a common center as circles or diameters.

CONTOUR: The outline of an object.

CONTROL: A device used to regulate the function of a unit.

COOLER: A heat exchanger used to remove heat from the hydraulic fluid.

COUNTERBORE: The enlargement of the end of a hole to a specified diameter and depth.

COUNTERSINK: The chamfered end of a hole to receive a flat head screw.

DASH NUMBER: A number preceded by a dash after the drawing number that indicates right- or left-hand parts as well as neutral parts and/or detail and assembly drawings. The coding is usually special to a particular industry.

DATUM: A point, line, surface or plane assumed to be exact for purposes of computation from which the location of other features are established.

DEDENDUM: The radial distance between the pitch circle and the bottom of the tooth.

DESIGN SIZE: The size of a feature after an allowance for clearance has been applied and tolerances have been assigned.

DETAIL DRAWING: A drawing of a single part that provides all the information necessary in the production of that part.

DIE: A tool used to cut external threads by hand or machine. Also a tool used to impart a desired shape to a piece of metal.

DIE-CASTING: A method of casting metal under pressure by injecting into metal dies of a die-casting machine. Also the part formed by die-casting.

DIE STAMPING: A piece cut out by a die.

DISPLACEMENT: The quantity of fluid which can pass through a pump, motor or cylinder in a single revolution or stroke.

DOWEL PIN: A pin which fits into a hole in an abutting piece to prevent motion or slipping, or to ensure accurate location of assembly.

DRAFT: The angle or taper on a pattern or casting that permits easy removal from the mold or forming die.

ECCENTRIC: Not having a common center. A device that converts rotary motion into reciprocating (back and forth) motion.

EFFECTIVITY: The serial number(s) of an aircraft, machine, assembly or part on which a drawing change applies. The change may be indicated as an effective date and would apply on that date forward.

ENCLOSURE: A rectangle drawn around a component or components to indicate the limits of an assembly. Port connections are shown on the enclosure line.

EXTRUSION: Metal which has been shaped by forcing it in the hot or cold state through dies of the desired shape.

FEATURE: A portion of a part, such as a diameter, hole, keyway or flat surface.

FEEDBACK (or feedback signal): The output signal from a feedback element.

FERROUS: Metals that have iron as their base material.

FILLET: A concave intersection between two surfaces to strengthen the area.

FILTER: A device whose primary function is the retention by a porous media of insoluble contaminants from a fluid.

FINISH: General finish requirements such as paint, chemical or electroplating rather than surface texture or roughness. (See surface texture.)

FIT: The clearance or interference between two mating parts.

FIXTURE: A device used to position and hold a part in a machine tool. It does not guide the cutting tool.

FLANGE: An edge or collar fixed at an angle to the main part or web as an I-beam.

FLAT PATTERN: A layout showing true dimensions of a part before bending. May be actual size pattern on polyester film for shop use.

FLUID:

1. A liquid or gas.
2. A liquid that is specially compounded for use as a power-transmitting medium in a hydraulic system.

FLUIDICS: A contraction of the words "fluid" and "logic," fluidics is a technology concerned with logical control functions and makes use of low pressure fluid interaction to produce control signals. Fluidic devices have no moving parts.

FORGING: Metal shaped under pressure with or without heat.

FORM TOLERANCING: Permitted variation of a feature from the perfect form indicated on the drawing.

FUSION WELD: The intimate mixing of molten metals.

GEOMETRIC DIMENSIONING AND TOLERANCING: A means of dimensioning and tolerancing a drawing with respect to the actual function or relationship of part features which can be most economically produced. It includes positional and form dimensioning and tolerancing.

GUSSET: A small plate used in reinforcing assemblies.

HARDNESS TEST: Techniques used to measure the degree of hardness of heat-treated materials.

HEAT EXCHANGER: A device which transfers heat through a conducting wall from one fluid to another.

HEAT TREATMENT: The application of

heat to metals to produce desired qualities of hardness, toughness and/or softness. (See anneal.)

HOBBING: A special gear cutting process. The gear blank and hob rotate together as in mesh during the cutting operation.

HONE: A method of finishing a hole or other surface to a precise tolerance by using a spring loaded abrasive block and rotary motion.

HORSEPOWER (HP): The power required to lift 550 pounds one foot in one second or 33,000 pounds one foot in one minute. A horsepower is equal to 746 watts or to 42.4 British thermal units per minute.

HYDRAULIC CONTROL: Control which is actuated by hydraulically induced forces.

HYDRAULICS: Engineering science pertaining to liquid pressure and flow.

INCREMENTAL SYSTEM: A system of numerically controlled machining that always refers to the preceding point when making the next movement. Also known as continuous path or contouring method of N/C machining.

INDICATOR: A precision measuring instrument for checking the trueness of work.

INTERCHANGEABILITY: The condition that assures the universal exchange or mutual substitution of units or parts of a mechanism or an assembly.

INVOLUTE: A spiral curve generated by a point on a chord as it unwinds from a circle or a polygon.

JIG: A device used to hold a part to be machined and positions and guides the cutting tool.

JOGGLE: A bend in a part to fit over other parts.

KERF: The slit or channel left by a saw or other cutting tool.

KEY: A small piece of metal (usually a pin or bar) used to prevent rotation of a gear or pulley on a shaft.

KNURL: The process of marking the surface of a part by rolling depressions in the surface.

LAP: To finish a surface with a very fine abrasive impregnated in a soft metal.

LIMITS: The extreme permissible dimensions of a part resulting from the application of a tolerance.

MAGNAFLUX: A nondestructive inspection technique that makes use of a magnetic field and magnetic particles to locate internal flaws in ferrous metal parts.

MAXIMUM MATERIAL CONDITION: When a feature contains the maximum amount of material, that is: minimum hole diameter and maximum shaft diameter. Abbreviated MMC.

MILL: To remove metal with a rotating cutting tool on a milling machine.

MISMATCH: The variance between depths of machine cuts on a given surface.

NEXT ASSEMBLY: The next object or machine on which the part or sub-assembly is to be used.

NOMINAL SIZE: A general classification term used to designate size of a commercial product.

NONFERROUS: Metals not derived from an iron base or an iron alloy base, such as aluminum, magnesium and copper.

NORMALIZING: A process in which ferrous alloys are heated and then cooled in still air to room temperatures to restore the uniform grain structure free of strains caused by cold working or welding.

ORTHOGRAPHIC PROJECTION: A multiview drawing that shows every feature of an object in its true size and shape.

PASSIVATION: Particularly applicable to stainless steel, it is a conditioning of the surface with a low strength nitric acid dip that develops the "stainless" property and prevents random staining due to "free iron" particles left from machining.

PICKLE: The removal of stains and oxide scales from parts by immersion in an acid solution.

PILOT: A protruding diameter on the end

of a cutting tool designed to fit in a hole and guide the cutter in machining the area around the hole.

PILOT HOLE: A small hole used to guide a cutting tool for making a larger hole. Also used to guide drill of larger size.

PILOT VALVE: An auxiliary valve used to control the operation of another valve. The controlling stage of a 2-stage valve.

PINION: The smaller of two mating gears.

PITCH: The distance from a point on one thread to a corresponding point on the next thread.

PLAN VIEW: The top view of an object.

PORT: An internal or external terminus of a passage in a component.

POSITIONAL TOLERANCING: The permitted variation of a feature from the exact or true position indicated on the drawing.

PROCESS SPECIFICATION: A description of the exact procedures, materials and equipment to be used in performing a particular operation such as a milling operation or spray painting.

PUMP: A device which converts mechanical force and motion into hydraulic fluid power.

QUENCHING: Cooling metals rapidly by immersing them in liquids or gases.

RAM: A single-acting cylinder with a single diameter plunger rather than a piston and rod. The plunger in a ram-type cylinder.

REAMING: To finish a drilled hole to a close tolerance.

RECIPROCATION: A straight line, back-and-forth motion or oscillation.

REFERENCE DIMENSION: Used only for information purposes and does not govern production or inspection operations.

REGARDLESS OF FEATURE SIZE (RFS): The condition where tolerance of position or form must be met irrespective of where the feature lies within its size tolerance.

RELEASE NOTICE: The authorization indicating the drawing has been cleared for use in production.

RELIEF VALVE: Pressure operated valve which bypasses pump delivery to the reservoir, limiting system pressure to a predetermined maximum value.

RESERVOIR: A container for storage of liquid in a fluid power system.

RESISTANCE WELDING: The process of welding metals by using the resistance of the metals to the flow of electricity to produce the heat for fusion of the metals.

RESTRICTION: A reduced cross-sectional area in a line or passage which produces a pressure drop.

ROTARY ACTUATOR: A device for converting hydraulic energy into rotary motion - - a hydraulic motor.

SANDBLAST: The process of removing surface scale from metal by blowing a grit material against it at very high air pressure.

SECTION: A cross-sectional view at a specified point of a part or assembly.

SENSOR: Devices which convert physical conditions into information which can be understood by the control system.

SEQUENCE:

1. The order of a series of operations or movements.
2. To divert flow to accomplish a subsequent operation or movement.

SERRATIONS: Condition of a surface or edge having notches or sharp teeth.

SERVO MECHANISM: A mechanism subjected to the action of a controlling device which will operate as if it were directly actuated by the controlling device, but capable of supplying power output many times that of the controlling device, this power being derived from an external and independent source.

SHIM: A piece of thin metal used between mating parts to adjust their fit.

SOLENOID: A coil of wire carrying an electric current possessing the characteristics of a magnet.

SPECIFICATION: A detailed description of a part or material giving all information not shown on the graphic part of the blueprint such as quality, size, quantity and manufacturer's name.

SPLINE: A raised area on a shaft (external) designed to fit into a recessed area of a mating part.

SPOT FACE: A machined circular spot on the surface of a part to provide a flat bearing surface for a screw, bolt, nut, washer or rivet head.

SPOT WELD: A resistance type weld that joins pieces of metal by welding separate spots rather than a continuous weld.

STRESS RELIEVING: To heat a metal part to a suitable temperature and hold that temperature for a determined time then cooled gradually in air. This treatment reduces the internal stresses induced by casting, quenching, machining, cold working or welding.

SUMP: A reservoir.

SUPERSEDENCE: The replacing of one part by another. A part that has been replaced is said to be superseded.

SURFACE TEXTURE: The lay, roughness, waviness and flaws of a surface.

TABULAR DIMENSION: A type of rectangular datum dimensioning in which dimensions from mutually perpendicular datum planes are listed in a table on the drawing instead of on pictorial portion.

TANGENT: A line drawn to the surface of an arc or circle so that it contacts the arc or circle at only one point.

TAP: A rotating tool used to produce internal threads by hand or machine.

TEMPERING: Creating ductility and toughness in metal by heat treatment process.

TEMPLATE: A pattern or guide.

TENSILE STRENGTH: The maximum load (pull) a piece can support without breakage or failure.

TOLERANCE: The total amount of variation permitted from the design size of a part.

TORQUE: The rotational or twisting force in a turning shaft.

TRANSDUCER (or feedback transducer): An element which measures the results at the load and sends a signal back to the amplifier.

TRUE POSITION: The basic or theoretically exact position of a feature.

TUMBLING: The process of removing rough edges from parts by placing them in a rotating drum that contains abrasive stones, liquid and a detergent.

TYPICAL (TYP): This term, when associated with any dimension or feature, means the dimension or feature applies to the locations that appear to be identical in size and configuration.

VERNIER SCALE: A small moveable scale attached to a larger fixed scale, for obtaining fractional subdivisions of the fixed scale.

WORKING DRAWING: A set of drawings which provide details for the production of each part and information for the correct assembly of the finished product.

Standard Abbreviations

Term	Abbreviation
A	
Abrasive	ABRSV
Accessory	ACCESS
Accumulator	ACCUMR
Acetylene	ACET
Actual	ACT
Actuator	ACTR
Addendum	ADD
Adhesive	ADH
Adjust	ADJ
Advance	ADV
Aeronautic	AERO
Alclad	CLAD
Alignment	ALIGN
Allowance	ALLOW
Alloy	ALY
Alteration	ALT
Alternate	ALT
Aluminum	AL
American Wire Gage	AWG
Anneal	ANL
Anodize	ANOD
Approved	APPD
Approximate	APPROX
Asbestos	ASB
As Required	AR
Assemble	ASSEM
Assembly	ASSY
Automatic	AUTO
Auxiliary	AUX
Average	AVG
B	
Babbit	BAB
Base Line	BL
Bend Radius	BR
Bevel	BEV
Blueprint	BP or B/P
Bolt Circle	BC
Bracket	BRKT
Brass	BRS
Brazing	BRZG
Brinell Hardness Number	BHN
Bronze	BRZ
Brown & Sharpe (Gage)	B&S
Burnish	BNH
Bushing	BUSH
C	
Calculated	CACL
Cancelled	CANC
Capacity	CAP
Carburize	CARB
Case Harden	CH
Cast Iron	CI
Center	CTR
Center to Center	C TO C
Centigrade	C
Centimeter	CM
Centrifugal	CENT
Chamfer	CHAM
Chamfer or Radius	C/R
Check Valve	CV
Chrome Vanadium	CR VAN
Circuit	CKT
Circular	CIR
Circumference	CIRC
Clearance	CL
Closure	CLOS
Coated	CTD
Cold-Drawn Steel	CDS
Cold-Rolled Steel	CRS
Color Code	CC
Concentric	CONC
Condition	COND
Contour	CTR
Control	CONT
Copper	COP
Counterbore	CBORE
Countersink	CSK
Coupling	CPLG
Cylinder	CYL
D	
Datum	DAT
Decimal	DEC
Decrease	DECR
Degree	DEG
Detail	DET

Standard Abbreviations

Developed Length	DL
Developed Width	DW
Deviation	DEV
Diagonal	DIAG
Diagram	DIAG
Diameter	DIA
Diameter Bolt Circle	DBC
Diametral Pitch	DP
Dimension	DIM
Disconnect	DISC
Dowel	DWL
Draft	DFT
Drafting Room Manual	DRM
Drawing	DWG
Drawing Change Notice	DCN
Drill	DR
Drop Forge	DF
Duplicate	DUP
E	
Each	EA
Eccentric	ECC
Effective	EFF
Electric	ELEC
Enclosure	ENCL
Engine	ENG
Engineer	ENGR
Engineering	ENGRG
Engineering Change Order	ECO
Engineering Order	EO
Equal	EQ
Equivalent	EQUIV
Estimate	EST
F	
Fabricate	FAB
Fillet	FIL
Finish	FIN
Finish All Over	FAO
Fitting	FTG
Fixed	FXD
Fixture	FIX
Flange	FLG
Flat Head	FHD
Flat Pattern	F/P
Flexible	FLEX
Fluid	FL
Forged Steel	FST
Forging	FORG
Furnish	FURN
G	
Gage	GA
Gallon	GAL
Galvanized	GALV
Gasket	GSKT
Generator	GEN
Grind	GRD
Ground	GRD
H	
Half-Hard	1/2 H
Handle	HDL
Harden	HDN
Head	HD
Heat Treat	HT TR
Hexagon	HEX
High Carbon Steel	HCS
High Frequency	HF
High Speed	HS
Horizontal	HOR
Hot-Rolled Steel	HRS
Hour	HR
Housing	HSG
Hydraulic	HYD
Hydrostatic	HYDRO
I	
Identification	IDENT
Impregnate	IMPG
Inch	IN
Inclined	INCL
Include, Including, Inclusive	INCL
Increase	INCR
Independent	INDEP
Indicator	IND
Information	INFO
Inside Diameter	ID
Installation	INSTL
Interrupt	INTER
J	
Joggle	JOG
Junction	JCT
K	
Keyway	KWY
L	
Laboratory	LAB
Lacquer	LAQ
Laminate	LAM
Left Hand	LH
Length	LG
Letter	LTR
Limited	LTD
Limit Switch	LS
Linear	LIN
Liquid	LIQ
List of Material	L/M
Long	LG
Low Carbon	LC
Lubricate	LUB
M	
Magnaflux	M
Magnesium	MAG
Maintenance	MAINT
Major	MAJ
Malleable	MALL
Malleable Iron	MI
Manual	MAN
Mark	MK
Master Switch	MS
Material	MATL
Maximum	MAX
Measure	MEAS

Mechanical	MECH
Medium	MED
Middle	MID
Military	MIL
Minimum	MIN
Miscellaneous	MISC
Modification	MOD
Mold Line	ML
Motor	MOT
Multiple	MULT
N	
Nickel Steel	NS
Nomenclature	NOM
Nominal	NOM
Normalize	NORM
Not to Scale	NTS
Number	NO
O	
Obsolete	OBS
Opposite	OPP
Ounce	OZ
Outside Diameter	OD
Over-All	OA
P	
Package	PKG
Parting Line (Castings)	PL
Pattern	PATT
Piece	PC
Pilot	PLT
Pitch	P
Pitch Circle	PC
Pitch Diameter	PD
Plan View	PV
Plastic	PLSTC
Plate	PL
Pneumatic	PNEU
Port	P
Positive	POS
Pounds Per Square Inch	PSI
Pounds Per Square Inch Gage	PSIG
Pressure	PRESS
Primary	PRI
Process, Procedure	PROC
Product, Production	PROD
Q	
Quality	QUAL
Quantity	QTY
Quarter-Hard	1/4 H
R	
Radius	RAD or R
Ream	RM
Receptacle	RECP
Reference	REF
Regular	REG
Regulator	REG
Release	REL
Required	REQD
Right Hand	RH
Rivet	RIV
Rockwell Hardness	RH
Round	RD
S	
Schedule	SCH
Schematic	SCHEM
Screw	SCR
Screw Threads	
American National Coarse	NC
American National Fine	NF
American National Extra Fine	NEF
American National 8 Pitch	8N
American Standard Taper Pipe	NTP
American Standard Straight Pipe	NPSC
American Standard Taper (Dryseal)	NPTF
American Standard Straight (Dryseal)	NPSF
Unified Screw Thread Coarse	UNC
Unified Screw Thread Fine	UNF
Unified Screw Thread Extra Fine	UNEF
Unified Screw Thread 8 Thread	8UN
Section	SECT
Sequence	SEQ
Serial	SER
Serrate	SERR
Sheet	SH
Silver Solder	SILS
Soft Grind	SO GR
Solenoid	SOL
Special	SPL
Specification	SPEC
Spot Face	SF
Stainless Steel	SST
Steel	STL
Stock	STK
Symbol	SYM
Symmetrical	SYM
System	SYS
T	
Tabulate	TAB
Tangent	TAN
Tapping	TAP
Teeth	T
Tensil Strength	TS
Thick	THK
Thread	THD
Tolerance	TOL
Tool Steel	TS.
Torque	TOR
Total Indicator Reading	TIR
True Involute Form	TIF
Tungsten	TU
Typical	TYP

Standard Abbreviations

V	
Vacuum	VAC
Variable	VAR
Vernier	VER
Vertical	VERT
Vibrate	VIB
Void	VD
Volt	V
Volume	VOL

W	
Washer	WASH
Watt	W
Weatherproof	WP
Wide, Width	W
Wrought Iron	WI
Y	
Yield Point (PSI)	YP
Yield Strength (PSI)	YS

Standard Tables and Symbols

Decimal and Metric Equivalents

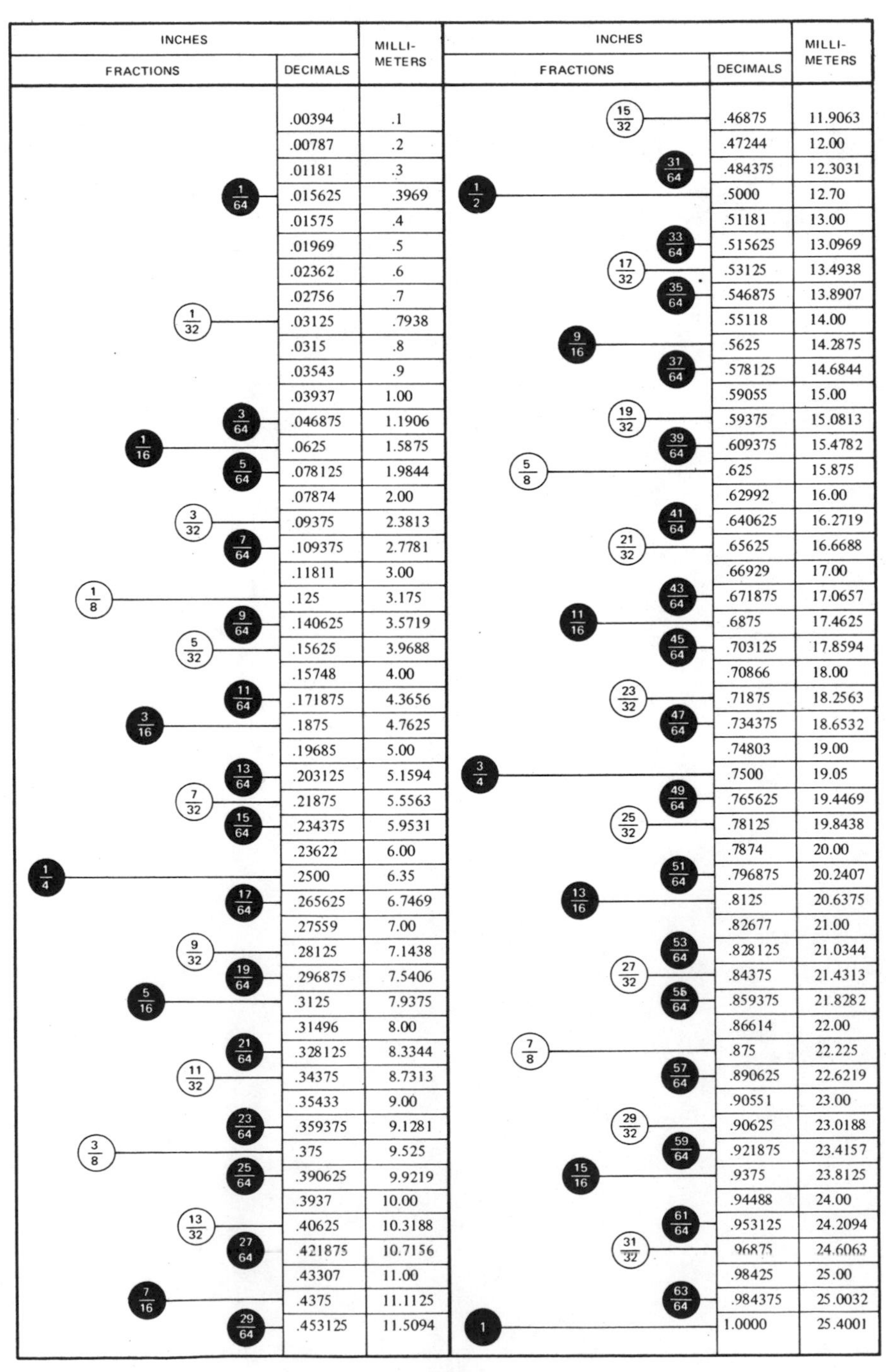

INCHES: FRACTIONS	INCHES: DECIMALS	MILLIMETERS
	.00394	.1
	.00787	.2
	.01181	.3
1/64	.015625	.3969
	.01575	.4
	.01969	.5
	.02362	.6
	.02756	.7
1/32	.03125	.7938
	.0315	.8
	.03543	.9
	.03937	1.00
3/64	.046875	1.1906
1/16	.0625	1.5875
5/64	.078125	1.9844
	.07874	2.00
3/32	.09375	2.3813
7/64	.109375	2.7781
	.11811	3.00
1/8	.125	3.175
9/64	.140625	3.5719
5/32	.15625	3.9688
	.15748	4.00
11/64	.171875	4.3656
3/16	.1875	4.7625
	.19685	5.00
13/64	.203125	5.1594
7/32	.21875	5.5563
15/64	.234375	5.9531
	.23622	6.00
1/4	.2500	6.35
17/64	.265625	6.7469
	.27559	7.00
9/32	.28125	7.1438
19/64	.296875	7.5406
5/16	.3125	7.9375
	.31496	8.00
21/64	.328125	8.3344
11/32	.34375	8.7313
	.35433	9.00
23/64	.359375	9.1281
3/8	.375	9.525
25/64	.390625	9.9219
	.3937	10.00
13/32	.40625	10.3188
27/64	.421875	10.7156
	.43307	11.00
7/16	.4375	11.1125
29/64	.453125	11.5094

INCHES: FRACTIONS	INCHES: DECIMALS	MILLIMETERS
15/32	.46875	11.9063
	.47244	12.00
31/64	.484375	12.3031
1/2	.5000	12.70
	.51181	13.00
33/64	.515625	13.0969
17/32	.53125	13.4938
35/64	.546875	13.8907
	.55118	14.00
9/16	.5625	14.2875
37/64	.578125	14.6844
	.59055	15.00
19/32	.59375	15.0813
39/64	.609375	15.4782
5/8	.625	15.875
	.62992	16.00
41/64	.640625	16.2719
21/32	.65625	16.6688
	.66929	17.00
43/64	.671875	17.0657
11/16	.6875	17.4625
45/64	.703125	17.8594
	.70866	18.00
23/32	.71875	18.2563
47/64	.734375	18.6532
	.74803	19.00
3/4	.7500	19.05
49/64	.765625	19.4469
25/32	.78125	19.8438
	.7874	20.00
51/64	.796875	20.2407
13/16	.8125	20.6375
	.82677	21.00
53/64	.828125	21.0344
27/32	.84375	21.4313
55/64	.859375	21.8282
	.86614	22.00
7/8	.875	22.225
57/64	.890625	22.6219
	.90551	23.00
29/32	.90625	23.0188
59/64	.921875	23.4157
15/16	.9375	23.8125
	.94488	24.00
61/64	.953125	24.2094
31/32	.96875	24.6063
	.98425	25.00
63/64	.984375	25.0032
1	1.0000	25.4001

DRILL SIZE DECIMAL EQUIVALENTS – METRIC AND INCH

NUMBER AND LETTER DRILLS.

Drill No.	Frac	Deci
80	—	.0135
79	—	.0145
	1/64	.0156
78	—	.0160
77	—	.0180
76	—	.0200
75	—	.0210
74	—	.0225
73	—	.0240
72	—	.0250
71	—	.0260
70	—	.0280
69	—	.0292
68	—	.0310
	1/32	.0313
67	—	.0320
66	—	.0330
65	—	.0350
64	—	.0360
63	—	.0370
62	—	.0380
61	—	.0390
60	—	.0400
59	—	.0410
58	—	.0420
57	—	.0430
56	—	.0465
	3/64	.0469
55	—	.0520
54	—	.0550
53	—	.0595
	1/16	.0625
52	—	.0635
51	—	.0670
50	—	.0700
49	—	.0730
48	—	.0760
	5/64	.0781
47	—	.0785
46	—	.0810
45	—	.0820
44	—	.0860
43	—	.0890
42	—	.0935
	3/32	.0938
41	—	.0960
40	—	.0980
39	—	.0995
38	—	.1015
37	—	.1040
36	—	.1065
	7/64	.1094
35	—	.1100
34	—	.1110
33	—	.1130
32	—	.116
31	—	.120
	1/8	.125
30	—	.129
29	—	.136

Drill No.	Frac	Deci
	9/64	.140
28	—	.141
27	—	.144
26	—	.147
25	—	.150
24	—	.152
23	—	.154
	5/32	.156
22	—	.157
21	—	.159
20	—	.161
19	—	.166
18	—	.170
	11/64	.172
17	—	.173
16	—	.177
15	—	.180
14	—	.182
13	—	.185
	3/16	.188
12	—	.189
11	—	.191
10	—	.194
9	—	.196
8	—	.199
7	—	.201
	13/64	.203
6	—	.204
5	—	.206
4	—	.209
3	—	.213
	7/32	.219
2	—	.221
1	—	.228
A		.234
	15/64	.234
B	—	.238
C	—	.242
D	—	.246
	1/4	.250
E	—	.250
F	—	.257
G	—	.261
	17/64	.266
H	—	.266
I	—	.272
J	—	.277
	9/32	.281
K	—	.281
L	—	.290
M	—	.295
	19/64	.297
N	—	.302
	5/16	.313
O	—	.316
P	—	.323
	21/64	.328
Q	—	.332
R	—	.339
	11/32	.344

Drill No.	Frac	Deci
S	—	.348
T	—	.358
	23/64	.359
U	—	.368
	3/8	.375
V	—	.377
W	—	.386
	25/64	.391
X	—	.397
Y	—	.404
	13/32	.406
Z	—	.413
	27/64	.422
	7/16	.438
	29/64	.453
	15/32	.469
	31/64	.484
	1/2	.500
	33/64	.516
	17/32	.531
	35/64	.547
	9/16	.562
	37/64	.578
	19/32	.594
	39/64	.609
	5/8	.625
	41/64	.641
	21/32	.656
	43/64	.672
	11/16	.688
	45/64	.703
	23/32	.719
	47/64	.734
	3/4	.750
	49/64	.766
	25/32	.781
	51/64	.797
	13/16	.813
	53/64	.828
	27/32	.844
	55/64	.859
	7/8	.875
	57/64	.891
	29/32	.906
	59/64	.922
	15/16	.938
	61/64	.953
	31/32	.969
	63/64	.984
	1	1.000

METRIC DRILLS.

MM	DEC.	MM	DEC.	MM	DEC.	MM	DEC.
1.	.0394	3.2	.1260	6.3	.2480	9.5	.3740
1.05	.0413	3.25	.1280	6.4	.2520	9.6	.3780
1.1	.0433	3.3	.1299	6.5	.2559	9.7	.3819
1.15	.0453	3.4	.1339	6.6	.2598	9.75	.3839
1.2	.0472	3.5	.1378	6.7	.2638	9.8	.3858
1.25	.0492	3.6	.1417	6.75	.2657	9.9	.3898
1.3	.0512	3.7	.1457	6.8	.2677	10.	.3937
1.35	.0531	3.75	.1476	6.9	.2717	10.5	.4134
1.4	.0551	3.8	.1496	7.	.2756	11.	.4331
1.45	.0571	3.9	.1535	7.1	.2795	11.5	.4528
1.5	.0591	4.	.1575	7.2	.2835	12.	.4724
1.55	.0610	4.1	.1614	7.25	.2854	12.5	.4921
1.6	.0630	4.2	.1654	7.3	.2874	13.	.5118
1.65	.0650	4.25	.1673	7.4	.2913	13.5	.5315
1.7	.0669	4.3	.1693	7.5	.2953	14.	.5512
1.75	.0689	4.4	.1732	7.6	.2992	14.5	.5709
1.8	.0709	4.5	.1772	7.7	.3031	15.	.5906
1.85	.0728	4.6	.1811	7.75	.3051	15.5	.6102
1.9	.0748	4.7	.1850	7.8	.3071	16.	.6299
1.95	.0768	4.75	.1870	7.9	.3110	16.5	.6496
2.	.0787	4.8	.1890	8.	.3150	17.	.6693
2.05	.0807	4.9	.1929	8.1	.3189	17.5	.6890
2.1	.0827	5.	.1968	8.2	.3228	18.	.7087
2.15	.0846	5.1	.2008	8.25	.3248	18.5	.7283
2.2	.0866	5.2	.2047	8.3	.3268	19.	.7480
2.25	.0886	5.25	.2067	8.4	.3307	19.5	.7677
2.3	.0906	5.3	.2087	8.5	.3346	20.	.7874
2.35	.0925	5.4	.2126	8.6	.3386	20.5	.8071
2.4	.0945	5.5	.2165	8.7	.3425	21.	.8268
2.45	.0965	5.6	.2205	8.75	.3445	21.5	.8465
2.5	.0984	5.7	.2244	8.8	.3465	22.	.8661
2.6	.1024	5.75	.2264	8.9	.3504	22.5	.8858
2.7	.1063	5.8	.2283	9.	.3543	23.	.9055
2.75	.1083	5.9	.2323	9.1	.3583	23.5	.9252
2.8	.1102	6.	.2362	9.2	.3622	24.	.9449
2.9	.1142	6.1	.2402	9.25	.3642	24.5	.9646
3.	.1181	6.2	.2441	9.3	.3661	25.	.9843
3.1	.1220	6.25	.2461	9.4	.3701		

TAP DRILL SIZES FOR UNIFIED STANDARD SCREW THREADS

Screw Thread		Tap Drill
Major Diameter	Threads Per Inch	Size Or Number
0	80	3/64
1	64	53
	72	53
2	56	50
	64	50
3	48	47
	56	45
4	40	43
	48	42
5	40	38
	44	37
6	32	36
	40	33
8	32	29
	36	29
10	24	25
	32	21
12	24	16
	28	14
1/4	20	7
	28	3
5/16	18	F
	24	I

Screw Thread		Tap Drill
Major Diameter	Threads Per Inch	Size Or Number
3/8	16	5/16
	24	Q
7/16	14	U
	20	25/64
1/2	13	27/64
	20	29/64
9/16	12	31/64
	18	33/64
5/8	11	17/32
	18	37/64
3/4	10	21/32
	16	11/16
7/8	9	49/64
	14	13/16
1	8	7/8
	12	59/64
1 1/8	7	63/64
	12	1 3/64
1 1/4	7	1 7/64
	12	1 11/64
1 3/8	6	1 7/32
	12	1 19/64
1 1/2	6	1 11/32
	12	1 27/64

TAP DRILL SIZES FOR ISO METRIC THREADS

Nominal Size mm	Series			
	Coarse		Fine	
	Pitch mm	Tap Drill mm	Pitch mm	Tap Drill mm
1.4	0.3	1.1	–	–
1.6	0.35	1.25	–	–
2	0.4	1.6	–	–
2.5	0.45	2.05	–	–
3	0.5	2.5	–	–
4	0.7	3.3	–	–
5	0.8	4.2	–	–
6	1.0	5.0	–	–
8	1.25	6.75	1	7.0

Nominal Size mm	Series			
	Coarse		Fine	
	Pitch mm	Tap Drill mm	Pitch mm	Tap Drill mm
10	1.5	8.5	1.25	8.75
12	1.75	10.25	1.25	10.50
14	2	12.00	1.5	12.50
16	2	14.00	1.5	14.50
18	2.5	15.50	1.5	16.50
20	2.5	17.50	1.5	18.50
22	2.5	19.50	1.5	20.50
24	3	21.00	2	22.00
27	3	24.00	2	25.00

Sizes: Primary	Sizes: Secondary	Basic major diameter	Threads per inch — Series with graded pitches: Coarse UNC	Fine UNF	Extra fine UNEF	Series with constant pitches: 4UN	6UN	8UN	12UN	16UN	20UN	28UN	32UN	Sizes
0		0.0600	—	80	—	—	—	—	—	—	—	—	—	0
	1	0.0730	64	72	—	—	—	—	—	—	—	—	—	1
2		0.0860	56	64	—	—	—	—	—	—	—	—	—	2
	3	0.0990	48	56	—	—	—	—	—	—	—	—	—	3
4		0.1120	40	48	—	—	—	—	—	—	—	—	—	4
5		0.1250	40	44	—	—	—	—	—	—	—	—	—	5
6		0.1380	32	40	—	—	—	—	—	—	—	—	UNC	6
8		0.1640	32	36	—	—	—	—	—	—	—	—	UNC	8
10		0.1900	24	32	—	—	—	—	—	—	—	—	UNF	10
	12	0.2160	24	28	32	—	—	—	—	—	—	UNF	UNEF	12
1/4		0.2500	20	28	32	—	—	—	—	—	UNC	UNF	UNEF	1/4
5/16		0.3125	18	24	32	—	—	—	—	—	20	28	UNEF	5/16
3/8		0.3750	16	24	32	—	—	—	—	UNC	20	28	UNEF	3/8
7/16		0.4375	14	20	28	—	—	—	—	16	UNF	UNEF	32	7/16
1/2		0.5000	13	20	28	—	—	—	—	16	UNF	UNEF	32	1/2
9/16		0.5625	12	18	24	—	—	—	UNC	16	20	28	32	9/16
5/8		0.6250	11	18	24	—	—	—	12	16	20	28	32	5/8
	11/16	0.6875	—	—	24	—	—	—	12	16	20	28	32	11/16
3/4		0.7500	10	16	20	—	—	—	12	UNF	UNEF	28	32	3/4
	13/16	0.8125	—	—	20	—	—	—	12	16	UNEF	28	32	13/16
7/8		0.8750	9	14	20	—	—	—	12	16	UNEF	28	32	7/8
	15/16	0.9375	—	—	20	—	—	—	12	16	UNEF	28	32	15/16
1		1.0000	8	12	20	—	—	UNC	UNF	16	UNEF	28	32	1
	1 1/16	1.0625	—	—	18	—	—	8	12	16	20	28	—	1 1/16
1 1/8		1.1250	7	12	18	—	—	8	UNF	16	20	28	—	1 1/8
	1 3/16	1.1875	—	—	18	—	—	8	12	16	20	28	—	1 3/16
1 1/4		1.2500	7	12	18	—	—	8	UNF	16	20	28	—	1 1/4
	1 5/16	1.3125	—	—	18	—	—	8	12	16	20	28	—	1 5/16
1 3/8		1.3750	6	12	18	—	UNC	8	UNF	16	20	28	—	1 3/8
	1 7/16	1.4375	—	—	18	—	6	8	12	16	20	28	—	1 7/16
1 1/2		1.5000	6	12	18	—	UNC	8	UNF	16	20	28	—	1 1/2
	1 9/16	1.5625	—	—	18	—	6	8	12	16	20	—	—	1 9/16
1 5/8		1.6250	—	—	18	—	6	8	12	16	20	—	—	1 5/8
	1 11/16	1.6875	—	—	18	—	6	8	12	16	20	—	—	1 11/16
1 3/4		1.7500	5	—	—	—	6	8	12	16	20	—	—	1 3/4
	1 13/16	1.8125	—	—	—	—	6	8	12	16	20	—	—	1 13/16
1 7/8		1.8750	—	—	—	—	6	8	12	16	20	—	—	1 7/8
	1 15/16	1.9375	—	—	—	—	6	8	12	16	20	—	—	1 15/16
2		2.0000	4 1/2	—	—	—	6	8	12	16	20	—	—	2
	2 1/8	2.1250	—	—	—	—	6	8	12	16	20	—	—	2 1/8
2 1/4		2.2500	4 1/2	—	—	—	6	8	12	16	20	—	—	2 1/4
	2 3/8	2.3750	—	—	—	—	6	8	12	16	20	—	—	2 3/8
2 1/2		2.5000	4	—	—	UNC	6	8	12	16	20	—	—	2 1/2
	2 5/8	2.6250	—	—	—	4	6	8	12	16	20	—	—	2 5/8
2 3/4		2.7500	4	—	—	UNC	6	8	12	16	20	—	—	2 3/4
	2 7/8	2.8750	—	—	—	4	6	8	12	16	20	—	—	2 7/8
3		3.0000	4	—	—	UNC	6	8	12	16	20	—	—	3
	3 1/8	3.1250	—	—	—	4	6	8	12	16	—	—	—	3 1/8
3 1/4		3.2500	4	—	—	UNC	6	8	12	16	—	—	—	3 1/4
	3 3/8	3.3750	—	—	—	4	6	8	12	16	—	—	—	3 3/8
3 1/2		3.5000	4	—	—	UNC	6	8	12	16	—	—	—	3 1/2
	3 5/8	3.6250	—	—	—	4	6	8	12	16	—	—	—	3 5/8
3 3/4		3.7500	4	—	—	UNC	6	8	12	16	—	—	—	3 3/4
	3 7/8	3.8750	—	—	—	4	6	8	12	16	—	—	—	3 7/8
4		4.0000	4	—	—	UNC	6	8	12	16	—	—	—	4
	4 1/8	4.1250	—	—	—	4	6	8	12	16	—	—	—	4 1/8
4 1/4		4.2500	—	—	—	4	6	8	12	16	—	—	—	4 1/4
	4 3/8	4.3750	—	—	—	4	6	8	12	16	—	—	—	4 3/8
4 1/2		4.5000	—	—	—	4	6	8	12	16	—	—	—	4 1/2
	4 5/8	4.6250	—	—	—	4	6	8	12	16	—	—	—	4 5/8
4 3/4		4.7500	—	—	—	4	6	8	12	16	—	—	—	4 3/4
	4 7/8	4.8750	—	—	—	4	6	8	12	16	—	—	—	4 7/8
5		5.0000	—	—	—	4	6	8	12	16	—	—	—	5
	5 1/8	5.1250	—	—	—	4	6	8	12	16	—	—	—	5 1/8
5 1/4		5.2500	—	—	—	4	6	8	12	16	—	—	—	5 1/4
	5 3/8	5.3750	—	—	—	4	6	8	12	16	—	—	—	5 3/8
5 1/2		5.5000	—	—	—	4	6	8	12	16	—	—	—	5 1/2
	5 5/8	5.6250	—	—	—	4	6	8	12	16	—	—	—	5 5/8
5 3/4		5.7500	—	—	—	4	6	8	12	16	—	—	—	5 3/4
	5 7/8	5.8750	—	—	—	4	6	8	12	16	—	—	—	5 7/8
6		6.0000	—	—	—	4	6	8	12	16	—	—	—	6

ISO METRIC SCREW THREAD STANDARD SERIES

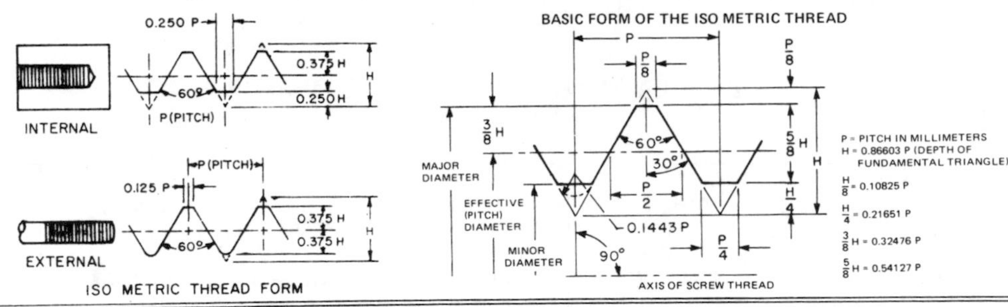

Nominal Size Diam. (mm) Column[a]			Pitches (mm)														Nom-inal Size Diam. (mm)
			Series With Graded Pitches		Series With Constant Pitches												
1	2	3	Coarse	Fine	6	4	3	2	1.5	1.25	1	0.75	0.5	0.35	0.25	0.2	
0.25			0.075	—	—	—	—	—	—	—	—	—	—	—	—	—	0.25
0.3			0.08	—	—	—	—	—	—	—	—	—	—	—	—	—	0.3
	0.35		0.09	—	—	—	—	—	—	—	—	—	—	—	—	—	0.35
0.4			0.1	—	—	—	—	—	—	—	—	—	—	—	—	—	0.4
	0.45		0.1	—	—	—	—	—	—	—	—	—	—	—	—	—	0.45
0.5			0.125	—	—	—	—	—	—	—	—	—	—	—	—	—	0.5
	0.55		0.125	—	—	—	—	—	—	—	—	—	—	—	—	—	0.55
0.6			0.15	—	—	—	—	—	—	—	—	—	—	—	—	—	0.6
	0.7		0.175	—	—	—	—	—	—	—	—	—	—	—	—	—	0.7
0.8			0.2	—	—	—	—	—	—	—	—	—	—	—	—	—	0.8
	0.9		0.225	—	—	—	—	—	—	—	—	—	—	—	—	—	0.9
1			0.25	—	—	—	—	—	—	—	—	—	—	—	—	0.2	1
	1.1		0.25	—	—	—	—	—	—	—	—	—	—	—	—	0.2	1.1
1.2			0.25	—	—	—	—	—	—	—	—	—	—	—	—	0.2	1.2
	1.4		0.3	—	—	—	—	—	—	—	—	—	—	—	—	0.2	1.4
1.6			0.35	—	—	—	—	—	—	—	—	—	—	—	—	0.2	1.6
	1.8		0.35	—	—	—	—	—	—	—	—	—	—	—	—	0.2	1.8
2			0.4	—	—	—	—	—	—	—	—	—	—	—	0.25	—	2
	2.2		0.45	—	—	—	—	—	—	—	—	—	—	—	0.25	—	2.2
2.5			0.45	—	—	—	—	—	—	—	—	—	—	0.35	—	—	2.5
3			0.5	—	—	—	—	—	—	—	—	—	—	0.35	—	—	3
	3.5		0.6	—	—	—	—	—	—	—	—	—	—	0.35	—	—	3.5
4			0.7	—	—	—	—	—	—	—	—	—	0.5	—	—	—	4
	4.5		0.75	—	—	—	—	—	—	—	—	—	0.5	—	—	—	4.5
5			0.8	—	—	—	—	—	—	—	—	—	0.5	—	—	—	5
		5.5	—	—	—	—	—	—	—	—	—	—	0.5	—	—	—	5.5
6			1	—	—	—	—	—	—	—	—	0.75	—	—	—	—	6
		7	1	—	—	—	—	—	—	—	—	0.75	—	—	—	—	7
8			1.25	1	—	—	—	—	—	—	1	0.75	—	—	—	—	8
		9	1.25	—	—	—	—	—	—	—	1	0.75	—	—	—	—	9
10			1.5	1.25	—	—	—	—	—	1.25	1	0.75	—	—	—	—	10
		11	1.5	—	—	—	—	—	—	—	1	0.75	—	—	—	—	11
12			1.75	1.25	—	—	—	—	1.5	1.25	1	—	—	—	—	—	12
	14		2	1.5	—	—	—	—	1.5	1.25[b]	1	—	—	—	—	—	14
		15	—	—	—	—	—	—	1.5	—	1	—	—	—	—	—	15
16			2	1.5	—	—	—	—	1.5	—	1	—	—	—	—	—	16
		17	—	—	—	—	—	—	1.5	—	1	—	—	—	—	—	17
	18		2.5	1.5	—	—	—	2	1.5	—	1	—	—	—	—	—	18
20			2.5	1.5	—	—	—	2	1.5	—	1	—	—	—	—	—	20
	22		2.5	1.5	—	—	—	2	1.5	—	1	—	—	—	—	—	22
24			3	2	—	—	—	2	1.5	—	1	—	—	—	—	—	24
		25	—	—	—	—	—	2	1.5	—	1	—	—	—	—	—	25
		26	—	—	—	—	—	—	1.5	—	1	—	—	—	—	—	26
	27		3	2	—	—	—	2	1.5	—	1	—	—	—	—	—	27
		28	—	—	—	—	—	2	1.5	—	1	—	—	—	—	—	28
30			3.5	2	—	—	(3)	2	1.5	—	1	—	—	—	—	—	30
		32	—	—	—	—	—	2	1.5	—	—	—	—	—	—	—	32
	33		3.5	2	—	—	(3)	2	1.5	—	—	—	—	—	—	—	33
		35[c]	—	—	—	—	—	—	1.5	—	—	—	—	—	—	—	35[c]
36			4	3	—	—	—	2	1.5	—	—	—	—	—	—	—	36
		38	—	—	—	—	—	—	1.5	—	—	—	—	—	—	—	38
	39		4	3	—	—	—	2	1.5	—	—	—	—	—	—	—	39
		40	—	—	—	—	3	2	1.5	—	—	—	—	—	—	—	40
42			4.5	3	—	4	3	2	1.5	—	—	—	—	—	—	—	42
	45		4.5	3	—	4	3	2	1.5	—	—	—	—	—	—	—	45

a Thread diameter should be selected from columns 1, 2 or 3; with preference being given in that order.
b Pitch 1.25 mm in combination with diameter 14 mm has been included for spark plug applications.
c Diameter 35 mm has been included for bearing locknut applications.
The use of pitches shown in parentheses should be avoided wherever possible.
The pitches enclosed in the bold frame, together with the corresponding nominal diameters in Columns 1 and 2, are those combinations which have been established by ISO Recommendations as a selected "coarse" and "fine" series for commercial fasteners. Sizes 0.25 mm through 1.4 mm are covered in ISO Recommendation R 68 and, except for the 0.25 mm size, in AN Standard ANSI B1.10.

(ANSI)

SHEET METAL AND WIRE GAGE DESIGNATION

GAGE NO.	AMERICAN OR BROWN & SHARPE'S A.W.G. OR B. & S.	BIRMING-HAM OR STUBS WIRE B.W.G.	WASHBURN & MOEN OR AMERICAN S.W.G.	UNITED STATES STANDARD	MANU-FACTURERS' STANDARD FOR SHEET STEEL	GAGE NO.
0000000	----	----	.4900	.500	----	0000000
000000	.5800	----	.4615	.469	----	000000
00000	.5165	----	.4305	.438	----	00000
0000	.4600	.454	.3938	.406	----	0000
000	.4096	.425	.3625	.375	----	000
00	.3648	.380	.3310	.344	----	00
0	.3249	.340	.3065	.312	----	0
1	.2893	.300	.2830	.281	----	1
2	.2576	.284	.2625	.266	----	2
3	.2294	.259	.2437	.250	.2391	3
4	.2043	.238	.2253	.234	.2242	4
5	.1819	.220	.2070	.219	.2092	5
6	.1620	.203	.1920	.203	.1943	6
7	.1443	.180	.1770	.188	.1793	7
8	.1285	.165	.1620	.172	.1644	8
9	.1144	.148	.1483	.156	.1495	9
10	.1019	.134	.1350	.141	.1345	10
11	.0907	.120	.1205	.125	.1196	11
12	.0808	.109	.1055	.109	.1046	12
13	.0720	.095	.0915	.0938	.0897	13
14	.0642	.083	.0800	.0781	.0747	14
15	.0571	.072	.0720	.0703	.0673	15
16	.0508	.065	.0625	.0625	.0598	16
17	.0453	.058	.0540	.0562	.0538	17
18	.0403	.049	.0475	.0500	.0478	18
19	.0359	.042	.0410	.0438	.0418	19
20	.0320	.035	.0348	.0375	.0359	20
21	.0285	.032	.0317	.0344	.0329	21
22	.0253	.028	.0286	.0312	.0299	22
23	.0226	.025	.0258	.0281	.0269	23
24	.0201	.022	.0230	.0250	.0239	24
25	.0179	.020	.0204	.0219	.0209	25
26	.0159	.018	.0181	.0188	.0179	26
27	.0142	.016	.0173	.0172	.0164	27
28	.0126	.014	.0162	.0156	.0149	28
29	.0113	.013	.0150	.0141	.0135	29
30	.0100	.012	.0140	.0125	.0120	30
31	.0089	.010	.0132	.0109	.0105	31
32	.0080	.009	.0128	.0102	.0097	32
33	.0071	.008	.0118	.00938	.0090	33
34	.0063	.007	.0104	.00859	.0082	34
35	.0056	.005	.0095	.00781	.0075	35
36	.0050	.004	.0090	.00703	.0067	36
37	.0045	----	.0085	.00624	.0064	37
38	.0040	----	.0080	.00625	.0060	38
39	.0035	----	.0075	-----	----	39
40	.0031	----	.0070	-----	----	40
41	.0028	----	.0066	-----	----	41
42	.0025	----	.0062	-----	----	42
43	.0022	----	.0060	-----	----	43
44	.0020	----	.0058	-----	----	44
45	.0018	----	.0055	-----	----	45
46	.0016	----	.0052	-----	----	46
47	.0014	----	.0050	-----	----	47
48	.0012	----	.0048	-----	----	48

Electronic Symbols

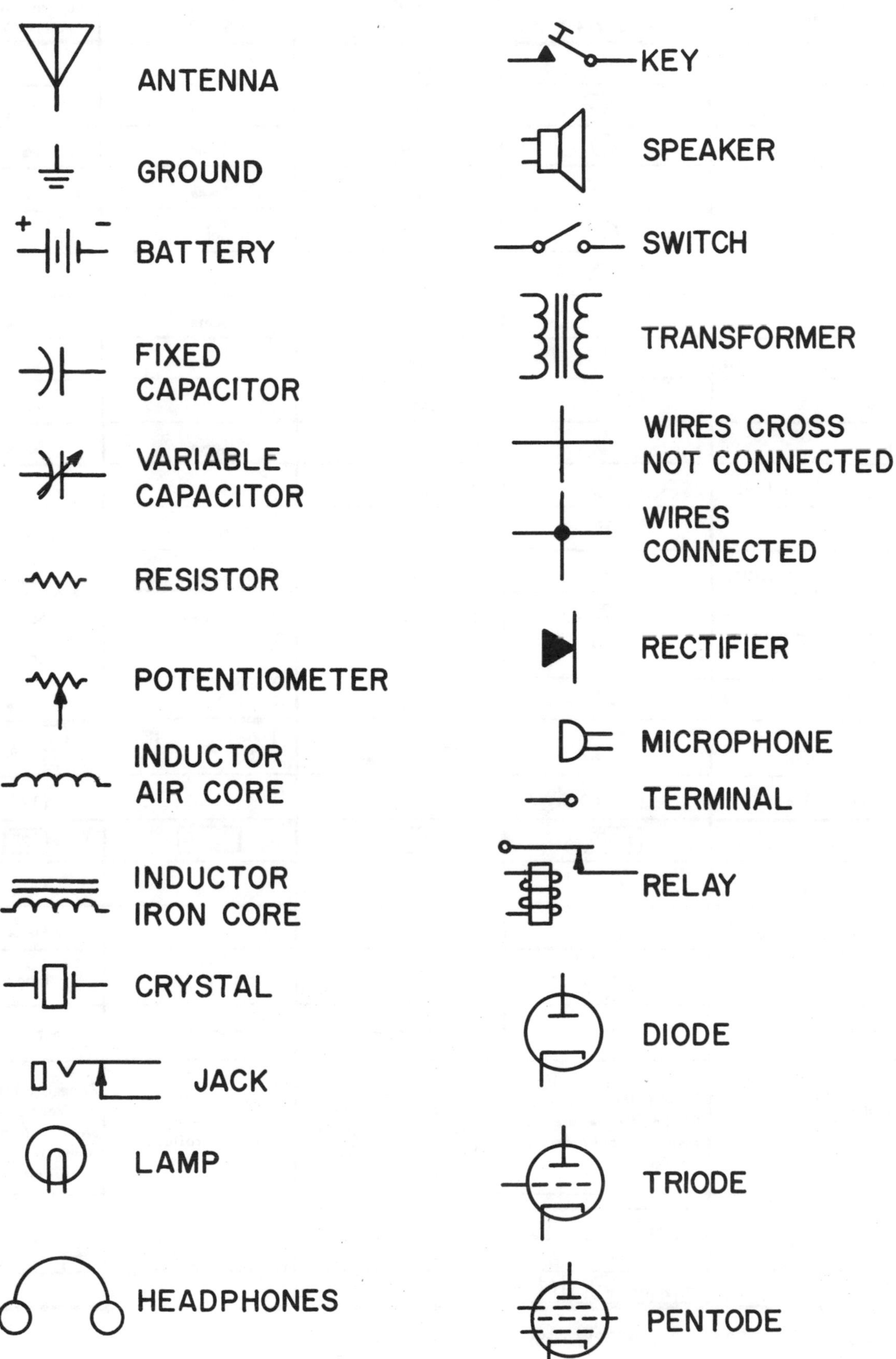

POSITIONAL AND FORM TOLERANCES

Characteristics	American ANSI Y14.5	British BS 308	Canadian CSA B78.2	International ISO R1101
Straightness	—	Same	Same	Same
Flatness	⏥	Same	Same	Same
Roundness (Circularity)	○	Same	Same	Same
Cylindricity	⌭	Same	Same	Same
Profile of a Line	⌒	Same	Same	Same
Profile of a Surface	⌓	Same	Same	Same
Parallelism	//	Same	Same	Same
Perpendicularity (Squareness)	⊥	Same	Same	Same
Angularity	∠	Same	Same	Same
Position	⌖	Same	Same	Same
Concentricity (Coaxiality)	◎	Same	Same	Same
Symmetry	⌯	Same	Same	Same
Maximum Material Condition	Ⓜ	Same	Same	Same
Diameter	Ø	Same	Same	Same
Circular Runout	↗	Same	Same	Same
Total Runout	⌰ *	None	None	None
Datum Identification	-A-	A or	-A- or	A or
Reference Dimension	(5.000)	(127)	(5.000)	(127)
Basic Dimension	[5.000]	[127]	[5.000]	[127]
Regardless of Feature Size	Ⓢ	None	None	None
Projected Tolerance Zone	Ⓟ	None	Ⓟ	None
Datum Target	A 1	A 1	A 1	None
Part Symmetry	None	⊣⊢—·—⊣⊢	⊣⊢—·—⊣⊢	⊣⊢—·—⊣⊢
Shape of the tolerance zone	Zone is total width. Ø specified where zone is circular or cylindrical.	Zone is a total width in direction of leader arrow. Ø specified where zone is circular or cylindrical.	Zone shape evident from chacteristic being controlled.	Zone is a total width in direction of leader arrow. Ø specified where zone is circular or cylindrical.
Sequence within the feature control symbol	⌖ | A | B | C | Ø.02 Ⓜ or ⌖ | Ø .02 Ⓜ | A | B | C	⌖ | Ø 0.5 Ⓜ | A | B | C	⌖ | .02 Ⓜ | A | B | C	⌖ | Ø 0.5 Ⓜ | A | B | C

* "TOTAL" specified under the feature control symbol.

(ANSI)

Precision Sheet Metal Set-Back Chart

MATERIAL THICKNESS

90 DEG. BEND RADIUS

	.016	.020	.025	.032	.040	.051	.064	.072	.078	.081	.091	.102	.125	.129	.156	.162	.187	.250
1/32	.034	.039	.046	.055	.065	.081	.102	.113	.121	.125	.139							
3/64	.041	.046	.053	.062	.072	.090	.108	.119	.127	.131	.145							
1/16	.048	.053	.059	.068	.079	.093	.110	.122	.134	.138	.152							
5/64	.054	.060	.066	.075	.086	.100	.117	.127	.138	.144	.158							
3/32	.061	.066	.073	.082	.092	.107	.124	.134	.142	.146	.160							
7/64	.068	.073	.080	.089	.099	.113	.130	.141	.148	.153	.167	.181						
1/8	.075	.080	.086	.095	.106	.120	.137	.147	.155	.159	.172	.186	.216	.221				
9/64	.081	.087	.093	.102	.113	.127	.144	.154	.162	.166	.179	.193	.223	.228	.263			
5/32	.088	.093	.100	.109	.119	.134	.150	.161	.169	.173	.186	.200	.230	.235	.270	.278		
11/64	.095	.100	.107	.116	.126	.140	.157	.168	.175	.179	.192	.207	.236	.242	.277	.284	.317	
3/16	.102	.107	.113	.122	.133	.147	.164	.174	.182	.186	.199	.213	.243	.248	.283	.291	.324	.405
13/64	.108	.114	.120	.129	.140	.154	.171	.181	.189	.193	.206	.220	.250	.255	.290	.298	.330	.412
7/32	.115	.120	.127	.136	.146	.161	.177	.188	.196	.199	.212	.227	.257	.262	.297	.305	.337	.419
15/64	.122	.127	.134	.143	.153	.167	.184	.195	.202	.206	.219	.233	.263	.269	.304	.311	.344	.426
1/4	.129	.134	.140	.149	.160	.174	.191	.201	.209	.213	.226	.240	.270	.275	.310	.318	.351	.432
17/64	.135	.141	.147	.156	.166	.181	.198	.208	.216	.220	.233	.247	.277	.282	.317	.325	.357	.439
9/32	.142	.147	.154	.163	.173	.187	.204	.215	.223	.226	.239	.254	.284	.289	.324	.332	.364	.446
5/16	.156	.161	.167	.176	.187	.201	.218	.228	.236	.240	.253	.267	.297	.302	.337	.345	.378	.459
11/32	.169	.174	.181	.190	.200	.214	.231	.242	.250	.253	.266	.281	.311	.316	.351	.359	.391	.473
3/8	.183	.188	.194	.203	.214	.228	.245	.255	.263	.267	.280	.294	.324	.329	.364	.372	.404	.486

X Y R (STOCK THICKNESS) T

DEVELOPED LENGTH = X+Y−Z

Z = SET-BACK ALLOWANCE FROM CHART

Standard Welding Symbols

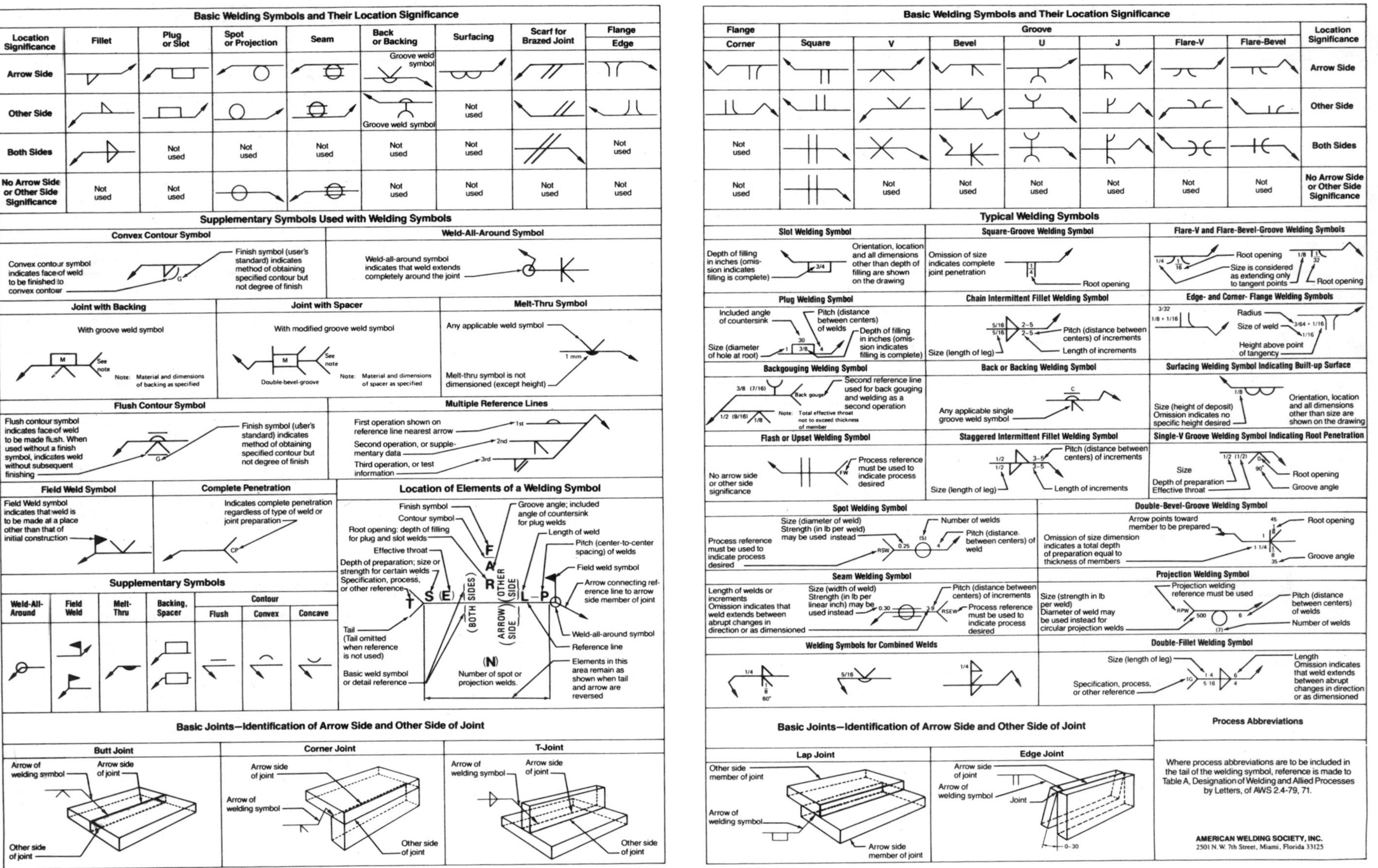

DESIGNATION OF WELDING PROCESSES BY LETTERS

	Welding Process	Letter Designation
Brazing	Infrared Brazing	IRB
	Torch Brazing	TB
	Furnace Brazing	FB
	Induction Brazing	IB
	Resistance Brazing	RB
	Dip Brazing	DB
	†Twin-Carbon Arc Brazing	TCAB†
	†Block Brazing	BB†
	†Flow Brazing	FLB†
Gas Welding	Oxyacetylene Welding	OAW
	Oxyhydrogen Welding	OHW
	Pressure Gas Welding	PGW
	†Air-Acetylene Welding	AAW†
Resistance Welding	Resistance-Spot Welding	RSW
	Resistance-Seam Welding	RSEW
	Projection Welding	RPW
	Flash Welding	FW
	Upset Welding	UW
	Percussion Welding	PEW
Arc Welding	Stud Welding	SW
	Plasma-Arc Welding	PAW
	Submerged Arc Welding	SAW
	Gas Tungsten-Arc Welding	GTAW
	Gas Metal-Arc Welding	GMAW
	Flux Cored Arc Welding	FCAW
	Shielded Metal-Arc Welding	SMAW
	Carbon-Arc Welding	CAW
	†Bare Metal-Arc Welding	BMAW†
	†Gas-Shielded Stud Welding	GSSW†
	†Atomic Hydrogen Welding	AHW†
	†Twin-Carbon Arc Welding	TCAW†
	†Gas Carbon-Arc Welding	GCAW†
	†Shielded Carbon-Arc Welding	SCAW†
Other Processes	Thermit Welding	TW
	Laser Beam Welding	LBW
	Induction Welding	IW
	Electroslag Welding	EW
	Electron Beam Welding	EBW
	†Nonpressure Thermit Welding	NTW†
	†Pressure Thermit Welding	PTW†
	†Flow Welding	FLOW†
Solid State Welding	Ultrasonic Welding	USW
	Friction Welding	FRW
	Forge Welding	FOW
	Explosion Welding	EXW
	Diffusion Welding	DFW
	Cold Welding	CW
	†Roll Welding	RW†
	†Die Welding	DW†
	†Hammer Welding	HW†

DESIGNATION OF CUTTING PROCESSES BY LETTERS

Cutting Process	Letter Designation
Arc Cutting	AC
Air Carbon-Arc Cutting	AAC
Carbon-Arc Cutting	CAC
Metal-Arc Cutting	MAC
Plasma-Arc Cutting	PAC
Oxygen Cutting	OC
Chemical Flux Cutting	FOC
Metal Powder Cutting	POC
Oxygen-Arc Cutting	AOC

(Used with the permission of the AMERICAN WELDING SOCIETY)

Standard Graphic Symbols

THE SYMBOLS SHOWN CONFORM TO THE AMERICAN NATIONAL STANDARDS INSTITUTE (ANSI) SPECIFICIATIONS. BASIC SYMBOLS CAN BE COMBINED IN ANY COMBINATION. NO ATTEMPT IS MADE TO SHOW ALL COMBINATIONS.

LINES AND LINE FUNCTIONS	
LINE, WORKING	
LINE, PILOT (L>20W)	
LINE, DRAIN (L<5W)	
CONNECTOR	
LINE, FLEXIBLE	
LINE, JOINING	
LINE, PASSING	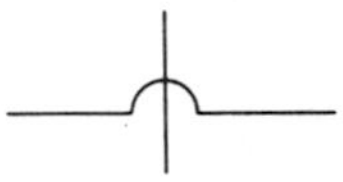
DIRECTION OF FLOW, HYDRAULIC PNEUMATIC	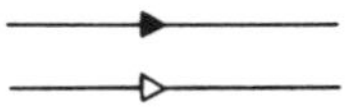
LINE TO RESERVOIR ABOVE FLUID LEVEL BELOW FLUID LEVEL	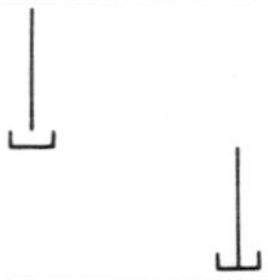
LINE TO VENTED MANIFOLD	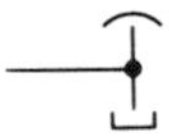
PLUG OR PLUGGED CONNECTION	
RESTRICTION, FIXED	
RESTRICITION, VARIABLE	

METHODS OF OPERATION	
PRESSURE COMPENSATOR	
DETENT	
MANUAL	
MECHANICAL	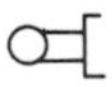
PEDAL OR TREADLE	
PUSH BUTTON	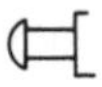

PUMPS	
PUMP, SINGLE FIXED DISPLACEMENT	
PUMP, SINGLE VARIABLE DISPLACEMENT	

MOTORS AND CYLINDERS	
MOTOR, ROTARY, FIXED DISPLACEMENT	
MOTOR, ROTARY VARIABLE DISPLACEMENT	
MOTOR, OSCILLATING	
CYLINDER, SINGLE ACTING	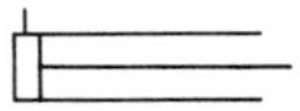
CYLINDER, DOUBLE ACTING	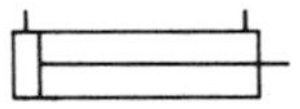
CYLINDER, DIFFERENTIAL ROD	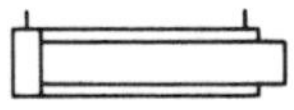
CYLINDER, DOUBLE END ROD	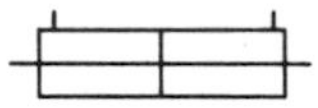
CYLINDER, CUSHIONS BOTH ENDS	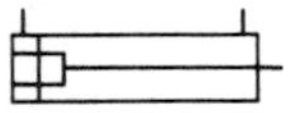

METHODS OF OPERATION	
LEVER	
PILOT PRESSURE	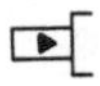
SOLENOID	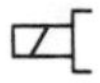
SOLENOID CONTROLLED, PILOT PRESSURE OPERATED	
SPRING	
SERVO	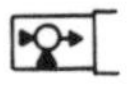

Standard Graphic Symbols

MISCELLANEOUS UNITS	
DIRECTION OF ROTATION (ARROW IN FRONT OF SHAFT)	
COMPONENT ENCLOSURE	
RESERVOIR, VENTED	
RESERVOIR, PRESSURIZED	
PRESSURE GAGE	
TEMPERATURE GAGE	
FLOW METER (FLOW RATE)	
ELECTRIC MOTOR	M
ACCUMULATOR, SPRING LOADED	
ACCUMULATOR, GAS CHARGED	
FILTER OR STRAINER	
HEATER	
COOLER	
TEMPERATURE CONTROLLER	
INTENSIFIER	
PRESSURE SWITCH	W
BASIC VALVE SYMBOLS	
CHECK VALVE	
MANUAL SHUT OFF VALVE	
BASIC VALVE ENVELOPE	
VALVE, SINGLE FLOW PATH, NORMALLY CLOSED	

BASIC VALVE SYMBOLS (CONT.)	
VALVE, SINGLE FLOW PATH, NORMALLY OPEN	
VALVE, MAXIMUM PRESSURE (RELIEF)	
BASIC VALVE SYMBOL, MULTIPLE FLOW PATHS	
FLOW PATHS BLOCKED IN CENTER POSITION	
MULTIPLE FLOW PATHS (ARROW SHOWS FLOW DIRECTION)	
VALVE EXAMPLES	
UNLOADING VALVE, INTERNAL DRAIN, REMOTELY OPERATED	
DECELERATION VALVE, NORMALLY OPEN	
SEQUENCE VALVE, DIRECTLY OPERATED, EXTERNALLY DRAINED	
PRESSURE REDUCING VALVE	
COUNTER BALANCE VALVE WITH INTEGRAL CHECK	
TEMPERATURE AND PRESSURE COMPENSATED FLOW CONTROL WITH INTEGRAL CHECK	
DIRECTIONAL VALVE, TWO POSITION, THREE CONNECTION	
DIRECTIONAL VALVE, THREE POSITION, FOUR CONNECTION	
VALVE, INFINITE POSITIONING (INDICATED BY HORIZONTAL BARS)	

(Vickers Industrial Division)

Acknowledgments

The author wishes to express his appreciation to colleagues in the Division of Technology: Dr. Louis J. Pardini, Marshall R. Minter, Russell G. Biekert and Donal L. Hay; and to John L. Peterson, Superintendent, Southwest Indian Polytechnic Institute, Albuquerque, New Mexico.

He would also like to recognize the following industries who have contributed to this text:

Aerojet-General Corporation, El Monte, Calif.
AiResearch Manufacturing Company, Phoenix, Ariz.
Albina Engine and Machine Works, Portland, Ore.
Allis-Chalmers, Milwaukee, Wis.
American Machine & Foundry Company, Richmond, Va.
American Society of Mechanical Engineers, New York, N.Y.
American Standard, Inc., Buffalo, N. Y.
American Welding Society, New York, N. Y.
Atlas Press Co., Kalamazoo, Mich.
Beech Aircraft Corporation, Wichita, Kan.
Bell Aerosystems Company, Buffalo, N. Y.
Bellows-Valvair, Akron, Ohio
The Boeing Company, Seattle, Washington
Brown & Sharpe Mfg. Co., North Kingstown, R. I.
Capital Machine Tool, Phoenix, Ariz.
Caterpillar Tractor Co., Peoria, Ill.
Cessna Aircraft Company, Wichita, Kan.
Chrysler Corporation, Detroit, Mich.
The Cincinnati Milling Machine Co., Cincinnati, Ohio
Clark Equipment Company, Battle Creek, Mich.
Corning Glass Works, Corning, N. Y.
Crescent Tool Co., Jamestown, N. Y.
Fairchild Hiller, Germantown, Md.
The Firestone Tire & Rubber Company, Akron, Ohio
G A F Corporation, La Habra, Calif.
General Dynamics, Fort Worth, Texas
General Motors Engineering Staff, Warren, Mich.
Gleason Works, Rochester, N. Y.
Goodyear Aerospace, Litchfield Park, Ariz.
Halstead & Mitchell Co., Zelienople, Pa.
Hamilton Manufacturing Company, Two Rivers, Wis.
Hamilton Watch Company, Lancaster, Pa.
Harley-Davidson Motor Co., Milwaukee, Wis.
Heald Corp., Worcester, Mass.
Hewlett-Packard, Palo Alto, Calif.
Hobbs Manufacturing Company, Worcester, Mass.
Honeywell Inc., Minneapolis, Minn.
Hughes Aircraft Company, Los Angeles, Calif.
Lennox Industries Inc., Marshalltown, Ia.
Lincoln Electric Company, Cleveland, Ohio
Lockheed Aircraft Corporation, Burbank, Calif.
Manitowoc Engineering Co., Manitowoc, Wis.
Martin Marietta Corporation, Denver, Colo.
McDonnell Douglas, St. Louis, Mo.
Mead Fluid Dynamics, Chicago, Ill.
Minnesota Mining and Manufacturing Co., St. Paul, Minn.
Moline Tool Company, Moline, Ill.
Motorola Inc., Semiconductor Products Div., Phoenix, Ariz.
The National Cash Register Company, Dayton, Ohio
National Fluid Power Association, Thiensville, Wis.
National Lock Hardware, Rockford, Ill.
North American Rockwell, El Segundo, Calif.
Norton Company, Worcester, Mass.
Oliver Machinery Co., Grand Rapids, Mich.
Outboard Marine Corporation, Waukegan, Ill.
Pacific Valves, Inc., Long Beach, Calif.
The Perkin-Elmer Corporation, Norwalk, Conn.
Frederick Post, Los Angeles, Calif.
Pratt & Whitney Aircraft, East Hartford, Conn.
Pratt & Whitney Inc., Machine Tool Div., West Hartford, Conn.
Precision Products Inc., Phoenix, Ariz.
Rheem Manufacturing Company, Linden, N. J.
Rockwell Manufacturing Company, Pittsburg, Pa.
Ross Heat Exchanger Division, Buffalo, N. Y.
Ryan Aeronautical Company, San Diego, Calif.
Schwinn Bicycle Company, Chicago, Ill.
Sheldon Machine Co., Inc., Chicago, Ill.
Skil Corporation, Chicago, Ill.
South Bend Lathe, South Bend, Ind.
Sperry Flight Systems Division, Phoenix, Ariz.
Stainless Steel Products, Burbank, Calif.
The L. S. Starrett Company, Athol, Mass.
Sterling Instrument Div., Designatronics Inc., Mineola, N.Y.
Stewart-Warner Corporation, Chicago, Ill.
Sunnen Products Company, St. Louis, Mo.
Talley Industries, Inc., Mesa, Ariz.
The Trane Co., La Crosse, Wis.
Unidynamics Phoenix, Inc., Phoenix, Ariz.
United Aircraft, East Hartford, Conn.
Vickers Industrial Division, Troy, Mich.
Western Electric, New York, N. Y.
Western Gear Corp., Precision Products Div., Lynwood, Cal.
Westinghouse Air Brake Company, Pittsburg, Pa.
Wilton Corporation, Schiller Park, Ill.

Index

Index

Industry Prints used in this write-in text.